The Cultural Nature of Attachment

Contextualizing Relationships and Development

Strüngmann Forum Reports

Julia R. Lupp, series editor

The Ernst Strüngmann Forum is made possible through
the generous support of the Ernst Strüngmann Foundation,
inaugurated by Dr. Andreas and Dr. Thomas Strüngmann.

This Forum was supported by the
Deutsche Forschungsgemeinschaft

The Cultural Nature
of Attachment

Contextualizing Relationships
and Development

Edited by

Heidi Keller and Kim A. Bard

Program Advisory Committee:

Kim A. Bard, Marjorie Beeghly, Nandita Chaudhary,
William D. Hopkins, Heidi Keller, Julia R. Lupp,
and Masako Myowa

The MIT Press

Cambridge, Massachusetts
London, England

© 2017 Massachusetts Institute of Technology and
the Frankfurt Institute for Advanced Studies

Series Editor: J. R. Lupp
Editorial Assistance: M. Turner, A. Ducey-Gessner, C. Stephen
Photographs: N. Miguletz
Lektorat: BerlinScienceWorks

The book was set in TimesNewRoman and Arial.
Printed and bound in the United States of America.

Library of Congress Cataloging-in-Publication Data is available.

ISBN 978-0-262-03690-0

Ernst Strüngmann Forum (22nd: 2016 : Frankfurt am Main, Germany)

10 9 8 7 6 5 4 3 2 1

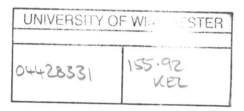

Contents

The Ernst Strüngmann Forum

Science is a highly specialized enterprise—one that enables areas of enquiry to be minutely pursued, establishes working paradigms and normative standards, and supports rigor in experimental research. Some issues, however, do not fall neatly into the purview of a single disciplinary field. Here, specialization can hinder conceptualization and limit the generation of potential problem-solving approaches. The Ernst Strüngmann Forum was created to explore these types of problems.

Founded on the tenets of scientific independence and the inquisitive nature of the human mind, the Ernst Strüngmann Forum is dedicated to the continual expansion of knowledge. Its activities promote interdisciplinary communication on high-priority issues encountered in basic science. Through its innovative communication process, the Ernst Strüngmann Forum provides a creative environment within which experts scrutinize high-priority issues from multiple vantage points.

This process begins with the identification of themes. By nature, a theme constitutes a problem area that transcends classic disciplinary boundaries, is of high-priority interest, and requires concentrated, multidisciplinary input to address the issues. Proposals are received from leading scientists active in their field and selected by an independent Scientific Advisory Board. Once approved, the Ernst Strüngmann Forum convenes a steering committee to refine the scientific parameters of the proposal and select participants. Approximately one year later, a central gathering, or Forum, is held to which circa forty experts are invited. Expansive discourse is employed to approach the problem. Often, this necessitates reexamining long-established ideas and relinquishing conventional perspectives. Yet when this is accomplished, new insights begin to emerge. As a final step, the resultant ideas and newly gained perspectives from the entire process are communicated to the scientific community for further consideration and implementation.

Preliminary discussion for this topic began in 2014, when the need was expressed to examine how attachment theory could be broadened to encompass cross-cultural and cross-species research. From January 5–7, 2015, the Program Advisory Committee (Kim A. Bard, Marjorie Beeghly, Nandita Chaudhary, William D. Hopkins, Heidi Keller, Julia Lupp, and Masako Myowa) met to refine the scientific framework for this Forum, which was held in Frankfurt am Main from April 3–8, 2016.

This volume synthesizes the discourse that transpired between a diverse group of experts and is comprised of two types of contributions. Background information is provided on key aspects of the overall theme. These chapters, drafted before the Forum, have been reviewed and subsequently revised. In addition, Chapters 4, 6, 8, and 10 summarize the extensive discussions of the

working groups. These chapters are not consensus documents nor are they proceedings; they transfer the essence of this multifaceted discourse, expose areas
where opinions diverge, and highlight topics in need of future enquiry.

An endeavor of this kind creates its own unique group dynamics and puts
demands on everyone who participates. Each invitee played an active role,
and for their efforts, I am grateful to all. A special word of thanks goes to the
Program Advisory Committee, to the authors and reviewers of the background
papers, as well as to the moderators of the individual working groups (Barbara
Finlay, William Hopkins, Nandita Chaudhary, and Marjorie Beeghly). The
rapporteurs of the working groups (Kristen Hawkes, Allyson Bennett, Gilda
Morelli, and Suzanne Gaskins) deserve special recognition, for to draft a report
during the Forum and finalize it in the months thereafter is no simple matter.
Finally, I extend my appreciation to Heidi Keller and Kim Bard: as chairper-
sons of this 22nd Ernst Strüngmann Forum, their engagement throughout the
process was crucial.

A communication process of this nature relies on institutional stabil-
ity and an environment that encourages free thought. The generous support
of the Ernst Strüngmann Foundation, established by Dr. Andreas and Dr.
Thomas Strüngmann in honor of their father, enables the Ernst Strüngmann
Forum to pursue its work in the service of science. In addition, the follow-
ing valuable partnerships are gratefully acknowledged: the Scientific Advisory
Board, which ensures the scientific independence of the Forum; the Deutsche
Forschungsgemeinschaft, for its supplemental financial support; and the
Frankfurt Institute for Advanced Studies, which shares its intellectual setting
with the Forum.

Long-held views are never easy to put aside. Yet, when this is achieved,
when the edges of the unknown begin to appear and the resulting gaps in
knowledge are able to be identified, the act of formulating strategies to fill such
gaps becomes a most invigorating activity. On behalf of everyone involved,
I hope this volume will convey a sense of this lively exercise and lead to an
inclusive conceptualization of attachment.

Julia R. Lupp, Director Ernst Strüngmann Forum
Frankfurt Institute for Advanced Studies (FIAS)
Ruth-Moufang-Str. 1, 60438 Frankfurt am Main, Germany
https://esforum.de/

List of Contributors

Bard, Kim A.   Department of Psychology, University of Portsmouth, Portsmouth PO1 2DY, U.K.

Beeghly, Marjorie   Department of Psychology, Wayne State University, Detroit, MI 48202, U.S.A.

Bennett, Allyson J.   Department of Psychology, University of Wisconsin Madison, Madison, WI 53715, U.S.A.

Bohr, Yvonne   LaMarsh Centre for Child and Youth Research, Department of Psychology, York University, Toronto, ON, Canada M3J 1P3

Butler, David L.   Graduate School of Education, Kyoto University, Yoshida-honmachi, Sakyo-ku, Kyoto 606-8501, Japan

Chaudhary, Nandita   Department of Human Development and Childhood Studies, Lady Irwin College, University of Delhi, New Delhi 110001, India

Chen, Stephen H.   Department of Psychology, Wellesley College, Wellesley, MA 02481

Chisholm, James S.   School of Anatomy, Physiology, and Human Biology, University of Western Australia, Crawley, WA 6009, Australia

Fairbanks, Lynn A.   Semel Institute, University of California at Los Angeles, Los Angeles, CA 91103, U.S.A.

Feldman, Ruth   Department of Psychology, Bar-Ilan University, 52900 Ramat-Gan, Israel

Finlay, Barbara L.   Department of Psychology, Cornell University, Uris Hall, Ithaca, NY 14853, U.S.A.

Gaskins, Suzanne   Department of Psychology, Northeastern Illinois University, Chicago, IL 60625, U.S.A.

Gazzola, Valeria   Social Brain Laboratory, Netherlands Institute for Neuroscience, Royal Netherlands Academy of Arts and Sciences, 1105 BA Amsterdam, The Netherlands

Gernhardt, Ariane   Department of Culture and Development, Osnabrück University, School of Human Sciences, 49076 Osnabrück, Germany

Giedd, Jay   Division of Child and Adolescent Psychiatry, University of California, San Diego, La Jolla, CA 92093

Gottlieb, Alma   Department of Anthropology, University of Illinois at Urbana-Champaign, 607 S. Mathews Avenue, Urbana, IL 61801, U.S.A.

Hawkes, Kristen   Department of Anthropology, University of Utah, Salt Lake City, UT 84112-0060, U.S.A.

Hopkins, William D.   Institute for Neuroscience, Georgia State University, Atlanta, GA 30303, U.S.A.

Johow, Johannes   Department of Philosophy, Justus-Liebig-Universität Gießen, 35394 Gießen, Germany

Kalcher-Sommersguter, Elfriede   Institute of Zoology, University of Graz, Graz, 8010, Austria

Keller, Heidi   Faculty of Human Sciences, Institute of Psychology, University of Osnabrück, 49069 Osnabrück, Germany

Lamb, Michael E.   Department of Psychology, School of Biology, Cambridge University, Cambridge, U.K.

Liebal, Katja   Department of Education and Psychology, Freie Universität Berlin, 14195 Berlin, Germany

Liu, Cindy H.   Department of Psychiatry, Beth Israel Deaconess Medical Center, Harvard Medical School, and Department of Psychology, University of Massachusetts Boston, Boston, MA 02125, U.S.A.

Lupp, Julia R.   Ernst Strüngmann Forum, Frankfurt Institute for Advanced Studies, Ruth-Moufang-Straße 1, 60438 Frankfurt am Main, Germany

Morelli, Gilda A.   Applied Developmental and Educational Psychology, Boston College, MA 02467

Murray, Marjorie   Anthropology Programme, Instituto de Sociologia, Pontificia Universidad Católica de Chile, correo 22, Chile

Myowa, Masako   Graduate School of Education, Kyoto University, Yoshida-honmachi, Sakyo-ku, Kyoto 606-8501, Japan

Quinn, Naomi   Department of Cultural Anthropology, Duke University, Durham, NC 27708, U.S.A.

Rosabal-Coto, Mariano   Instituto Investigaciones Psicológicas, Universidad de Costa Rica, San José, Costa Rica

Scheele, Dirk   Department of Medical Psychology and Psychiatry, University of Bonn, 53105 Bonn, Germany

Scheidecker, Gabriel   Department of Political and Social Sciences, Institute of Social and Cultural Anthropology, SFB 1171 Affective Societies, Freie Universität Berlin, 14195 Berlin, Germany

Sheridan, Margaret A.   Department of Psychology and Neuroscience, University of North Carolina, Chapel Hill, NC, 27599-3270, U.S.A.

Sommer, Volker   Department of Anthropology, University College London, WC1E 6BT London, U.K.

Suomi, Stephen J.   Laboratory of Comparative Ethology, NICHD, National Institutes of Health, Suite 8030, MSC 7971, Bethesda, MD 20892, U.S.A.

Takada, Akira   Graduate School of Asian and African Area Studies, Kyoto University, Sakyo-ku, Kyoto, 606-8501, Japan

Teti, Douglas M. Human Development and Family Studies, Pennsylvania State University, PA 16802, U.S.A.

Thierry, Bernard Department of Ecology, Physiology, and Ethology, Hubert Curien Pluridisciplinary Institute, CNRS University of Strasbourg, Strasbourg 67087, France

Thompson, Ross A. Department of Psychology, University of California, Davis, CA 95616, U.S.A.

Tomoda, Akemi Child Development Research Center, University of Fukui, Fukui, Japan

Tottenham, Nim Department of Psychology, Columbia University, New York, NY 10025, U.S.A.

Tronick, Ed Department of Psychology, University of Massachusetts Boston, Boston, MA 02125, U.S.A.

Vicedo, Marga Institute for the History and Philosophy of Science, University of Toronto, Toronto M5S 1K7, Canada

Wang, Leslie Department of Sociology, University of Massachusetts Boston, Boston, MA 02125, U.S.A.

Weisner, Thomas S. Department of Psychiatry and Department of Anthropology, UCLA, Los Angeles, CA 90024, U.S.A.

Yovsi, Relindis D. Avenue Léon Debatty 32, 1070 Brussels, Belgium

1

Introduction

Heidi Keller, Kim A. Bard, and Julia R. Lupp

Abstract

Science, and by extension society, requires a comprehensive theory of attachment to guide research and practice—one grounded in a contextualized conception of attachments and their development, which encompasses knowledge from diverse disciplines engaged in the study of human development. To improve on the current paradigm, this volume embraces the diversity of attachment systems across cultures and primate species, and assesses the core assumptions and methods of attachment theory. Resultant understanding is used to project an updated version of attachment theory—one that can be applied across cultures. Suggestions for more culturally sensitive research methods are proposed and ideas applicable to current practice and policies discussed. A reconceptualized theory of attachment is presented based on principles that are generalizable, valid, and reliable across diverse primates and diverse human cultures. In addition, the need to make adjustments in attachment philosophy is stressed, and strategies are discussed to communicate and work with researchers, policy-makers, practitioners, and other stakeholders.

Background

With his formulation of attachment theory, John Bowlby (1958, 1969, 1973, 1980) initiated a paradigm that defined how children's development and the evolutionary functions of primate parenting would be understood for decades to come. At the time of its conception, basic science was dominated by a mechanistic understanding of human development (seen as a chain of stimulus-response patterns), while a psychoanalytic worldview (with its strong emphasis of unconscious processes) dominated application, especially clinical practice. In primatology, scientific debate oscillated between operant conditioning explanations and psychodynamic accounts of the origins of infant-caregiver bonds. Bowlby's great achievement was to enlist diverse scientific traditions (e.g., ethology and evolutionary approaches, systems theory, psychoanalysis) in formulating a foundation by which the development of socioemotional processes could be understood. His resulting theory, later complemented by the work of Mary Ainsworth (1967; Ainsworth et al. 1978), was compelling and attracted much attention, albeit with a time delay typical for new approaches in science.

The initial enthusiasm that greeted his seminal contribution, however, cannot offset the fact that Bowlby's understanding was incomplete, at best, and sometimes simply wrong. Although Bowlby was receptive to many new ideas, he never embraced core principles of evolutionary theory or cultural variation in parenting and children's development. He also did not acknowledge the existence of variability in caregiving practices across primate species. His cultural blindness was pointed out by Margaret Mead as early as 1954, yet her critique went unheeded (Mead 1954; Vicedo, this volume). Even today, attachment researchers have not integrated the core principles of evolution, culture, and cross-species variation into attachment theory, although they do acknowledge that different conceptions and strategies of parenting exist across cultures. They continue to claim, though, that attachment theory has strong cultural roots, due to the observational studies that Ainsworth conducted in Uganda (Ainsworth 1967). These Ugandan observations, however, are not representative of all rural non-Western cultures. Indeed, they conflict with many anthropological and cultural psychological observations that have been conducted in similar sub-Saharan villages. Differences in child-rearing philosophies, practices, and children's developmental trajectories exist and have been extensively documented for different subsistence-based communities, urban families in non-Western countries, as well as migrants and refugees in Western societies (Harwood et al. 1995; Gottlieb 2004; Keller 2007; Quinn and Mageo 2013; Vicedo 2013; Otto and Keller 2014; LeVine and LeVine 2016; Gottlieb and DeLoache 2017). Yet this decisive body of evidence has not been able to change prevailing views on attachment theory.

As a result, we (HK and KAB) approached the Ernst Strüngmann Forum to request support in examining attachment theory against the backdrop of current knowledge in cultural psychology, anthropology, evolutionary theory, primatology, and neuroscience. Our contention was that science, and by extension society, requires a comprehensive theory to guide its work—one grounded in a contextualized understanding of attachments and their development, encompassing knowledge from the many disciplines that engage in the study of human development.

The Ernst Strüngmann Forum is dedicated to the expansion of knowledge in science. It facilitates open discourse on problems faced in research—topics that require the input of multiple areas of expertise to generate greater understanding. These topics reflect real problems encountered by researchers and often indicate areas where existing paradigms may need to shift or where new ones are required. Throughout the process fostered by the Forum, "gaps in knowledge" are exposed and potential ways forward are collectively pursued. Consensus is never forced nor is it necessarily the goal. Instead, results of these multifaceted discussions are synthesized and disseminated to permit testing of emergent ideas, to support further debate on contentious issues, and to stimulate future research.

The overarching aim of this Forum on the cultural nature of attachment was to reconceptualize what is meant by attachment. It was not set up to scrutinize the underlying philosophical or ideological assumptions of current attachment theory, or to rehash the contributions of Bowlby and Ainsworth, or to revisit the basic mammalian biology of bonding (e.g., Carter et al. 2005). Instead, we sought to scrutinize the concept of evolution upon which attachment theory is grounded and to enlist multiple perspectives—from cross-cultural and cross-species research as well as new information from epigenetics and neuroscience (e.g., Jablonka and Lamb 2005; Suomi 2008)—to create a novel, expanded view of infant attachment(s). In doing so, we embraced the diversity of attachment systems across cultures and primate species, and used an inclusive perspective to evaluate the core assumptions and methods of attachment theory.

This volume summarizes the results of our extended discussions. In it you will find proposals for an inclusive theory of attachment, suggestions for more culturally sensitive research methods, and novel ideas applicable to current practice and policies. Attachment theory and philosophy need to adjust to the dynamic nature of science, to reflect the research that informs it and stay valid. To this end, we discuss ways to communicate and apply resultant understanding to researchers, policy-makers, practitioners, and other stakeholders.

Perspectives

Science does not take place in a vacuum and is also not static. It plays out within a context driven by worldviews and broad philosophical frameworks. The methodologies, theories, and research questions that emerge reflect this setting. Influenced by changing ideas and informed by new data, science needs to be a dynamic process, as the following examples illustrate.

Many years ago, in collaboration with Irenäus Eibl-Eibesfeldt, one of us (HK) explored the universal nature of face-to-face contact between infants and adult caregivers. Using film footage of Yanomami Indians and Trobriand villagers, clear evidence of face-to-face contact was found in these two groups, both of which differed significantly from Western middle-class families (Keller et al. 1988). This work also documented enormous quantitative differences in the amount of face-to-face contact exhibited across cultures. At the time, this variability did not arouse HK's interest like it does today (see Keller and Chaudhary, this volume).

Working within a different context, one of us (KAB) investigated the extent to which face-to-face contact between infants and adult caregivers typifies a human unique engagement system. Her initial finding was that chimpanzees engage in some face-to-face contact, but not a lot, which suggested a species difference (Bard 1994). Later though, based on increased knowledge of intergroup diversity in chimpanzees as well as humans, this conclusion was revised: In chimpanzees, just as in humans, there is significant variation in the amount of face-to-face contact exhibited by mother-infant pairs across two groups (Bard et al. 2005). Similar developmental processes apply to the infant-caregiver engagement system of chimpanzees and of humans; specifically, the amount of face-to-face contact is inversely related to the amount of physical contact experienced by very young infants.

Looking for phenomena defined in accordance to Western standards or mores may reveal the existence of these concepts in other, diverse environments. Such definitions, however, cannot address the validity of these phenomena within local meaning or value systems and may inhibit the examination of other, potentially more important dimensions of the same construct. Cross-cultural studies of cognition, intelligence, and the "big five" personality traits provide many examples of how an incomplete understanding can be falsely interpreted as proof of universality (e.g., Nisbett and Norenzayan 2002). There is enormous variability in the types of child caregiving arrangements and socialization strategies practiced across human cultures and primate groups, and attachment mechanisms differ as a result. A theory of attachment must account for this variability.

Attachment theorists claim that a strong evolutionary foundation is embedded into attachment theory due to Bowlby's interest in studies of rhesus macaques and perspectives from Robert Hinde (an ethologist) and Harry Harlow (an experimental psychologist) (e.g., Suomi et al. 2008). This work did provide a complement to human studies on exploration/secure base and reinforced the importance of early attachments. However, Bowlby reached conclusions based on a single nonhuman primate species living in captivity and on a limited understanding of evolutionary processes. What resulted was, at best, inaccurate and, at worst, biased.

Attachment theory assumes that an adaptive behavioral system underpins the evolutionary foundation of attachment, since attachment relationships help infants survive and thrive. Further, it holds that a specific way of mothering—one in line with Western middle-class childcare philosophy—is best for the healthy development of all infants. Neither, however, is supported by evidence.

Genetic fitness is at the core of evolutionary thinking: reproductive success is the ultimate goal, both in terms of physical and psychological development. Well-being does not drive genetic fitness. Ever since Trivers (1972) differentiated r- and K-selection strategies, reproductive styles have been correlated to different contextual conditions. Had attachment theory incorporated such a contextual view of developmental processes, a more differential understanding of parenting qualities and child development would have been the norm (Myowa and Butler; Hawkes et al.; and Chisholm, this volume; see also Lamb et al. 1984a).

How, then, can a popular but incomplete theory be brought in line with current understanding? What aspects need to be altered, and how might this be accomplished? To approach these questions, the following working groups were formed at the Forum:

- Evolution and attachment across primate groups (Hawkes et al., Chapter 4, this volume)
- Neural foundations of variability in attachment (Bennett et al., Chapter 10, this volume)
- Cultural evidence for different conceptions of attachment (Morelli et al., Chapter 6, this volume)
- Meaning and methods in the study and assessment of attachment (Gaskins et al., Chapter 8, this volume)

Their discussions benefited from the input of primatologists, evolutionary biologists, cultural anthropologists, cultural psychologists, neuroscientists, attachment theorists, and developmental psychologists. Based on findings from current research in their fields, participants worked collectively to expand the concept of attachment, using context-specific definitions of infant-caregiver attachments and their development.

Reconceptualizing Attachment

Evolution and Attachment across Primate Groups

In his early writings, Bowlby explicitly stressed the contextual nature of attachment, yet he focused primarily on the social environment, which in his view was defined by the mother. He did not recognize that the social environment is embedded within an ecological setting; that adaptive behaviors (including mothering) depend on ecosocial contexts, affordances, and constraints. Maternal investment does indeed vary according to context, and it can differ as a function of the caregiving arrangements of the social system in which the mother lives, which must be systematically considered (for a detailed discussion, see Vicedo 2013). What is currently lacking in theoretical accounts of attachment, but crucial for evolutionary theory, is the *contextual embeddedness* of the child or infant in his/her respective social and ecological environment (Harkness and Super 1996; Jablonka and Lamb 2007; Quinn and Mageo 2013; Otto and Keller 2014).

Bowlby used the rhesus macaque caregiving system as the evolutionary model for attachment. He did not acknowledge the enormous variability in caregiving arrangements that exist in nonhuman primate species. For instance, cotton-top tamarins rely on distributed caretaking; capuchin monkeys behave in similar ways toward their mothers as they do toward siblings or unrelated adults; and monogamy is rare among nonhuman primates (e.g., Sommer 2000; Bard 2018). Moreover, the very cultural context of a researcher can influence which type of behavior is chosen for study. For instance, Japanese primatologists primarily study cooperation, whereas Euro-American primatologists are predominantly interested in competition (de Waal 2001).

Research in evolution, as a complex process, has expanded greatly over the last decades and now reveals the inadequacies of Bowlby's concept of the environment of evolutionary adaptedness (EEA) to explain the origin of attachment. In Chapter 3, Masako Myowa and David Butler address the inadequacy of using rhesus macaques as a model for infant caregiving systems in all primates. They provide a phylogenetic history of attachment among primates and identify features of attachment that are shared or which differ between humans and nonhuman primates. Importantly, they consider possible cognitive, social, and ecological factors associated with these similarities and/or differences in attachment among primates.

From a complex adaptive systems perspective, James Chisholm posits in Chapter 11 that the human capacity for culture emerged with the evolution of human attachment by means of selection for increased mother-infant cooperation in the resolution of parent-offspring conflict. After outlining the evolutionary-developmental logic of attachment, parent-offspring conflict, and the view of culture as "extended embodied minds," he describes how the embodied mind and its attachments might have been extended beyond the mammalian mother-infant dyad to include expanding circles of cooperative individuals and groups. Since attachment came before and gave rise to culture, no culture could exist for long that did not accommodate the attachment needs of its infants.

In Chapter 4, Kristen Hawkes et al. extend the evolutionary perspective with a critical look at the causes and consequences of varying care in primates. Interactions between infants, mothers, and others in a range of species are used to assess variations and commonalities, as well as to explore how development in human infants can be understood in terms of maturational state at birth and weaning compared to other primates. They conclude with a consideration of the long-term effects of infant experience in primates other than humans. Interactions between particular chimpanzee mothers and infants are described and show that trust relationships between mothers and human researchers reveal variations in mothering style that appear to result from early life events, recent experience, and social context.

Neural Foundations of Variability in Attachment

Neuroscience offers novel insight into processes that support the integration of the social brain, cultural contexts, and development of attachment relationships beyond the human case. For instance, the cortical organization of adult chimpanzees is differentially influenced by early-rearing experiences (Bogart et al. 2014); laterality in the posterior superior temporal gyrus has been implicated in the processing of social information in chimpanzees (Hopkins et al.

2014b); and genetic variation in the arginine vasopressin V1a receptor gene is significantly associated with receptive joint attention in adult chimpanzees (Hopkins et al. 2014a). Previously, these correlations of brain structure and function with social behaviors and polymorphisms in receptor gene were associated with pair bonding in humans and voles (e.g., Phelps 2010). A scientifically valid theory of attachment must include knowledge of the neurobiological mechanisms that support plasticity in attachment outcomes (Bennett et al., this volume; Panksepp 1998; Coan 2008; Bogart et al. 2014).

Margaret Sheridan and Kim Bard look at the neural consequences of infant attachment in Chapter 9, using evidence from nonhuman primates and institutionalized infants. Whereas attachment theory suggests that the function of attachment primarily relates to the regulation of negative affect, they argue that neurobiological evidence illustrates the impact of attachment relationships on two neural systems not typically considered: the neural substrates of reward learning and the neural substrates that support complex cognitive function, such as executive function.

In Chapter 10, Allyson Bennett et al. continue the discussion into the neural foundations for variability in attachment, posing critical questions on how relationships are initiated. Instead of conceptualizing attachment as a single type of relationship or a rigid developmental channel, they propose that attachment is necessary to understand the neural foundations of multiple infant-caregiver relationships, and the role these play in developing competence across the life span. They suggest that this approach will help identify common neurobiological elements of attachment as well as the remarkable plasticity and diversity within and across individuals, cultures, and species.

Cultural Evidence for Different Conceptions of Attachment

Attachment theory is based on a particular conceptualization of infants, which we now refer to as WEIRD (Western, educated, industrialized, rich, democratic—a term coined by Henrich et al. 2010), where the infant is regarded as a separate, active, independent, and autonomous person who acts on its own wishes and desires. This view of the infant necessitates complete responsiveness from the environment, in particular the mother. In the anthropological and psychological literature, however, significant work has revealed variance in human caregiving as well as different cultural conceptions of relationships. The processes by which attachment forms and the involvement of different social partners differ substantially according to the environment in which children grow and develop. Attachment researchers, however, have not incorporated principles from this work or used it to update attachment theory.

In Chapter 2, Marga Vicedo places attachment research in its historical context and details how early attachment theorists have ignored cultural diversity. She examines various challenges to the ethological attachment theory and frames the discussion around two of its fundamental tenets: the universality of attachment patterns and the biological foundations of the attachment system. She demonstrates how these challenges have not yet been successfully addressed and calls for better models of the coevolution of culture and biology.

The types of caregiving arrangements experienced by infants in non-WEIRD settings are vast, and there is an array of arrangements and responsibilities for caretakers (Quinn and Mageo 2013; Otto and Keller 2014). The mother can be the primary caregiver within a network of others, as in the Aka. The mother can be the primary caregiver for a short period of time, as in the Beng community, after which care is complemented and substituted by other caregivers. The mother can be a primary caregiver who holds intensive caregiving relationships with other infants, and even animals, as in the Pirahã. Distributed caregiving arrangements are also possible, where the mother is not necessarily the primary caregiver, as in the Brazilian favelas and the Cameroonian Nso.

In Chapter 5, Heidi Keller and Nandita Chaudhary present evidence of diverse childcare arrangements in cultures outside of Western norms and argue that these arrangements are normative in their respective cultural contexts. They stress that infant care, in all environments, is far more than just an isolated, biopsychological phenomenon: it is an activity deeply imbued with cultural meanings, values, and practices. Challenging the core assumptions that attachment is dyadic and mother-oriented, they propose a "cradles of care" model to address different possibilities in child-rearing conditions, independent of geographical place and age group of caregivers.

Gilda Morelli et al. (Chapter 6) take a pluralistic approach to attachment and present an alternative view to classic attachment theory. Because children develop attachment relationships that are locally determined, they argue that the study of child development must be informed by a systematic, ethnographical approach—one that involves observing, talking with, and listening to local people as they go about living their lives. They hold that a child's social network is of paramount importance.

Both WEIRD and non-WEIRD perspectives have implications for the very definition of attachment. The affectional bond—a relatively long-enduring tie in which the singular partner is important as a unique, noninterchangeable individual (WEIRD perspective)—is just one possible solution. To explore this further, Cindy Liu and colleagues (Chapter 7) examine the concept of monotropy, a basic component of attachment theory, by looking at the practice of transnational separation in Chinese immigrant families. Prolonged separation between parents and children is a common occurrence for many families in the United States and China. Thus it provides a cultural exemplar to extend and situate the meaning of attachment.

Meaning and Methods in the Study and Assessment of Attachment

To assess the quality of attachment relationships, researchers have relied on the Strange Situation Procedure (Ainsworth et al. 1978), a systematic observational procedure whose validity is limited to middle-class white Americans living in the 1950s and 1960s in an urban U.S. setting. The assumptions underlying this procedure are that infants have dyadic attachments with adult partners, that infants encounter strangers moderately often but should be wary of them, that infants engage in daily independent exploration and are frequently in a room alone, and that a sensitive adult caregiver is available to respond quickly should an infant vocalize distress (Ainsworth et al. 1978). These types of experiences, however, are not common to many infants around the world: In many cultures, infants have multiple caregivers and infants are never alone. In some cultures, infants are not encouraged to express emotion, especially negative emotions. In others, strangers are not perceived with wariness (Keller and Chaudhary as well as Morelli et al., this volume). Thus serious issues of validity arise when the Strange Situation Procedure is used in cultures other than the one for which it was designed. In part, due to a reliance on the Strange Situation Procedure, infants well adapted to their culture-specific attachment system can be erroneously labeled as atypical or even pathological from a Western/urban perspective, and vice versa (Keller 2007).

The issues surrounding how attachment should be measured and assessed across diverse cultures were addressed by Suzanne Gaskins and colleagues. In Chapter 8, they propose that attachment systems fulfill two universal functions: they provide socially organized resources for the infant's protection and psychobiological regulation as well as a privileged entry point for social learning. Based on this consideration of the functions of attachment that could be applied universally, Gaskins et al. suggest ways to understand the nature of the cultural and ecological contexts that organize attachment systems, and propose a wide range of research strategies to facilitate the extension and contextual validity of measures of attachment across cultures and species.

Emergent Issues

As one might suspect, many issues emerged during the Forum that could not be resolved; these topics have been highlighted in the individual chapters for future attention. Two particular topics, however, were written up after the Forum to help direct future discourse: (a) the current status of attachment theory and (b) the implications of attachment-related research for policy and practice.

Current attachment theory. In our evaluation of the core assumptions and methods of attachment, it became apparent that we needed to have a coherent account of the state of current attachment theory. This would also enable us to assess claims that attachment theory has already been substantially revised. In response, Ross Thompson prepared an overview of twenty-first century attachment theory (Chapter 12) and, importantly, highlighted points of contention that remain:

- To whom do infants become attached?
- How should differences in attachment relationships be characterized?
- What influences lead to differences in attachment relationships?
- What are the outcomes of differences in attachment?

His characterization of contemporary attachment theory underscores an ongoing tension between a monotropic view of attachment and the recognition of the importance of multiple attachments. Contemporary attachment researchers do recognize that a much wider range of normative attachments develop in early childhood than was previously acknowledged. However, they also consider much of the evidence of cultural variability to be largely irrelevant to attachment. As Thompson writes (p. 318, this volume):

> While culturally oriented researchers ask for greater *culturally informed attachment research,* attachment researchers sometimes wonder where they can find greater *attachment-informed cultural studies.* When they survey the research literature on culture and attachment, attachment researchers find relatively few studies that address the central claims of attachment theory in an informative way: as indicated above, research that might be relevant is often not focused on the developmental experience of young children.

Yet ethnographic and cross-cultural studies of non-Western societies over the last fifty years have explicitly addressed children's relational networks, emotional regulation, separation, and other issues that are central to attachment theory (e.g., Whiting 1963; Weisner and Gallimore 1977; Sorenson 1979; Tronick et al. 1987; Rogoff et al. 1993; Rothbaum et al. 2002; Konner 2005; Quinn and Mageo 2013; Otto and Keller 2014; Lancy 2015; LeVine and LeVine 2016). Does this reveal an interdisciplinary disconnect and, if so, what can be done to resolve it?

It is important to acknowledge and understand disciplinary differences. Ethnographic researchers, for example, do not like to apply methods developed for one cultural context to measure outcomes in another, distinct context. For instance, in their work with Aka foragers, Meehan and Hawks (2013:108) hold, that "the Strange Situation Procedure is not appropriate in all cultural contexts." One could go a step further and argue that imposing conditions that are deemed to be grossly inappropriate in a cultural context (e.g., separating children from others or leaving them alone in a room) is unethical. Equally, evaluating beliefs and behaviors in one culture according to the standards of another (e.g., the sensitivity scale, the Attachment Q sort) may be grossly misleading and also unethical (see, e.g., the discussion of warmth in Keller and Chaudhary, this volume). As has been repeatedly affirmed (e.g., Marvin et al. 1977; Takahashi 1986; LeVine et al. 1994; Harwood et al. 1995; Hewlett and Lamb 2005; Lancy 2008; Quinn and Mageo 2013; Otto and Keller 2014), some of the original and core assumptions of attachment theory are not applicable to many cultures around the world. Still, there are issues that appear to result from disciplinary differences, and these await resolution.

Implications for policy and practice.   Over the past fifty years, attachment theory has permeated a wide range of professions that serve children and families, impacting policy at multiple levels. Although it may be perfectly appropriate to provide therapeutic interventions to infants and caregivers who are not adjusting well within their cultural setting, it is problematic to make diagnoses of pathology due to a lack of understanding of alternative cultural norms. As this volume illustrates, what qualifies as normal, atypical, abnormal, and/or pathological varies substantially across cultures and primate species. How, then, should application paradigms be altered to reflect current knowledge and culturally informed perspectives on attachment?

In Chapter 13, Suzanne Gaskins et al. examine how current understanding of the cultural nature of attachment can be integrated into policy and practice. They address the development of policy on multiple levels and the methods used to implement the results. They also discuss the process of translating research into policy and practice and propose an inclusive process that involves including all relevant stakeholders to minimize bias.

In Chapter 14, Mariano Rosabal-Coto et al. offer a critical appraisal of current applications and approaches to draw attention to the importance of program designs for the future. Because child-rearing practices vary across cultures, they stress that the value systems which motivate different practices must be recognized and accounted for when applications are developed and implemented. They issue a call for researchers to become proactive in rectifying misuses of attachment theory and in designing new applications that reflect cultural variation.

Further steps.   To change the prevailing paradigm of attachment—both in theory and practice—will not be easy. It requires viewing the phenomenon as an evolved universal developmental task: one that has to be solved in context-specific, culture-sensitive ways to have adaptive value. Because the concept of attachment impacts the lives of so many individuals, we view this not as an academic exercise but rather as a moral imperative.

We hope this volume will spur further discourse and guide future study based on truly universal principles: ones that are generalizable, valid, and reliable across all primates and all human cultures. Importantly, we hope that you will join us in furthering the understanding necessary to complete and apply a contextualized and culturally informed attachment theory.

Acknowledgments

We thank the Ernst Strüngmann Foundation for enabling this important discourse as well as the Frankfurt Institute for Advanced Studies and Deutsche Forschungsgemeinschaft for their additional support. We wish to extend a special word of gratitude to all of the contributors, for it is never easy to find time, among life's many responsibilities, to engage in such comprehensive discussions. This commitment, however, was crucial to further existing understanding into the foundations of young children's socioemotional development.

2

The Strange Situation of the Ethological Theory of Attachment

A Historical Perspective

Marga Vicedo

Abstract

This chapter examines the history of some challenges to John Bowlby's and Mary Ainsworth's ethological attachment theory (EAT). Bowlby and Ainsworth argued that the mother-infant relationship is a natural dyad designed by evolution in which the instinctual responses of one party activate instinctual responses in the other, and that secure attachment is an adaptation. This chapter focuses on EAT's two fundamental tenets: the universality of attachment patterns and the biological foundations of the attachment system. It shows that several scholars have challenged those tenets over the years and argues that attachment researchers have not addressed those challenges successfully.

Introduction

Commissioned by the World Health Organization (WHO) to write a report about the effects of maternal separation, British psychiatrist and psychoanalyst John Bowlby published *Maternal Care and Mental Health* (Bowlby 1951), followed two years later by its popular version, *Child Care and the Growth of Love* (Bowlby 1953). Bowlby argued that maternal care and love are essential for a child's psychological development. Although highly influential, Bowlby's conclusions also incited great controversy. In 1962, the WHO published a new report with the revealing title *Deprivation of Maternal Care: A Reassessment of Its Effects* (World Health Organization 1962). In this report a number of scholars criticized Bowlby mainly for unduly extrapolating from observations of sick children in hospitals and children in severely deprived conditions to infants growing up in standard circumstances. By that time, however, Bowlby had turned to biology to ground his views about the determinant role of maternal care.

In his 1958 paper, "The Nature of the Child's Tie to His Mother," Bowlby introduced his ethological attachment theory (EAT), which he later expanded in his 1969 book *Attachment*. According to this theory, natural selection designed a system to attach infants to their mothers, and a correct integration of this mother-infant dyad is necessary for a child's adequate emotional development. North American psychologist Mary Ainsworth and some of her students presented observational and experimental work in support of this theory (Ainsworth 1967; Ainsworth et al. 1978).

Since then, EAT has become one of the most prominent and influential theories of child development. It has informed a wide range of policies and practices in education, law, and child care (see Chapters 13 and 14, this volume). It has also expanded into a theory of personality, with scholars analyzing attachment issues in practically all of an individual's relationships.

Since its inception, EAT has also received serious criticisms. However, this important point about its contested status is little known because most historical presentations to date are celebratory accounts written by attachment theorists or supporters (Ainsworth and Bowlby 1991; Bretherton 1991, 1992; van der Horst 2011), or by others who rely on their accounts (Karen 1998).

My own historical examination of post-World War II American views about children's emotional needs has revealed that scholars in a variety of fields raised a series of powerful criticisms of attachment theory that have not been adequately addressed to this day (Vicedo 2013). According to Bowlby, the Austrian ethologist Konrad Lorenz's studies of imprinting, the experiments with rhesus monkeys conducted by the American psychologist Harry Harlow and British primatologist Robert Hinde, as well as the observations of children by British social worker James Robertson and North American psychologist Mary Ainsworth all supported his ethological attachment theory. But I have shown that this was not so (Vicedo 2013). Comparative psychologists and animal researchers convincingly disproved many of Lorenz's assertions about imprinting and successfully challenged his conception of instincts. Harlow emphasized the existence of different affectional systems and showed that the lack of maternal care alone did not cause irreversible effects. Hinde highlighted the need to understand mother-infant relations in the context of family and social relations in each primate group and showed that the effects of early separations per se do not determine the infant's future behavior. Robertson denied that his observations in hospitals supported Bowlby's views, and he opposed Bowlby's extrapolation from cases of long separations in hospitals to the context of everyday child-rearing practices. Several psychologists and psychoanalysts, including Anna Freud, also criticized Bowlby's use of data about children suffering from profound and extended deprivations to explain development in standard circumstances (for references, see Vicedo 2013). Each of these criticisms is significant on its own. Taken together, they amount to a serious challenge to EAT's standing as good science.

In my view, a combination of social factors and methodological/rhetorical strategies explains the great appeal of Bowlby's theory in the American context. In the United States, scientific and social interest in the development of emotions grew considerably after World War II. Specifically, the question of the effects of maternal separation on a child's emotional development acquired poignant relevance in discussions about the consequences of women's increasing presence in the workforce. In the context of Cold War debates about gender roles and working mothers, Bowlby's views helped justify the patriarchal family with its separate, clearly defined parental roles. In addition, Bowlby and Ainsworth used several strategies which helped boost their theory's popularity: they presented a united front despite their differences; they cited mostly work that supported their views; and they repeatedly claimed that biological science validated their theory.

My book, *The Nature and Nurture of Love* (Vicedo 2013), focuses on the use of ethological ideas to support attachment theory until the late 1970s. In this chapter, I examine another fertile source of criticisms not addressed in my book: the challenges to the theory from cross-cultural studies of childhood and child-rearing conducted by cultural anthropologists. This line of criticism goes back to the 1950s and has acquired impressive momentum since the 1990s. I will argue that cultural psychologists and anthropologists have presented a powerful challenge to the uniformity of attachment behaviors and the normativity of secure attachments. In addition, since attachment researchers appeal to the biological basis of attachment to defend the universality of attachment behaviors, I analyze EAT's evolutionary framework in Bowlby's work and in the work of his followers. I argue that, in fact, there is no adequate scientific foundation for that appeal.

Recently, social anthropologist Sara Harkness (2015:196) wrote:

> …attachment research at present finds itself in a strange situation, still committed to a vision of infant-caregiver relationships based on a (probably idealized) model of middle-class Anglo-American family life in the 1950s that does not correspond to the circumstances of care for most of the world's babies.

By exploring the challenges to EAT from cultural anthropology and psychology, and the shortcomings in EAT's evolutionary framework, I aim to illuminate in this chapter how EAT came to find itself in that strange situation.

Mead: Challenging the Focus on the Mother-Infant Dyad

The famous American anthropologist Margaret Mead launched the first challenge to Bowlby's universalistic claims. As I will show below, neither Bowlby nor Ainsworth took this challenge seriously.

As one of the first anthropologists to make childhood central to ethnographic studies, Mead became a leader of the Culture and Personality Studies movement. This movement, which reached its most fertile years in the 1930s and 1940s, called for an interdisciplinary effort to understand child-rearing and its role in different societies. Mead shared the conviction that childhood studies required input from diverse fields and areas of expertise (Mead 1955). Among the many efforts to move in this direction, the Six Cultures Study of Socialization (SCSS) deserves special mention. Initiated in 1954 by Harvard anthropologist John W. M. Whiting, with the collaboration of Yale psychologists Irvin L. Child and William W. Lambert, SCSS included studies carried out in Mexico, New England, the Philippines, India, and southwestern Kenya (Whiting and Whiting 1975; LeVine 2001, 2007, 2010).

Mead welcomed the post-World War II surge of interest in child development, but worried because social prescriptions and policies were being justified by ideas that lacked adequate support (Mead 1954:474). In a review of the work done from the mid-1940s to the mid-1950s, including Bowlby's influential 1951 report for the WHO, Mead pointed out two problematic trends: the inflated statements about the role of any single factor in child-rearing, and the "exaggerated and poorly supported claims of the importance of the mother as a single figure in the infant's life" (Mead 1954:476).

Like Bowlby, Mead was interested in the biological aspects of caretaking, but she was wary of possible misconceptions. She noted that Lorenz's and Tinbergen's ethological work opened the "possibility of tackling from a new point of view…the whole question of the instinctual elements in parent-child relationships" (Mead 1954:476). However, she also called attention to an important confusion:

> At present, the specific biological situation of the continuing relationship of the child to its biological mother and its need for care by human beings are being hopelessly confused in the growing insistence that child and biological mother, or mother surrogate, must never be separated, that all separation even for a few days is inevitably damaging, and that if long enough it does irreversible damage.

In her opinion, there was no anthropological evidence to support "the value of such an accentuation of *the tie between mother and child*." In fact, cross-cultural studies suggested that "adjustment is most facilitated if the child is cared for by many warm, friendly people" (Mead 1954:477, italics added for emphasis).

Bowlby did not view caretaking that way, as he made clear in the title of his 1958 paper "The Nature of the Child's Tie to His Mother." In this first presentation of EAT, he appealed to biology to defend the instinctual nature of the ties between mother and child. Further, he argued that without a proper functioning of those ties, children would develop emotional pathologies. Bowlby also defended the child's need for monotropy, the existence of a hierarchy in attachments, with the infant's main attachment centered on a single figure, usually the biological mother (Bowlby 1958). He did not include, however, information about mother-infant relations in different cultures.

In 1962, when the WHO published an evaluation of the conclusions of Bowlby's 1951 report, Mead noted again in her contribution the ethnocentric character of Bowlby's ideas. She reiterated her view that studies of other societies showed that a large number of nurturing figures could provide the security children need for healthy emotional development. She concluded that by giving the mores of his own culture the status of universal behavior and by positing a biological underpinning, Bowlby had committed the sin of "reification." He had taken "a set of ethnocentric observations on our own society, combined with assumptions of biological requirements which are incompatible with *Homo sapiens*," and turned them into "a set of universals" (Mead 1962:58). No doubt, these were serious charges.

Bowlby left the task of responding to his critics to Mary Ainsworth. An American-Canadian psychologist, Ainsworth had worked in Bowlby's group in London before moving to Uganda in 1954 and again, a year later, to the United States. At the time, she was planning to write a book with Bowlby and James Robertson. In writing her response for the WHO 1962 volume, Ainsworth consulted with Bowlby and followed his suggestions (Vicedo 2013).

In her response to Mead, Ainsworth claimed that Mead had misunderstood Bowlby. The view that Bowlby sponsored "an exclusive mother-child pair as the ideal" was a misunderstanding for several reasons:

1. Bowlby "argued for the desirability of a major mother-figure, not necessarily the biological mother, whose care is supplemented by other figures, including a father-figure."
2. Dispersion of maternal care was "not likely to be the norm in any primitive society."
3. "It seems entirely likely that the infant himself is innately monotropic."

So, "a situation (whether brought about by an 'experimental society' or through some individual variation in a traditional society) which impedes monotropic attachment will distort the normal course of development" (Ainsworth 1962:146–147). In this strange text, Ainsworth first denied that Bowlby defended monotropy, but then went on to argue that monotropy was probably the norm in primitive societies, that it was probably innate, and that departures from it would lead to pathology. Thus, after claiming that Mead's critique of Bowbly rested on the false assumption that he supported monotropy, Ainsworth literally presented a defense of monotropy rather than a rejection.

Ainsworth never addressed Mead's challenge seriously. The limited nature of Ainsworth's own research, which became a major reference point in the attachment literature, is telling. During her 1954–1955 stay in Uganda, Ainsworth conducted a study of the relationships between mothers and their infant children in that culture. She followed the first fifteen months of life of 28 babies during home visits to about 20 Ganda households, observing mother-infant interactions while interviewing the mothers with the help of an interpreter. This resulted in her 1967 book *Infancy in Uganda*. She focused on three basic aspects of an infant's behavior to assess attachment: the use of mother as a secure base for exploration; the infant's distress in brief separations from mother; and the infant's fear when encountering strangers. She classified the infants she had observed into three groups: secure-attached (16 children); insecure-attached (7 children); and non-attached (5 children). Ainsworth identified numerous factors that played a role in the development of infant-mother attachment, but she concluded that the determinant ones were the conduct and feelings of the mother, mainly her sensitivity in responding to her baby's signals (Ainsworth 1967:400). Although most of these children were taken care of by several caretakers, Ainsworth did not investigate this. Ainsworth's book was important in expanding the limited database upon which Bowlby was relying, but it did not provide an in-depth ethnographic analysis of Ganda's childcare practices. Though Ainsworth discussed Mead's views on multiple caregivers, she did not examine the cooperative caretaking of the Ganda nor did she explore the significance of the Ganda family practices in their socio-cultural context.

Bowlby did not address Mead's challenge either. In 1969, two years after the publication of Ainsworth's book, Bowlby published *Attachment* in which he mentioned the word "anthropological" only once and referred to Mead only once, in a footnote, and just to reiterate that she had misunderstood him (Bowlby 1969:303). Although he presented a theory of child development that aimed to have a universal character, Bowlby did not engage the existing anthropological literature on children from diverse cultures.[1] This did not escape attention. At least one reviewer found Bowlby's small data sample and selective referencing regarding other cultures problematic: "Of observations in non-European cultures, repeated use is made only of Ainsworth's research in Uganda" (Maas 1970:414–415).

The lack of serious engagement with Mead's points is surprising since her critique was directed at three fundamental levels: empirical, conceptual, and methodological. First, she did not think that Bowlby had sufficient evidence to claim that (a) maternal care and love were *sine qua non* for a child's emotional development and (b) separation from the mother would result in catastrophic

---

[1] He also made no mention of Mead in any of his other books. Failure to address Mead's critique has persisted for decades, in particular by sympathizers of Bowlby's views in their attempts to define the history of the field (Karen 1998; van der Horst 2011).

consequences for both the individual and the species. Second, she pointed out that Bowlby was mixing up two issues: child-rearing practices needed to support a specific sociocultural arrangement versus practices essential for human survival. Third, she challenged Bowlby's approach to studying children's basic emotional needs by focusing on those aspects of child-rearing practices that were important in his own society.

How was it possible to have ignored Mead's critique, given her intellectual stature in American social science? At least four factors played a role: First, in general, Bowlby did not address criticisms of his work, and neither did Ainsworth. Second, despite its influence in the postwar years, by the mid-1960s the Culture and Personality movement had practically disappeared. In addition, the lack of a unified theoretical framework and the problems of operationalizing research and translating between cultures made it difficult for the different ethnographic studies to cohere into a general account of child development (LeVine 2001). Third, the disciplinary goals of anthropology and psychology propelled studies of children in different directions. Much of psychological research has aimed to establish universal generalizations about the human mind, focusing on experimentation to reach knowledge of assumed universal capacities that explain human behavior. In contrast, anthropological studies of childhood have focused on understanding specific practices in a variety of cultures (Vicedo 2017). Fourth, ethology was successful in arguing that social behavior is a matter of instincts. In the mid- to late-1960s, biological explanations of human behavior held great appeal in the social sciences and the wider society (Degler 1991). Books such as Lorenz's *On Aggression* (Lorenz 1966) and Desmond Morris's *The Naked Ape* (Morris 1967) became instant bestsellers. In addition, other writers who were sympathetic to the aims of ethology published highly successful books on the animal roots of human conduct (Ardrey 1961, 1966; Tiger 1969). The search for the biological roots of human behavior also fueled interest in animal research. Lorenz's studies of imprinting in ducks and Harry Harlow's experiments with rhesus monkeys made it into *Life Magazine* and *The New York Times*.

Though Mead along with other anthropologists, such as Ashley Montagu (1968), remained critical of biological reductionism, biological explanations of human behavior enjoyed tremendous popularity. These conditions worked in favor of Bowlby's approach in presenting EAT as an account of universal affects and behaviors based on biological knowledge. This stood in stark contrast to anthropology's emphasis on the diversity of emotional experiences and on the complexity of studying emotions in different societies. The romance of child development with biology dates back to the origin of the field and is fraught with bitter disappointments (Morss 1990). In the wake of ethology's success and with advances in genetics and evolution grabbing the headlines, Bowlby's proposal gained momentum.

Bowlby: The Stone Age Child and the Paleo Mother

Given the centrality of Bowlby's claims about the evolutionary support for
EAT and the cursory manner in which they are often presented, it is worth examining Bowlby's main ideas in detail. My analysis helps to clarify Bowlby's
position on some of the points that have become controversial in the attachment literature. It also shows that Bowlby's presentation of the attachment
system as an adaptation was a hypothesis for which he did not provide sufficient evidence.

In his writings from the 1950s, Bowlby imported several concepts from the
discipline of ethology. Starting in the 1930s, the European animal researchers
Konrad Lorenz and Niko Tinbergen developed ethology as the biological study
of animal behavior and argued that much social behavior is instinctual. After
World War II, ethology as a field of scientific inquiry became highly successful. For their foundational contributions, Tinbergen and Lorenz would share the
Nobel Prize in Physiology or Medicine with Karl von Frisch in 1973. Bowlby
interacted with Lorenz and Tinbergen at several workshops and corresponded
with them regularly.

In 1958, Bowlby aimed to provide a synthesis of psychoanalytical and ethological ideas through his paper, "The Nature of the Child's Tie to His Mother."
He argued that infants are born with instinctual responses which "tie" them to
their mothers; the behavior of mothers, in turn, is preprogrammed to connect
them to their infants. Five instinctual behaviors help the child create this bond
with the mother: sucking, clinging, following, crying, and smiling. Though
Bowlby did not specify which maternal behaviors are instinctual, he claimed
that the proper functioning of the mother-infant tie is necessary for the adequate emotional and psychological development of the infant.

In support of his views, Bowlby appealed to the authority of biology: "The
theory of Component Instinctual Responses, it is claimed, is rooted firmly in
biological theory and requires no dynamic which is not plainly explicable in
terms of the survival of the species" (Bowlby 1958:369). Bowlby emphasized
that he was rejecting the psychoanalytic concept of instinct as energy that needs
to be released and was adopting in its place the ethological concept of instinct.
Lorenz had defined instincts as species-specific patterns of behavior that are
innate, are the result of natural selection, and are impervious to experience.
For Bowlby, the behaviors of infant toward mother and mother toward child
are instances of instinctual behavior. Studying the literature on animal behavior, especially primatology, he developed these ideas further over the next ten
years and presented them in his 1969 book *Attachment* (Bowlby 1969).

Attachment was the first volume of his influential trilogy on the effects of
maternal deprivation (the three titles were *Attachment, Separation,* and *Loss*),
and the major work that delineated his ethological theory of attachment. After
a brief introductory discussion about his "point of view" (i.e., psychoanalysis
informed by ethology), Bowlby presented the "observations to be explained"

(i.e., the data that his theory needed to explain). He listed the work of Dorothy Burlingham and Anna Freud at the Hampstead Nurseries in London, who showed that children separated from their families during World War II often regressed in their development. He also listed the work of René Spitz and Katherine Wolf, which looked at the detrimental effects of multiple caretakers for children who lived in institutions. He relied especially on the observations provided by James Robertson (Bowlby 1982:26). Robertson had documented the profound distress experienced by children separated from their families during lengthy hospital stays. It is, however, important to note that by 1969, Spitz's work had been thoroughly discredited (for a devastating critique of the work on maternal deprivation that Bowlby relied upon in his 1951 WHO report, see Pinneau 1955; Casler 1961), and Anna Freud had criticized Bowlby's extrapolations of the experiences of children during the war to the home setting in normal circumstances (Freud 1960).

The book's second part is devoted to the analysis of "instinctive behavior." Bowlby considered that some patterns of human behavior, including the care of babies and the attachment of young to parents, "seem best considered as expressions of some common plan and, since they are of obvious survival value, as instances of instinctive behaviour" (Bowlby 1969:39).

In *Attachment*, Bowlby revised his earlier views about instincts, reflecting the influence of Tinbergen and Hinde. Over the years, ethologists had modified their goals and main ideas as a result of criticisms from other researchers (Burkhardt 2005; Vicedo 2013). In 1953, American comparative psychologist Daniel Lehrman criticized Lorenz for not recognizing the significant role of development in behavior and for wrongly assuming that the inheritance of a trait implies its developmental fixity (Lehrman 1953). Hinde, a close friend of Lehrman, also criticized the ethological concept of drive (Hinde 1956). Bowlby now considered instinctive behavior to be "the result of integrated control systems operating within a certain kind of environment" (Bowlby 1969:44). He emphasized that instinctive behavior is not inherited; rather "what is inherited is a potential to develop certain sorts of system, termed here behavioural systems," that are then shaped by the particular environment of development (Bowlby 1969:45). But there are limits. And those limits are essential to an understanding of the biological and psychological consequences of changes in the environment (Bowlby 1969:46):

> The recognition that behavioural equipment, like anatomical and physiological equipment, can contribute to survival and propagation only when it develops and operates within an environment that falls within prescribed limits is crucial to an understanding of both instinctive behaviour and psychopathology.

For Bowlby, there is always an environment to which a system, man-made or biological, is adapted. He called this the system's "environment of adaptedness" (Bowlby 1969:47). In a man-made system, one would refer to the system's "environment of designed adaptedness." In living organisms, one should talk about the "environment of evolutionary adaptedness" (EEA). According to Bowlby, the EEA is different for each system in each species. To avoid confusion from the polysemy of the term "adaptation," he proposed using this word when referring to the process and "adaptedness" when referring to the state or condition of being adapted (Bowlby 1969:51).

Bowlby also asserted that "only within its environment of adaptedness can it be expected that a system will work efficiently" (Bowlby 1969:47). To explain this, he put forth an analogy with man-made systems. These work best in the "environment" for which they were designed. For example, a small car built to run on English roads will work best there. "Whether or not it will also suit other environments, however, is unknown," Bowlby wrote. Given the difficulty of knowing whether the car would work in the Arctic Circle or the Sahara, for example, he noted that "until it is shown to be more extended, it is wise to assume that the car's environment of adaptedness is limited to London streets" (Bowlby 1969:52).

Focusing then on humans, Bowlby claimed that most human behaviors are adaptations resulting from selection in "man's environment of evolutionary adaptedness" (Bowlby 1969:58). He argued that the rate of man-made environmental change had "far outstripped the pace at which natural selection is able to work." So, he noted, one could be "fairly sure that none of the environments in which civilised, or even half-civilised, man lives today conforms to the environment in which man's environmentally stable behavioural systems were evolved and to which they are intrinsically adapted" (Bowlby 1969:59). Bowlby concluded that "the environment in terms of which the adaptedness of man's instinctive equipment must be considered is the one that man inhabited for two million years until changes of the past few thousand years led to the extraordinary variety of habitats he occupies today." According to Bowlby, this original environment "presented the difficulties and hazards that acted as selective agents during the evolution of the behavioral equipment that still is man's today. This means that man's primeval environment is, almost certainly, also his environment of evolutionary adaptedness." Further, he claimed that if his conclusion was correct, "*the only relevant criterion by which to consider the natural adaptedness of any particular part of present-day man's behavioral equipment is the degree to which and the way in which it might contribute to population survival in man's primeval environment*" (Bowlby 1969:59). In sum, human behavioral systems were "designed" in the Pleistocene by natural selection and could only be understood by knowing their function in that ancestral environment.

The question naturally follows: Does the human behavioral equipment, "designed" by natural selection in the EEA, still work in today's environment? Bowlby said that this could not be known without empirical research. Furthermore, he claimed (Bowlby 1969:60–61) that he was not concerned with such questions:

...enormously important though they are, all questions as to whether man's present behavioural equipment is adapted to his many present-day environments, especially urban environments, are not strictly relevant to this book, which is concerned only with elemental responses originating in bygone times. What matters here is that, if man's behavioural equipment is indeed adapted to the primeval environment in which man once lived, it is only by reference to that environment that its structure can be understood....It is impossible to understand man's instinctive behaviour until we know something of the environment in which it evolved.

This is a surprising statement, given that Bowlby presented his book as an exploration of instinctive attachment behavior and its consequences for mental health.

Having established this general framework for understanding the human mind, Bowlby also translated his earlier 1958 paper into the new language of instinctual control systems. He proposed that babies and mothers are born with a repertoire of attachment behaviors that evolved through natural selection in the EEA. The infant's attachment system comprises those behaviors that lead the infant to seek proximity to the mother—behaviors that were selected by evolution because they enhanced the infant's protection from predators. In turn, the behavior of parents "that is reciprocal" to infants' attachment behavior conforms the caregiving system (Bowlby 1969:182). As noted above, in his 1958 paper Bowlby had not specified which maternal behaviors are instinctual. In 1969, his account of maternal caregiving was also underdeveloped. Even chapter thirteen, where Bowlby covered caregiving behavior, was devoted primarily to the child's behavior. The mother is portrayed as the one reacting or responding to the child's proximity-seeking actions: signaling behavior (crying, smiling, babbling) or approach behavior (approaching and following).

Strongly influenced by Lorenz's studies of imprinting, Bowlby viewed the tie between the mother and infant as an instance of that phenomenon. Imprinting is the process whereby the newborns in some species (e.g., ducks and geese) follow the first object they see upon hatching. Normally, that object would be their mother. According to Lorenz, if newborns do not follow their mother, they will not develop the social and sexual responses typical of their species in adulthood. For Lorenz, imprinting occurs during a critical period and has determinant effects on the social behavior of the animal. Other animal researchers, however, showed that imprinting is a more flexible phenomenon and that its effects can be reversed (Bateson 1966). In *Attachment*, Bowlby accepted some of the criticisms made of Lorenz's work and defended a more generic concept of imprinting. For him, this meant "the development of a clearly defined preference," which develops "fairly quickly...during a limited phase" and then remains "comparatively fixed" (Bowlby 1969:168). He asserted that the attachment between infant and mother is sufficiently similar to the process Lorenz described to be conceptualized as imprinting (Bowlby 1969:223).

In short, for Bowlby, the infant's attachment system and the caregiving system were configured by natural selection in the EEA. The child becomes imprinted on the mother. The mother's appropriate responses to the child during a crucial period of development are necessary for the survival and proper social development of the child.

But what would happen if there was a deviation from the environment to which those systems were adapted? As noted earlier, Bowlby claimed that it is not possible to know in advance how a system would perform outside its environment of adaptedness. Now, however, he asserted that a change in environment would likely imply a malfunctioning of the system. Bowlby (1969:166) was clear about this:

> …it is wise to be cautious and to assume that the more the social environment in which a human child is reared deviates from the environment of evolutionary adaptedness (which is probably father, mother, and siblings in a social environment comprising grandparents and a limited number of other known families), the greater will be the risk of his developing maladaptive patterns of social behaviour.

What constituted these "maladaptive" patterns of social behavior? For Bowlby, these were behaviors that do not promote survival and reproduction. Problems of attachment put at risk the survival of the species, although as long as the system fulfills its functions in enough individuals within a population, the species would likely survive. In addition, he suggested that at the level of individual mental health, problems of attachment would lead to individual psychopathology.

The question naturally arose: Is having an attachment to a person other than the biological mother a deviation that would lead to "maladaptive" patterns of behavior? That is, does a child need his/her biological mother (see Keller and Chaudhary, this volume)? As presented above, the question of whether Bowlby believed that a child attaches only to one person, and the question of whether that person is or should be the biological mother, became highly controversial. They remain so today. Let's address them.

First, regarding monotropy, Bowlby believed that although an infant can attach to more than one person, there is always one principal attachment figure. As he put it (Bowlby 1969:309):

> Because the bias of a child to attach himself especially to one figure seems to be well established and also to have far-reaching implications for psychopathology, I believe it merits a special term. In the earlier paper I referred to it as "monotropy."

In fact, in his original definition, monotropy was "the tendency for instinctual responses to be directed toward a particular individual or group of individuals and not promiscuously toward many" (Bowlby 1958:370). Clearly that definition was self-contradictory and needed clarification. So, in 1969, Bowlby made clear that he believed the child was biased to attach mainly to one person.

Second, as for the main figure, in many, though not all, of his writings, Bowlby included a footnote stating that the word "mother" was to be understood as "mother-figure"; that is, the person who provided care for the infant. In *Attachment*, Bowlby (1969:304) asked: "Can a woman other than a child's natural mother fill adequately the role of principal attachment-figure?" He answered in the affirmative, but qualified this as follows (Bowlby 1969:306):

> Though there can be no doubt that a substitute mother can behave in a completely mothering way to a child, and that many do so, it may well be less easy for a substitute mother than for a natural mother to do so. For example, knowledge of what elicits mothering behaviour in other species suggests that hormonal levels following parturition and stimuli emanating from the newborn baby himself may both be of great importance. If this is so for human mothers also, a substitute mother must be at a disadvantage compared with a natural mother. On the one hand, a substitute cannot be exposed to the same hormonal levels as the natural mother; on the other, a substitute may have little or nothing to do with the baby to be mothered until he is weeks or months old. In consequence of these limitations, a substitute's mothering responses may well be less strong and less consistently elicited than those of a natural mother.

In brief, Bowlby believed that biology had prepared the biological mother for the caregiver role. Thus, she would be the most adequate main attachment figure for her child.

Bowlby indeed saw the mother-child pair as the ideal. He considered monotropy a proven fact of nature, he considered the mother the natural attachment figure, and his work focused on the effects of deprivation or separation from the biological mother. Further, he believed that attachment problems resulted from lack of mothering, not from lack of caregiving (Bowlby 1969:357):

> Disturbances of attachment behavior are of many kinds. In the Western world much the commonest, in my view, are the results of too little mothering, or of mothering coming from a succession of different people.

He also condemned the existence of childcare centers on several occasions.

Last, but not least, Bowlby's specific use of words made his defense of the biological mother's central role clear. There is no reason to say "mother" if you mean "caretaker" or to refer to "mothering" if you mean "caregiving." There is no reason to say mother, but add a footnote to state that what you mean is "mother-figure." In the period during which he wrote, Bowlby was aware that most people were interpreting his writings to mean mother. He often cited authors who believed that lack of maternal love pushed their children into deep pathologies. For example, to support his views he cited the work of Bruno Bettelheim and Margaret Mahler, both of whom were well known for defending the idea that rejecting mothers cause their children's autism (Bowlby 1969:345–346; 1982:346–347).

It is also telling that Bowlby did not revise his views about the importance of the mother when he published a second edition of *Attachment* in 1982. He still presented the same "observations to be explained" that he had included in

1969, though by then most of those studies had been thoroughly criticized, as I pointed out above. By 1982, even Robertson had rejected in print the view that his observations of children in hospitals could support the generalizations made by Bowlby (Robertson and Robertson 1971).

In the 1982 edition, Bowlby mainly updated some assertions about primatology and biology, adopting ideas that further restricted who could be an attachment figure. He said one should not talk about the welfare of the "species," as he had done following Lorenz. Instead, he now referred to the genetic theory of natural selection as described by George C. Williams and popularized by Richard Dawkins and E. O. Wilson. Neo-Darwinian theory took "the individual gene" as the central unit of selection in evolution (Bowlby 1982:55). Adopting this view, Bowlby noted that "the ultimate outcome to be attained is always the survival of the genes an individual is carrying" (Bowlby 1982:56). In the gene-centric model proposed by Bowlby, only the biological mother or another kin member could have been selected by natural selection to assume a caretaker role in the EEA.

It must be made clear that the presentation of the human attachment system as an adaptation was a hypothesis advanced by Bowlby, but he did not present the necessary evidence to sustain that claim. First, most of the empirical work discussed in *Attachment* is about nonhuman animals. Second, in the neo-Darwinian, gene-centric evolutionary model, natural selection always selects among alternatives and can only lead to evolutionary change if a behavior has a genetic basis. Thus, to show that the secure attachment system (proximity-seeking infant and sensitive-responding mother) is in fact an adaptation, Bowlby would have needed to demonstrate:

- that there are genes for proximity-seeking behavior,
- that natural selection favored babies with genes for proximity seeking over babies without them,
- that sensitive mothering is genetically based, and
- that sensitive mothering provided higher fitness for mothers than its alternatives.

Bowlby, however, lacked evidence for any of these basic points. In fact, to date there are no studies which show the existence of genes for proximity-seeking and sensitive-mothering behaviors. In addition, we do not know how to assign fitness values to those behaviors. It is thus not possible to say that they are adaptations. I will return to these points later.

Given that Bowlby was presenting the evolutionary framework as a hypothesis, his ethological theory of attachment might have remained merely a grand theoretical scheme. Thanks to Ainsworth's research and the data generated by many scholars using the laboratory procedure she introduced to test attachment, that empirical foundation became much larger. Whether it is strong enough to support the theory in a convincing manner is, however, another issue.

Ainsworth: The Experimental Support from the Strange Situation

After leaving Uganda, Ainsworth obtained a job in the United States at Johns Hopkins University. In the mid-1960s, she attempted to replicate her Ganda study in Baltimore, Maryland. Ainsworth and three assistants observed 26 infants from white middle-class families in their homes. They took notes on mother-infant interactions at three-week intervals from 3–54 weeks of age. In Baltimore, however, Ainsworth did not observe the behaviors that she had used to assess attachment in Uganda. Thus, in 1964 she devised a procedure to elicit those behaviors: the Strange Situation.

The goal of the Strange Situation, a twenty-minute laboratory procedure, was to assess (a) whether the infants used the mother as a secure base to explore, (b) how they reacted when their mothers left them in an unfamiliar environment with a stranger, and (c) how they behaved when their mothers returned. It was a laboratory procedure that involved seven episodes of about three minutes each in the following sequence:

- The child and mother are alone.
- A female stranger enters the room and engages the infant.
- The mother leaves; the mother returns and the stranger leaves.
- The mother leaves again; the stranger returns.
- The mother returns and the stranger leaves.

Twenty-three of the infants from the Baltimore study (ranging in age from 9–24 months) were brought to the lab in their fifty-first week of study. Depending on the child's behavior, the children were divided in three main groups and eight subgroups.

In *Patterns of Attachment*, Ainsworth et al. (1978) presented the Strange Situation as a reliable test to categorize not only behaviors, but also children and their mothers. Depending mainly on the behavior of the infants upon their mothers' return, infants were classified into different categories: 65% as securely attached (B), those children who greeted the mother and sought contact with her; 20% as anxious-avoidant (A), those children who appeared to reject the mother upon her return; and 13% as anxious-resistant or ambivalent (C), those children who displayed both attachment behavior but also signs of being upset with the mother. In addition, the children's reactions were taken to be an index of the quality of their attachment to their mothers (Ainsworth et al. 1978:xi). According to Ainsworth et al. (1978:144–146), the securely attached children experienced sensitive maternal care; insecurely attached children had mothers who were inconsistent, rejecting, or unresponsive.[2]

Clearly, the maternal style is not observed in the Strange Situation since the mother only enters and leaves the room following a set script. In the original

---

[2]   Later, Main and Solomon (1990) introduced another category: disorganized (D) for children who did not fit any of the previous three categories.

study, the correlation between the child's behavior and the mother's behavior was established from the analysis of the observations of the mothers at home. Designed as a complement to the Baltimore study, the interpretation of the Strange Situation is tied to the results of that study. As Ainsworth et al. (1978:321) clearly stated:

> We must emphasize that individual differences in strange-situation behavior would have been well-nigh uninterpretable without extensive data about correlated individual differences in other situations, and especially without the naturalistic data that we collected in regard to Sample 1 at home throughout the first year of life.

In the Baltimore study, maternal behavior had been categorized along four dimensions: acceptance-rejection, cooperation-interference, accessibility-ignoring, and sensitivity-insensitivity. The analysis showed that infants of group B had the most sensitive mothers, while mothers in group A were more rejecting than those of infants in group C.

Despite the small sample and the fact that this study could at most establish a correlation between maternal care and child's behavior, Ainsworth presented it as showing a causal relation. Thus, the quality of the care provided by the mother was simply inferred from the child's behavior observed in the lab. Since the authors claimed that maternal sensitivity was the main factor in shaping the attachment, the categorization of children automatically entailed a categorization of the mothers. As a way of identifying sensitive mothers, this categorization also implied a moral judgment about mothers (LeVine and Norman 2001).

Furthermore, *Patterns of Attachment* presented the results noted above as proof of Bowlby's ethological attachment theory and its prescriptive implications. Ainsworth et al. (1978:95) argued that these were "normative findings" depicting "certain features of the species-characteristic organization of attachment behavior in the human 1-year-old and its interplay with other behavioral systems." They felt that these results confirmed Bowlby's theory: "We consider that the normative findings substantially support Bowlby's (1969, 1973) descriptions of the organization and function of infant attachment behavior" (Ainsworth et al. 1978:95). However, they did not explain how the behavior of children in a laboratory setting could provide empirical backing for Bowlby's claim about attachment being an adaptation, which would have required evidence about the course of historical developments many thousands of years ago.

Ainsworth had earlier followed Bowlby in adopting an evolutionary explanation for attachment and in viewing the biological mother as the primary attachment figure. In *Infancy in Uganda*, Ainsworth acknowledged the "new ethological view of instinct" as a main influence on her perspective (Ainsworth 1967:432). The same year that Bowlby published *Attachment*, Ainsworth endorsed Bowlby's position (Ainsworth 1969). Here, she addressed the questions regarding the deviations from the behavioral system constructed in the EEA and the role of the biological mother. Given that these two issues have been very controversial in the attachment literature to this day, it is worth citing Ainsworth at length.

Ainsworth (1969:1003) noted that Bowlby's conception of attachment took into account "both (a) the intraspecies uniformities in attachment and its development and (b) the deviations which are nonadaptive and which form the basis for a variety of pathologies." As she had stated earlier (Ainsworth 1969:1000):

> The function of a system is the one that gave it species-survival advantage in the "environment of evolutionary adaptedness"—the original environment in which the species first emerged. The biological function of the behavior may or may not give special advantage in one or another of the various environments in which present-day man lives, but this is a quite distinct consideration. Genetic programming continues to bias the infant to behave in ways adapted to the original environment of evolutionary adaptedness, and, similarly, under all the layers of individual learning and cultural acquisition, there is still a bias for mothers to behave reciprocally—a bias which may have been more or less sharpened or blunted by learning in any individual mother.

In this passage Ainsworth claimed that because attachment was an adaptation in the EEA, the infant is still genetically inclined to attach to its mother, and the mother is genetically inclined to attach to her infant. Therefore, for Ainsworth, the mother is the ideal attachment figure for the infant. Indeed, regarding the role of mothers, she claimed (Ainsworth 1969:995–996) that:

> …ethologists hold that those aspects of the genetic code which regulate the development of attachment of infant to mother are adapted to an environment in which it is a well-nigh universal experience that it is the mother (rather than some biologically inappropriate object) who will be present under conditions which facilitate the infant's becoming attached to her.

It is, however, unclear to whom she was referring, since she did not provide a citation to any work by ethologists.

In later years, Ainsworth continued to appeal to biology in order to support her views, but she only provided a handful of references to the biological foundations of attachment. Thus, in 1970, Ainsworth and Bell claimed that because of the risks involved in the helpless infancy of the human species and its need for protection, "it is inferred, therefore, that the genetic code makes provision for infant behaviors which have the usual (although not necessarily invariable) outcome of bringing infant and mother together" (Ainsworth and Bell 1970:51). However, from the fact that a certain feature would increase the fitness of an organism, we cannot infer that the genetic code made a "provision" for such a feature.

Finally, as mentioned above, Ainsworth presented her work on the Strange Situation as a contribution to Bowlby's evolutionary model of attachment. Thus, her 1978 book with Blehar, Waters, and Wall about patterns of attachment presented their work as throwing "important light on the concept of infant-mother attachment as viewed from an evolutionary-ethological standpoint" (Ainsworth et al. 1978:322), although what little they said about evolution is only a summary of Bowlby's views. They referred to the survival uses of protection and how attachment patterns are consistent with them. In a contemporary paper, Ainsworth (1979:37) claimed that "the behavior of the securely attached infant and his responsive mother...may be recognized as the expected evolutionary outcome of infant attachment and attachment behavior and of a reciprocal maternal behavior system which are preadapted to each other." In addition, "to the extent that the present environment of rearing departs from the environment to which a baby's behavior is preadapted, behavioral anomalies may be expected to occur" (Ainsworth 1979:5; for the same point, see Ainsworth et al. 1978:9).

Although these are rather confusing statements, they make one thing clear: if Ainsworth believed that the infant's attachment system and the maternal behavior system were "preadapted" to each other, then the ideal attachment figure for a child must be its mother. For all the talk in the works by Ainsworth and by Bowlby about attachment "figures," their emphasis on the genetic basis for these systems, the view that they are adaptations, and the notion that deviations from the EEA produced pathologies, all lead inevitably to seeing the mother-infant dyad as the ideal. Thus, Ainsworth's work did not help overcome the monotropic, mother-based positions of Bowlby.

Nevertheless, Ainsworth's twenty-minute procedure to categorize infants and mothers undoubtedly became central to the successful spread of Bowlby's views. In the context of post-World War II views of science, the operationalization of attachment research via the Strange Situation played a key role in the rise and expansion of attachment theory. Having an easy, short, and cheap laboratory tool to study attachment was crucial to EAT's enormous appeal in American psychology. As historians of science have documented, after World War II, psychology and the social sciences turned increasingly toward a model of science that aimed to emulate the natural sciences in their reliance on experiments, reproducibility of results, and search for universal generalizations (Herman 1995; Solovey 2013). The Strange Situation fit well within that model: it was a laboratory procedure that promised causal explanations of behavior and also afforded predictions. In addition, it was presented as confirming a theory that made universal claims about emotions backed by biological science. Thus, it was in line with the methodological and epistemological goals of mainstream postwar psychology.

However, serious problems remained. Not only had Mead's powerful challenge gone unanswered, but some developmental psychologists in the 1970s were calling attention to the reductionistic stance implicit in this vision of science, children, and mothers. In addition, Ainsworth's adoption of Bowlby's evolutionary framework to interpret her results encouraged criticisms of the uses of evolutionary ideas to support specific views on attachment.

Critical Assessments of the Strange Situation and the Ethological Attachment Theory: From the Child in the Lab to General Views about Child Development and Human Nature

Despite the key role of the Strange Situation in leading to an explosion of work on attachment, the procedure encountered serious criticisms. Several major developmental psychologists challenged the simplistic assumptions in this type of laboratory work. Others criticized specific implementations of it.

By the late 1970s, some child psychologists criticized the increasing reliance on simple experiments to understand child development. At Cornell, Urie Bronfenbrenner (1977:513) put it with graphic clarity: "Much of contemporary developmental psychology is *the science of the strange behavior of children in strange situations with strange adults for the briefest possible periods of time*."

So did Yale researcher William Kessen. In a 1978 address to the developmental psychology division of the American Psychological Association entitled "The American child and other cultural inventions," Kessen found it problematic to believe in a "free-standing" and "self-contained" child with its instincts, traits, and attachments as essential components of its nature. According to him, this vision of children also made it possible for psychologists to take the child as the unit of study rather than to consider the child as part of a more complex system of relations and network of influences. The conception of the "isolable" child had important and troubling consequences: "basically, we have observed those parts of development that the child could readily transport to our laboratories or to our testing sites" (Kessen 1979:819). This atomistic view of children led psychologists to ignore the *role of context* in child development.

Some scholars proposed an integration of experimental work in psychology with ethnographic research in order to get a fuller and more realistic view of child development. For example, developmental psychologists Charles M. Super and Sara Harkness put forward the concept of the "developmental niche" to refer to an individual's experience of culture at any developmental stage (Super and Harkness 1986). Michael Cole and his colleagues at the UCSD Laboratory of Comparative Human Cognition argued for the significance of the local interpersonal context to understand cognitive development (Cole 1985).

In addition, an exhaustive and careful review of Ainsworth's implementation of the Strange Situation Procedure led to substantial criticisms. In 1984, Michael E. Lamb, Ross A. Thompson, William P. Gardner, Eric L. Charnov, and David Estes presented a detailed analysis of the Strange Situation (Lamb et al. 1984a, b) which Lamb, Thompson, Gardner, Charnov, and Connell extended into book form a year later (Lamb et al. 1985). They found serious methodological problems with the procedure: the data were not sufficiently reliable and the samples were too small to support the generalizations put forth by Ainsworth's group. Furthermore, "observer reliability was never assessed in the homes and was inadequately assessed in the Strange Situation" (Lamb et al. 1985:65). Their review was devastating, leading to the conclusion that the results from the Baltimore study should "be viewed with great caution" (Lamb et al. 1984b:131).

Furthermore, the biological basis of attachment theory came under trenchant criticism in the early 1980s, soon after the publication of *Patterns of Attachment*. Hinde penned the first critical examination of EAT's biological claims. In 1982, he expressed concern about the appeal to natural selection in order to argue that "any one mothering style is necessarily best" (Hinde 1982:72). Hinde pointed out that it was unlikely that natural selection would have produced stereotypy, given that the optimal mothering behavior varies with a number of factors. He suggested that it was more likely that selection would have favored conditional maternal strategies (Hinde 1982:71). Finally, he also questioned "the assumption that infant behavior and maternal behavior are adapted to mesh with each other" (Hinde 1982:73).

Lamb and colleagues also provided a critical analysis of the use of biology in attachment theory (Lamb et al. 1984a, b, 1985). First, they rejected the adequacy of the imprinting model for conceptualizing an infant's attachment to an adult, since imprinting only operates in some species. In addition, they noted that there was no evidence for the existence of critical or sensitive periods in human social development (Lamb et al. 1985:24).

They also criticized how Ainsworth et al. (1978) interpreted the different patterns found in the Strange Situation by reference to Bowlby's evolutionary framework. The most common pattern of behavior, B, which involved seeking proximity and contact with the mother was labeled "secure attachment" because it was the pattern that Bowlby argued would have led to higher fitness in the EEA. The behavioral patterns of children in the A and C categories were considered "maladaptive." Lamb and coauthors agreed with Hinde's point that it was unlikely that natural selection had favored a single parental strategy, as behavioral ecologists supported the view that behavioral strategies are often conditional and frequency dependent. They thus challenged the "ideal single pattern" model underlying EAT and, specifically, the view that secure attachment, as evaluated by the Strange Situation, is adaptive, while other types of attachment behaviors are maladaptive (Lamb et al. 1985:52).

Lamb and his coauthors raised additional problems with the biological claims invoked to support EAT as well. They questioned the assumption that human parents and infants had evolved to form intimate, harmonious relationships, especially given the work of Robert Trivers on parent-offspring conflict (Trivers 1974). They called for caution in using the EEA as a heuristic tool since any hypothesis based on assumptions about it are "inevitably speculative" (Lamb et al. 1985:49). They also noted that unless researchers were able to establish the fitness consequences of the behavioral patterns observed in the Strange Situation, speculations about the evolutionary meaning of different patterns of infantile behavior were "of limited value" (Lamb et al. 1984a:164).

In addition, they called for clarification of the different meanings of the term "adaptation" as used in the attachment literature. Sometimes it was used to refer to developmental mental health and sometimes to biological fitness, and there was no reason to assume that these two were related. Lamb et al. noted as well that it was unclear "whether 'adaptive' attachment behavior" was thought to "bring fitness advantages to infants in contemporary times" (Lamb et al. 1984b:143). Again, what was adaptive in the EEA might not be biologically adaptive in the present, and might be of no consequence for contemporary psychological well-being either. Moreover, they pointed out that the conflation between the biological and psychological senses of adaptive was biasing the interpretation of empirical research. In their words: "Researchers… have tended to look for associations between B group behavior and a broad range of optimal outcomes, and links between A- or C-group behavior with more negative or maladaptive outcomes" (Lamb et al. 1985:55).

These assessments had far-reaching implications for the status of EAT and the Strange Situation. The charges were numerous and raised severe doubts about the core of the theoretical and empirical foundations of attachment studies. However, attachment researchers failed to respond to them. Lamb had been a student with Ainsworth, and she took his criticisms as a betrayal of the cause of attachment research (Ainsworth 1998; Karen 1998:265). As the lead author of the critical publications, he took the brunt of the unfavorable replies from insiders. Although Lamb went on to a brilliant career at the University of Cambridge, many attachment researchers effectively ostracized him.

In subsequent years the challenges to the foundations of attachment studies continued to mount. In the early 1990s, Robert Hinde and Joan Stevenson-Hinde questioned other claims about the biological foundations of attachment. They criticized the extrapolations from suppositions about the EEA to the view that secure attachment behavior is normal and, therefore, desirable. The title of their paper, "Attachment: Biological, Cultural and Individual Desiderata" highlighted the need to separate different claims made in the attachment literature, which often confused these levels. At the individual level, what matters is psychological well-being; at the biological level it is maximizing an individual's inclusive fitness; and at the cultural level it is the goals and norms of a society. They called for clarification of the claim that secure attachment is best (Hinde and Stevenson-Hinde 1990).

Despite the fact that Bowlby always appealed to Hinde's work to support his position, how far Hinde's views were from Bowlby's can be clearly appreciated in a paper that Hinde prepared as a tribute to Bowlby. Here, Hinde emphasized first that his research with rhesus monkey troops had shown that the effects of maternal separation depend on "diverse factors that interact in a complex way" (Hinde 1991:157). He saw those results as agreeing with what other researchers had shown in macaques (Stephen Suomi) and in human infants (Jim and Joyce Robertson, Michael Rutter). Thus, rather than focus on dyads, he stressed "the importance of seeing the individual as set within a network of relationships" (Hinde 1991:158). Further, he noted, in humans one needed to add the effect of "the sociocultural structure" (Hinde 1991:158).

Summarizing points he had made in earlier writings, Hinde strongly objected to the interpretation that the secure attachments identified in the Strange Situation implied that "what is natural is best, and that it is natural for children to have sensitive, responsive mothers" (Hinde 1991:160). After enumerating many of the points noted earlier, he concluded again that "the desideratum of 'psychological health' may depart from those of either biology or culture or both" (Hinde 1991:161). Noting that this was perhaps an issue "on which John Bowlby and I would not have seen exactly eye-to-eye," he concluded that he was "hesitant about describing a particular type of child-mother relationship as optimal." Condensed among nice words of praise for Bowlby's leadership and gratefulness for how much he owed him, Hinde's paper presented a group of key points against EAT.

Still, neither criticisms about the implementation of the Strange Situation nor criticisms of EAT's evolutionary claims had a serious impact on attachment research. As had already happened with Mead's earlier challenge to Bowlby, attachment scholars forged ahead by ignoring the mounting criticisms. In subsequent years, this field grew mostly by focusing on the Strange Situation Procedure. In fact, both supporters (Bretherton 1991:25) and critics (Quinn and Mageo 2013:4) have noted how the explosion of work using the Strange Situation has led to a confounding of attachment theory with the instrument. By facilitating the training of graduate students and the establishment of research networks, the Strange Situation led to an explosion of literature on attachment.

However, the interpretation of the role of maternal sensitivity in attachment that is at the center of many studies using the Strange Situation Procedure rests on very shaky foundations. The Strange Situation can be used for many purposes, but its use to infer that children's behavior is mainly due to the relationship with their mothers derives from a flawed study. Despite Lamb et al.'s criticisms, attachment theorists continue to present the Baltimore study as a model of careful research, asserting that the observers took running notes scored by time markers every five minutes, and translated their notes into audiorecorded narratives immediately after each visit (Bretherton 2013:462). However, my review of the archived original data in the summer of 2015 revealed that this was not the case. Although confidentiality prevents quoting directly from these data, the narrative reports from these observations that I have seen cannot be considered trustworthy scientific reports. Several of them are permeated with subjective evaluations of the mothers' personality from day one, including moral judgments. Other reports reveal tensions between the observer and the observed mothers. In addition, the reports from the different observers vary substantially in nature and quality, and most of them do not include notes taken every five minutes. In fact, one observer did not write up the observations until months later. Since these observations were key to Ainsworth's interpretation of children's behavior in the Strange Situation, the quality of their observations calls into question the validity of the Baltimore study and throws doubt on the links it supposedly helped to establish between the children's behavior and the conduct of mothers.

Attachment researchers also continued to assert that biological science supported EAT. In a 1991 paper, Ainsworth (1991:33) wrote: "The great strength of attachment theory in guiding research is that it focuses on a basic system of behavior—the attachment system—that is biologically rooted and thus species-characteristic." But to say that a behavior is "biologically rooted" is a vacuous statement. Most human behaviors are "biologically rooted," since we are living beings and the products of evolution.

As we saw, Ainsworth et al. (1978:322) presented their book on the patterns of attachment that are identified in the Strange Situation as contributing to the "evolutionary-ethological standpoint" of attachment theory. Other researchers make stronger claims about the significance of Ainsworth's work for EAT: "Ainsworth's work has paved the way to (1) a deeper understanding of the phylogenetic foundations of the socioemotional development of infants in the context of their relationships with their significant caregivers" (Sagi-Schwartz 1990:11). However, this widespread idea that the results of the Strange Situation provided confirmation for EAT is mistaken for the simple reason that the behavior of children in a lab is incapable of illuminating the phylogenetic foundations of children's emotions. Ainsworth's Strange Situation laboratory procedure may well measure how a child reacts after being left alone with a stranger, but research of this type cannot provide empirical support for the view that a particular form of mother-infant attachment is an adaptation. Claims about behavioral adaptations are claims about the history of the behaviors. However, the observation of current behaviors in a laboratory cannot provide evidence about their phylogenetic history. Phylogeny is about reconstructing evolutionary history. Ainsworth observed how children acted at home as well as under specific laboratory conditions. These observations can detect patterns of behavior. However, patterns of behavior per se do not provide any evidence about phylogeny. Phylogenetic reconstruction requires other kinds of data: about the history of the character (behavioral, morphological, or physiological trait) under study, about its origin, comparative data from different groups to enable decisions as to whether the character under consideration is a homology or analogy, etc. To reconstruct evolutionary history, we need to put together demographic data with paleontological evidence, primatology, archeological results, as well as knowledge about present-day hunter-gatherer societies.

Therefore, it is incorrect to present Ainsworth's work as providing evidence for the evolutionary foundations of attachment theory. Even assuming the validity of her results and her interpretation of them, her work is unable to confirm or refute claims about the phylogenetic history of human behaviors and emotions. As Bowlby himself once noted, to show that attachment is an adaptation, one must turn to the archaeological record, anthropological studies of hunter-gatherer societies, and studies of other primates (Bowlby 1969:61).

Recent Proposals: Saving Attachment in the Environment of Evolutionary Adaptedness

Much of the attachment literature has failed to address the challenges discussed above. However, some attachment researchers have understood the need for a different approach and tried to present better evidence for the evolutionary ideas that Bowlby and Ainsworth defended. In what follows I review the major proposals put forward since the 1990s to develop the "biological basis" of EAT. I conclude that those proposals have not presented sufficient evidence to accept EAT.[3]

The Universality of Human Social Attachment as an Adaptive Process

In several writings, German psychologists Klaus and Karin Grossmann have presented attachment as a "biological program" (Grossmann and Grossmann 1990:41, 45). In their 2005 paper, "Universality of Human Social Attachment as an Adaptive Process," they assert: "Attachment is the phylogenetically programmed propensity of a human child to form a special relationship with responsive caregivers as part of the infant-parent bond" (Grossmann and Grossmann 2005:199). It is not clear whether they are claiming that infants have a propensity to attach to parents, or to any caregiver. They emphasize the genetic basis of the attachment system: "We view attachments as the developmental process during which infants' genetic programs become phenotypically manifest, observable, and testable as a function of caretakers' responsiveness" (Grossmann and Grossmann 2005:202). They also present attachment as "a universal genetic program valid in all cultures, despite clearly observable variations in parental caregiving behaviors between existing cultures and within cultures in different epochs" (Grossmann and Grossmann 2005:202–203).

These strong pronouncements are surprising because the Grossmanns also recognize that attachment theory has yet to fulfill its aspiration to provide an ethological explanation of attachment behavior. They note the need to address four questions that Tinbergen had argued were central to a biological account of behavior: causation, development, function, and evolution (Tinbergen 1963). But the few paragraphs they provide under each category contain general assertions about behavior, with few specifics about attachment (Grossmann and Grossmann 2005:218–222). For example, under causation they state (Grossmann and Grossmann 2005:218): "The general assumption

---

[3] I do not examine James S. Chisholm's work because he presents his most recent views in this volume (see Chapter 11).

is that physiological structures of the central nervous system and hormonal processes 'cause' human behavioral adaptation at the proximate level." After some comments about the role of hormones in close bonds, they note that some "unanswered questions" remain. On the specifics about attachment, they admit that they can only speculate (Grossmann and Grossmann 2005:219, italics added for emphasis):

> Comparison of previously experienced maternal sensitivity of the insecurely with the securely attached toddlers suggests that the differences in physiological stress *may have been* a consequence of suboptimal attachment experiences, which, in turn, *may have influenced* the toddler's physiological reactions to stress. Hypo- or hyperactivation of physiological systems *might be a response* to past inappropriate management of the infant's homeostasis by the mothering figure…"

What the Grossmanns write about development is of limited value as well (Grossmann and Grossmann 2005:219): "In attachment theory, the developmental processes are assumed to be basic and thus universal for human infants." By this they mean that infants "are born prepared to become social" and ready to learn from those around them. After noting these facts, which are facts that no developmental theory would deny, they say (Grossmann and Grossmann 2005:219):

> As an open genetic program, infants' attachment development accommodates a limited variety of caregiving practices and styles if they comply at least minimally with an infants' (sic) basic attachment needs to be protected and cared for by at least one reliably available individual.

To support this, they cite studies in communist countries and Israeli *kibbutzim* where children were raised with interchangeable caregivers. These studies showed that children developed less competently, according to attachment measures of competency, but they tell us nothing about how the "open genetic program" operates. Nor do they present any research showing how this program develops in different individuals.

The sections on the function and evolution of attachment are equally disappointing. The section on function repeats Bowlby's view that attachment conferred protection from wild animals and dangers in the EEA, and simply adds that sociobiology would include other nonkin humans among the dangers. The rest of the section does not address function. The last section, evolution, also does not say anything about the evolution of attachment.

Not surprisingly, the Grossmanns conclude that further research is needed about the evolutionary aspects of attachment. Indeed, all the points they raise in their conclusion for further study show that one could not yet talk about the "universality of human social attachment as an adaptive process," despite the fact that this is the title of their paper. Last, and very importantly, the Grossmanns never clarify what they mean by adaptive in their paper.

The Nature of the Child's Ties

In an explicit recognition of the significance of the evolutionary framework for attachment theory, the leading chapter in all the editions of the *Handbook of Attachment* (the major compendium of advances in the field) is entitled "The Nature of the Child's Ties." The title pays homage to Bowlby's 1958 foundational paper, while dropping the controversial "mother" from the title. Still, it is not clear here to whom the child is supposed to attach, except to the generic "attachment figure." In this widely cited paper, June Cassidy (1999:4) claims: "The most fundamental aspect of attachment theory is its focus on the biological bases of attachment behavior." After summarizing Bowlby's views, Cassidy (1999:5) concludes:

> In a basic Darwinian sense, then, the proclivity to seek proximity is a behavioral adaptation in the same way that a fox's white coat on the tundra is an adaptation. Within this framework, attachment is considered a normal and healthy characteristic of humans throughout the life span.

Nowhere in the article, however, does she present the evidence needed to show that. Whereas we have a pretty good understanding of the genetics of coat colors in animals and the environments in which foxes live, we still do not know anything about the genetics for attachment behaviors and very little about the environments in which they were supposed to develop in the EEA. We also do not know which alternatives selection would have to choose from or their fitness values.

The lack of evidence to support evolutionary explanations of attachment behaviors is made evident in Cassidy's proposals to explain the evolution of monotropy. She offers the following scenario (Cassidy 1999:15, italics added for emphasis):

> First, the infant's tendency to prefer a principal attachment figure *may* contribute to the establishment of a relationship in which that one attachment figure assumes principal responsibility for the child....Second, monotropy *may* be the most efficient for the child as the child does not have to assess who would be the best person to help. Third, monotropy *may* be the child's contribution to a process I term "reciprocal hierarchical bonding," in which the child matches an attachment hierarchy to the hierarchy of the caregiving in his or her environment."

Yes, monotropy may have evolved in any (or all) of those ways. Then again, it may not have.

In addition to the lack of evidence for this just-so story about the evolution of monotropy, two other aspects of Cassidy's paper are noteworthy. First, it is confusing to see her explanation of monotropy given that numerous attachment theorists have claimed that monotropy is not part of attachment theory. Second, note the conflation between the biological, the normal, and the healthy. Is she suggesting that attachment is normal and healthy because it is an adaptation?

As Hinde and others have pointed out in several publications, one does not follow from the other.

Despite these issues, in the new 2016 edition of the *Handbook*, Cassidy (2016:17) concludes: "Bowlby and Ainsworth's original ideas have held up well…nothing that has emerged from the thousands of studies produced over the past 40 years has led to a serious challenge to the core theory" (Cassidy 2016:17). But, as we have seen here, the challenges are many and deep. At this point, it is difficult to understand what, if anything, those who are committed to this paradigm would consider a challenge to it.

The Innate Attachment System

Jeffry A. Simpson and Jay Belsky have published several papers aimed at developing an evolutionary framework for attachment, both independently (Belsky 1997a; Simpson 1999) and later in a joint chapter in the second and third editions of the *Handbook of Attachment* (Simpson and Belsky 2010, 2016).

In their joint paper, Simpson and Belsky (2010:131) claim that "attachment theory is one of a handful of major middle-level evolutionary theories." On the "innate attachment system" they note: "Guided by Darwin, Bowlby believed that the attachment system was genetically 'wired' into many species through intense directional selection during evolutionary history" (Simpson and Belsky 2010:132). However, since views about evolution have changed since Bowlby's early writings, they aim to "place attachment theory in a modern (neo-Darwinian) evolutionary perspective" (Simpson and Belsky 2010:132). According to Simpson and Belsky (2010:132–133), that perspective would include a series of views that "share a central premise: that much of the human mind and human social behavior reflect adaptations to the major obstacles to inclusive fitness that humans repeatedly faced throughout evolutionary history."

After one section about those central evolutionary views, they examine the "stable features of the social EEA" (Simpson and Belsky 2010:135): "For thousands of generations, our ancestors lived in small, cooperative groups. Most people within a tribe were biologically related to one another, and strangers were encountered rather infrequently." Later, in the section devoted to individual differences in attachment, Simpson and Belsky (2010:138) claim: "Ainsworth's Strange Situation is well suited to detect different patterns of attachment because it presents infants with two common cues to danger in the EEA: being left alone, and being left with a stranger." But how could situations that infants hardly ever, if at all, encountered in the EEA become "common" cues to danger? And if they were not common cues to danger, then what is the value of the Strange Situation Procedure for understanding better the EEA for attachment?

It is also unclear how one would evaluate what the best strategy might have been in situations in which the infant did not see the mother nearby in the EEA. Following Bowlby, Simpson and Belsky (2010:131) say that crying was the first best strategy to attract the mother. If that failed, the second-best strategy was to remain aloof so as not to attract predators. But if crying was likely to attract a predator, why was it the best strategy? In a social environment where babies were hardly ever left alone, would it not be the best strategy for a baby to wait calmly and quietly for the mother's return? Then, if the mother did not return, as a second strategy the baby could cry to alert another adult of the group. This ordering of strategies makes more sense and it has as much evidence—or as little—as the one assumed in attachment theory.

Following Hinde's suggestion that natural selection probably would not have led to stereotypy in parental care, Simpson and Belsky note, as others before them, that each attachment pattern (secure, anxious-ambivalent, and anxious-avoidant") *could be* a different "strategy" that "*could have solved* adaptive problems presented by different kinds of rearing environments" (Simpson and Belsky 2010:138, italics for emphasis). But note the highly speculative nature of the proposed evolutionary explanations, as in the case of avoidance behavior (Simpson and Belsky 2010:139, italics for emphasis):

> During evolutionary history, this behavioral strategy *would have increased* survival among infants who, *if they placed* too many demands on their parents, *might have been* abandoned.…*If* maternal rejection served as a proximal cue of the severity of future environments, avoidant tendencies *might have allowed* children not only to move away from their parents earlier, but to become more opportunistic and advantage-taking, thereby facilitating survival and early reproduction in such arduous environments.

Again, I do not deny that all of this may have happened that way. But what evidence do we have to substantiate that it did, or that this explanation is superior to others?

In the last part of their paper, Belsky and Simpson present different evolutionary models of social development across the life span. It is not possible to review them here. However, the existence of several different proposals attests to the fact that these are theoretical models, based on various imaginative scenarios, none of which have sufficient empirical support.

Mothers plus Others

Sociobiologist Sarah Hrdy has also presented attachment theory as being part of evolutionary theory, claiming that it is "arguably evolutionary theory's most important contribution to human well-being." Although in several writings she has challenged attachment theory's focus on the mother-infant dyad, Hrdy presents her views to "correct an underlying assumption about the universality of exclusive maternal care in primates, not to challenge Bowlby's fundamental insights" (Hrdy 2009:82). Thus, her own proposal for the evolution of attachment, the cooperative breeding hypothesis, is put forward to support attachment theory rather than to challenge it.

In Hrdy's view, Bowlby was right about attachment being an adaptation in the early part of primate evolution. Indeed, Hrdy (2009:114) claims that: "As attachment theorists have long assumed, all primate infants evolved to seek contact with a warm and nurturing mother. There is no questioning Bowlby's insight on this point." During this early period, according to Hrdy, ancestral primate babies had little to worry about since the mother's urges to stay attached to her babies were equally strong, and thus these mothers were in constant contact with their offspring. Yet Hrdy does not tell us why we cannot question something that she recognizes is an assumption. Neither does she tell us what evidence supports her reconstruction of primate evolution.

In Hrdy's evolutionary scenario, "at some point in the emergence of the genus *Homo*, however, mothers became more trusting, handing even quite young infants over to others to temporarily hold and carry" (Hrdy 2009:114). She argues that allomaternal assistance was essential for children's survival during the EEA. In turn, alloparents enhanced "their inclusive fitness by helping kin" (Hrdy 2005c:11). Furthermore, Hrdy sees cooperative breeding as the motor of the capacities for mind reading and cooperation that eventually led humans beyond other primates. She proposes that maternal separations caused the development of the little apes' increasing ability to read the facial expressions of others, leading eventually to the human's greater ability to read minds and to empathize, compared to other apes. Thus, in this account, cooperative breeding not only explains why human babies can attach to several caretakers but it also explains the extraordinary evolution of the cognitive and emotional capacities of our human ancestors.

In several places, however, Hrdy recognizes the speculative nature of her proposal. She asks, "as critical as a grandmother (or a great-aunt) might be, how likely were they to be present? The answer is not knowable" (Hrdy 2005c:16). At another point she acknowledges writing in a "speculative vein" and notes that her model "relies on a number of assumptions" (Hrdy 2005c:25, 28). Despite these qualifications, however, in other places she makes assertions for which we do not have—and probably cannot have—any good evidence. For example: "Back in the Pleistocene, any child who was fortunate enough to grow up acquired a sense of emotional security by default" (Hrdy 2009:290). I am not even sure we can understand what a "sense of emotional security" would have meant in the Pleistocene, let alone how to evaluate who had it and why.

To support her cooperative breeding hypothesis, Hrdy places strong weight on studies showing that in some contemporary societies with high child mortality, alloparents affect child survival. Although clearly this is important information, it is not sufficient to extrapolate to evolutionary scenarios in the EEA for two main reasons: (a) we do not have any information about child mortality rates in the Pleistocene or their causes; (b) from the fact that something would have been useful, we cannot conclude that it was selected. In addition, there are competing evolutionary explanations of human sociality (Sober and Wilson 1998; Richerson and Boyd 2005; Wilson 2012; Hawkes 2014).

Where Are We Today?

Some attachment theorists today provide sophisticated accounts of the possible evolutionary scenarios for attachment patterns that often take into account different ecological contexts and adopt a life history perspective. Some of them postulate the existence of different strategies of parental care and recognize that diverse attachment patterns can be adaptive. But the models put forth to defend the evolutionary basis of attachment rest on substantive assumptions about the human mind and about evolution that are still the subject of much debate (Gangestad and Simpson 2007). In addition, these models often rely on very different views. Thus, it is not clear which particular evolutionary framework is accepted in the attachment community today. The result is a number of competing frameworks, each of which is highly speculative.

In reviewing critically the proposals for an evolutionary account of attachment, I do not reject the general attempt to understand better the evolutionary history of the human mind and behavior. My point here is epistemological: Given the speculative nature of their proposals, scientists today cannot justifiably present EAT as part of evolutionary science or as grounded on accepted evolutionary views. Claims, such as "attachment theory is a special branch of Darwinian evolution theory" (van IJzendoorn et al. 2006:108), are unjustified.[4] While there is nothing wrong with using the EEA as a heuristic tool to help us think about the possible scenarios in the evolution of behavior, it is not legitimate to conclude that any one of these scenarios confers evidence for EAT. To advance in this area, attachment researchers need to clarify the diverse claims made in the literature, identify which of them are still largely speculative, determine what sorts of evidence each of them would require to move beyond the realm of speculation, and then pursue research that could produce evidence for or against the various claims.

To do this, I doubt that the best way is to reconstruct the shape of attachment in the EEA because, to begin, the concept of the EEA has been controversial (Laland and Brown 2002; Plotkin 2004:150–152; Buller 2005). For some authors, including Bowlby, the EEA refers to a period, the Pleistocene, from about 2 million years ago to 10,000 years ago. In contrast, other evolutionary psychologists claim that the EEA is not a concrete historical period, but the statistical composite of the combined set of selection pressures that led to a given adaptation. Regardless, we do not know how to reconstruct either of those scenarios. We know little about the social and psychological makeup of our ancestors or the selective pressures they encountered. Although evolutionary

---

4   I do not address their evolutionary model because it is basically a diagram that presents attachment as an innate mechanism, but they do not elaborate or present evidence to support this.

psychologists believe that we have sufficient data to derive some conclusions, in this case we do not know enough to favor one hypothesis over the others (monotropy, allomothering, conditional strategies, no attachment system).

Over the years, leading figures in evolutionary biology and evolutionary anthropology have emphasized the difficulties in reconstructing the evolutionary past of the human species (Lewontin 1995/1982:163–171; Tattersall 2012:200). We must recognize that the reconstruction of behavioral characteristics still is—and will remain—incredibly difficult. To do this in the area of attachment, we would need to know about migration patterns, family size and structure, diseases, nutrition, work and social habits, and social conditions that affect child-rearing: fertility rates, marriage patterns, variation in family size, etc. Using our fertile imaginations, we can construct many such scenarios, some more plausible than others. Yet, we are still far from being able to make well-supported scientific claims about child development in our deep past.

Equally important, even if we could assert that secure attachment was an adaptation in the EEA, what would be its relevance for us today? Bowlby, Ainsworth, and other attachment theorists argued that it would be relevant for us because our minds have not changed much since the EEA. As many evolutionary psychologists put it, we have Stone Age minds (Cosmides and Tooby 1997). Yet, this is highly controversial. How likely is it that we are still behaving and feeling like our Paleolithic ancestors? In fact, how likely is it that the way they behaved and felt remained static during two million years, a long time period that saw tremendous changes at many levels?

In my view, the assumption of stasis in the evolution of our emotional and psychological makeup is at odds with scientific work in different areas. Given new discoveries about epigenetics, neural plasticity, and developmental biology, why should one assume that our emotional makeup has remained fixed since the Stone Age? We know that evolutionary change can occur in relatively short periods of time. Peter and Rosemary Grant have documented evolution by natural selection in the Galapagos finches in a period of merely a few decades. They demonstrated that alterations in body and beak size resulted from changes in the food supply (Grant 1986; Grant and Grant 2008). Evolutionary changes in humans since the rise of domesticated animals have also been documented (Richerson and Boyd 2005:191–192). Other scholars argue that rates of adaptive evolution in the Neolithic and later periods were much higher than in earlier periods in human evolution (Hawks et al. 2007). What we know about epigenetics shows that evolutionary changes could be much faster than we had assumed as well (Jablonka and Lamb 2014).

In short, we need to have much more data about the genetics, the epigenetics, and the fitness of specific attachment behaviors as well as about the ecological and cultural circumstances of our ancestors. Until then, we can construct many "just-so stories" about evolutionary scenarios, but it would be premature to assert that babies have stone-age minds, as EAT claims. Ainsworth (2013/1983:459) said:

I have sometimes been accused of being out of touch with current changes in lifestyles, but I believe that the problem is that infants are perhaps a million or so years out of touch with them. Their inbuilt evolutionary adaptations tend not to match new lifestyles, much as we would like to believe infants to be infinitely adaptable.

But nobody is defending, as she suggests, the idea that infants are "infinitely adaptable." Instead, the point is that in order to show that the mind of an infant was set in stone "a million or so years" ago, as she believed, one needs to provide some evidence. So far, we do not have such evidence. (Additionally, note that Ainsworth's use of "adaptable" conflated the biological and psychological senses of adaptation).

Furthermore, even if the secure attachment pattern were proven to be an adaptation also to be adaptive now, it would still not follow that deviations from it are pathological. When the environment changes, a previous behavior may or may not be adaptive. It could be less or more adaptive than it was in the previous environment, or the change in environment could have no effect on the fitness of the trait.

We need to separate biological facts from social goals. When we say that secure attachment is optimal, we are not only making a claim about biological evolution, we are also making a value claim. What forms of child-rearing are valuable to us cannot be settled only by knowledge of biological and psychological processes. In child-rearing, what is best depends on one's views about what a good society is and on what counts as being a good child and a good caretaker in that good society. That is why to understand attachments we need to know how different societies integrate those values into child-rearing, which leads us back to where we started: culture.

New Insights from Ethnographic Research on Child-Rearing

While attachment workers focused mainly on (a) administering the Strange Situation Procedure in a variety of contexts and (b) testing children's competencies, many cross-cultural psychologists and anthropologists kept raising the point Mead had put forward: there is more diversity in child-rearing than attachment theorists have acknowledged. After interest in cultural psychology and psychological anthropology increased in the 1990s, studies of child-rearing revealed important new insights. It would be impossible to review them all here; I will summarize some with special relevance to attachment theory.

Some ethnographic studies provided much needed research on mothers. In her 1992 book, *Death Without Weeping: The Violence of Everyday Life in Brazil*, Nancy Scheper-Hughes demonstrated the complexity of maternal sentiments. She observed three generations of mothers living in dismal economic and social conditions in the slums of the sugar plantation town of Timbauba, Brazil, during 1964–1984 and 1985–1992. The average woman had

9.5 pregnancies, 8 births, and 3.5 infant deaths. Scheper-Hughes argued that the high infant death rate shaped the attitudes and emotions of mothers toward their infants. Aware that their babies might not survive, the mothers did not immediately "personalize" their infants, did not name them right after birth, and did not mourn their deaths as is done in Western societies. She concluded that maternal love is not a uniform, universal, and natural monolithic affect. Challenging attachment theory, she called for a "pragmatics of motherhood": a recognition that mother love "represents a matrix of images, meanings, sentiments, and practices that are everywhere socially and culturally produced" (Scheper-Hughes 1992:341; 2014).

That parental caregiving is not simply a system of innate responses is also made clear by David Lancy, who has shown that parents in many cultures focus on "not-attaching" to the infant until the infant can be considered a person in that community (Lancy 2014, 2015). Lancy compiled information on over 200 cases from the ethnographic and archaeological records representing all areas of the world and historical periods, from the Mesolithic to the present. He reports that in many societies, infants are not given status as a person at birth. For example, the Wari' native Amazonian communities compare an infant to unripe fruit because they consider that the infant is still in the process of being made, and the Nankani in northern Ghana delay judging whether a baby is human until they are sure it is not a spirit or bush-child. Since babies are not considered real persons, the parents resist attaching to them (Lancy 2015:41, 49).

Ethnographic studies also found that mothers have different goals and follow different norms in child-rearing (LeVine et al. 1994; Harwood et al. 1995; Weisner 2005; Mageo 2013; Otto 2014). For example, in Kenya, Gusii mothers, who bear an average of ten children, focus first on their survival and then on teaching them compliance, contrary to mothers in the United States who do not aim for compliance in child-rearing (LeVine et al. 1994). In her studies of Samoan family relations, Mageo (2013) found that socialization is not oriented to develop feelings of security, but to encourage separation.

Other important studies have demonstrated the need for research to go beyond the mother-infant dyad since mothers are not the only caregivers in many societies; indeed, cooperative child care is a widespread practice. Weisner and Gallimore (1977) presented a thorough review of anthropological studies showing the importance of children as caretakers. Since then, much more ethnographic work has confirmed that children and other family and community members are an intrinsic part of caretaking networks in many societies (Gottlieb 2004; Meehan and Hawks 2013; Gottlieb 2014; Röttger-Rössler 2014; Weisner 2014). Psychologists have also called for studying children within social networks (Lewis 2005).

Cultural anthropologists and psychologists have also argued that attachment researchers' views about children's competence have been shaped by a particular set of values prominent in Western societies. For example, creativity, autonomy, self-reliance, and independent exploration are all important personality traits for middle-class Westerners who educate their children to embrace those values. Other cultures, however, do not share those values. In some societies, parents educate their children to comply and subordinate their desires to the group's needs. If children in those cultures are measured by Western standards, their development will seem inadequate.

In sum, cross-cultural studies have shown that child-rearing goals, caretaking practices, conceptions of the child and childhood, and children's social relations and attachments are varied and multifaceted. They all acquire specific significance within the cultural context of a given society (Rogoff 2003; Keller 2007, 2013a; Everett 2014; Röttger-Rössler 2014; Keller and Chaudhary as well as Morelli et al., this volume).

Historians have also demonstrated that views about children, mothers, and child-rearing have changed over time, often in profound ways. Ecological, socioeconomic, and cultural factors have shaped different visions of childhood and child development throughout history (deMause 1975; Zelizer 1985; Elder et al. 1994; Koops and Zuckerman 2003; Stearns 2003; Kessel 2009; Mintz 2009; Everett 2014; Mayes and Lassonde 2014). Historical analysis also reveals deep changes in visions of motherhood and what counts as a good mother (Dally 1982; Apple and Golden 1997; Ladd-Taylor and Umansky 1998; Plant 2010). Additionally, different scientific perspectives have shaped varied, and sometimes contradictory, notions of caretaking and child-rearing (Hulbert 2003; Apple 2006).

On the basis of these results, many scholars have called for a rejection of the central core tenets of EAT. Some view EAT as a cultural ideological product of a specific historical context (Lancy 2014; Vicedo 2013; LeVine 2014). Other cultural anthropologists and cross-cultural psychologists call for further work on child-rearing in different cultural contexts and exploration of the significance of cultural variations (Rothbaum et al. 2000b; Keller 2008; Morelli and Henry 2013; Quinn and Mageo 2013; Keller 2014b; Otto and Keller 2014). None of these critics deny that children form attachments and that some of those influence their psychological growth. However, they reject the view that child development follows a universal and uniform pattern.

For a long time, attachment scholars remained reluctant to incorporate insights, provided by ethnographic work and psychological research, which fell outside their paradigm. Recently, some attachment supporters have written about the cultural dimensions of attachment, feeling frustrated by the criticisms of their cultural blindness (e.g., Mesman et al. 2016b). In addressing the cultural critiques, however, these authors often explain discrepancies away or find that studies which do not support their position are methodologically flawed (Mesman et al. 2016b:799). Most importantly, although they incorporate some ideas from cultural studies (e.g., the importance of multiple caretaking), they always end up concluding that all is well with the EAT paradigm as developed by Bowlby and Ainsworth. Thus, it is perhaps fair to say that the frustration is mutual, as students of other cultures believe that attachment researchers just pay lip service to the significance of culture, while they continue without sound justification to postulate a uniform attachment "behavioral system."

Supporters of the traditional EAT paradigm still believe there is a "preprogrammed" attachment system that is species-characteristic. Many of them talk about a biological program that is expressed in a universal and uniform manner with small cultural variations. They adopt, therefore, a vision of human nature in which biology is a fundamental stratum upon which culture superimposes new deposits through history. Ainsworth, for example, talked about the genetic programming "under all the layers of individual learning and cultural acquisition" (Ainsworth 1969:1000).

In 1966, however, that conception of the relation of biology to culture was brilliantly critiqued by the anthropologist Clifford Geertz. In "The Impact of the Concept of Culture on the Concept of Man," Geertz argued that anthropological studies and new studies in biology were already putting to rest a stratigraphic vision of human nature inherited from the eighteenth century, which saw man as "a composite of 'levels,' each superimposed upon those beneath it and underpinning those above it." This led to an intellectual strategy that aimed to "strip off the motley forms of culture" in order to reveal the "regularities of social organization," by peeling off the different layers superimposed on the biological foundations (Geertz 1973/1966:37). Geertz thought the human sciences were finally giving up the hopeless search for a universal and uniform human nature.

EAT is one of the areas in which, *pace* Geertz, the search for the universal and uniform human nature rooted (or determined) in our ancestral past has remained alive. Cultural critics argue, however, that we still need to carry out empirical studies about social relations in a large number of cultures before we are ready to establish generalizations about child-rearing and child development. Moreover, anthropological work has shown that no single factor can explain a child's emotional development and that we need to understand the cultural context in order to be able to interpret actions and emotions. Acts, behaviors, and feelings are part of cultural and psychological processes that acquire meaning within a large network of affects, other people's behavior and expectations, social norms, and individual goals.

That is why the relationship between anthropology and psychology cannot be a mere affair or marriage of convenience. Anthropologists cannot work without some assumptions about the human mind, and developmental psychologists, including attachment researchers, cannot make sound claims about the universal nature of the human mind or the human child without investigating real children in real situations in diverse societies. Child development studies need to incorporate ethnographic research in different cultures. By engaging psychological insights about the human mind, anthropologists could overcome what historian of science Peter Galison has called "the limits of localism" (Galison 2016). Whether scholars aim for a cultural psychology that might be able to offer "universalism without the uniformity" (Shweder and

Sullivan 1993:517) or for a program of "developmental contextualism" (Lamb 2005:110), they need to carry out interdisciplinary research that integrates the results from the field with work from the laboratory.[5] Toward this end, Morelli et al. (this volume) present a new paradigm to understand children's attachments in Chapter 6.

Conclusion

Bowlby and Ainsworth claimed that the evolutionary framework they adopted was the defining characteristic of the theory of attachment they jointly developed, and that is what other supporters of attachment theory have emphasized: "One of the great strengths of attachment theory is that it is based firmly within an evolutionary, ethological and ultimately a biological framework" (Byng-Hall 1991:6). Appealing to the biological basis of attachment, Bowlby, Ainsworth, and their many followers today have presented the ethological attachment theory as having universal application. They make two central claims: (a) the mother-infant dyad is a biological system that became an adaptation in the EEA, and (b) secure attachment is the path for an infant's optimal emotional development.

In this chapter, I have shown that those claims have had a contested history. A review of that history reveals that both claims remain unsubstantiated. From Mead to the present, ethnographic studies have provided extensive evidence for the diversity of child-rearing in different cultures. Those studies challenge the view that deviations from the secure-attachment pattern lead to pathological outcomes and that maternal sensitivity, as defined by attachment theorists, is necessary for a child's adequate emotional development. From Hinde's 1982 critique to the present, a number of researchers have also challenged the claim that evolutionary biology supports the notion that there is an ideal way of child-rearing.

Attachment theorists also continue to defend the uniformity and universality of attachment because in their view it is an adaptation. In turn, they believe that by being an adaptation, attachment must be universal and uniform. For example, Ainsworth (1991:33) wrote: "The great strength of attachment theory in guiding research is that it focuses on a basic system of behavior—the attachment system—that is biologically rooted and thus species-characteristic." Recently, Mesman et al. (2016b:804) claimed: "The cross-cultural studies included here support Bowlby's (1969/1982) idea that attachment is indeed a universal phenomenon, and an evolutionary explanation seems to be warranted."

But the logic tying biology and universality rests on a mistaken assumption. A feature of an organism can be an adaptation without being universal, and universality is not sufficient to show that a feature is an adaptation or adaptive.

---

[5] For an example of solid experimental work on how children learn to trust, see Harris (2012).

An evolutionary explanation is warranted when we have evidence for the evolution of the characteristic in question. In the case of the attachment system, it is still lacking.

In a very basic sense, EAT assumes a view of human nature in which diversity, variability, and historical change do not matter. By seeing variation as superficial, at best, and pathological, at worst, EAT denies the transformative power of historical change and cultural context.

What anthropology, cultural psychology, evolutionary biology, and history teach us is that diversity is the norm across time and space. Infants and mothers are part of sociocultural networks that vary and change. It is not simply that culture "has a role," provides the "context," or "adds" something to the developing mind of the child. Culture is much more than that: it is transformative at the individual and the species level. Psychological development is a process affected by many factors that interact in numerous ways. Evolution is also a historical process in which ecological, cultural, and biological factors interact in complex ways that affect ontogenetic and phylogenetic development. Perhaps what is needed are new metaphors and new conceptual frameworks to analyze the interrelation between all the factors that contribute to the dynamic processes of life histories at different levels.

As we become more aware of the intricate ways in which biology and culture influence each other, genetic-centric models do not seem the most appropriate vehicles for understanding the evolution of processes like child-rearing, for these are processes in which the role of cultural evolution is key. Better models of the coevolution of culture and biology are necessary before we can present a general theory of social attachments.

Bowlby adopted a conception of the evolution of mind that is still under considerable debate (Gangestad and Simpson 2007). He also relied on a neo-Darwinian view of evolution that has received substantial criticisms. As Denis Walsh has argued, we need to move beyond the conception of adaptation as design and of organisms as designed machines. He urges us to see organisms "as agents, making a place for themselves in the world" (Walsh 2015:185). As he reminds us, evolution is an ecological phenomenon, and organisms' plasticity to engage with the ecological conditions which they encounter and create themselves is central to adaptive change. If this is true for all organisms, how much more so it must be for human beings who have been actively creating their own environments at an unparalleled pace. As Tattersall (2012:148) reminds us: "We differ today far more from our earliest Pleistocene ancestors than do any other of the creatures with which we share the planet." That fact has much to do with our ability to modify the diverse environments we inhabit, an ability in which our biological and cultural features have interacted in complex ways for thousands of years.

Understanding that intertwined evolution in a way that does not simplify the biological and the cultural aspects of our nature still requires our concerted efforts. As Evelyn Fox Keller (2014a:2428) argues:

> If much of what the genome "does" is to respond to signals from its environment, then the bifurcations of developmental influences into the categories of genetic and environmental, or nature and nurture, makes no sense. Similarly, if we understand the term environment as including cultural dynamics, perhaps neither does the division of biological from cultural factors.

Reflecting elsewhere upon the conflicted history of biology and anthropology during the last century, Keller finds that going beyond the biological/cultural divide "is not going to be so easy" (Keller 2016a:38). Yet, what other good alternative do we have? Understanding the complexity of humans, who are integrated biological–socio–cultural systems, will require a better integration of the natural and human sciences.

Until we make further progress along these lines, we should be wary of scientific hype in the area of child development. As some scholars have recently pointed out, science hype (i.e., exaggerating the state of scientific knowledge in one area) hurts science, as the public becomes distrustful of the scientists who make claims beyond the evidence (Caulfield and Condit 2012). In areas in which scientific pronouncements influence matters of social policy, to go beyond the evidence is not only unscientific, it is socially irresponsible. In the area of child development, to avoid science hype, unsound policies based on unsupported claims, and ethnocentrism, it would behoove us to present our scientific findings in a way that acknowledges their often tentative and limited nature.

From the truism that neglect and abuse have serious detrimental effects for children, attachment theorists moved to erect an enormous theoretical edifice (LeVine and Miller 1990:79). But the occasional lip service paid to the existence of other types of caregiving is no longer sufficient. Critics are demanding a change in research practices through:

1. the decentering of WEIRD (western, educated, and from industrialized, rich, and democratic countries; Henrich et al. 2010) samples as the basis for theories about universal patterns of behavior, and
2. a healthy modesty about the use of speculations regarding adaptations and the role of these speculations in constructing a theory about human nature that has enormous implications for the lives of people all over the world.

Both scientific standards and social responsibility require attachment researchers and all of us to be more humble.

Reviewing Bowlby's 1969 book shortly after it was published, Maas (1970:414) already noted:

> What we need is not more dogma but verifiable patternings (sic) of influences upon man's early—and later—development. Ill-validated explanations may be seized upon all too quickly for prescriptive application. Theory, especially in this field of inquiry, should be developed, tested, presented, and reviewed critically.

Almost 50 years later, those words still hold true.

Acknowledgments

I am grateful to Juan Ilerbaig, Mark Solovey, Gilda Morelli, Robert LeVine, and Naomi Quinn for suggestions to improve this paper. I also benefited from discussions at the Forum and from comments sent by various participants.

3

The Evolution of Primate Attachment

Beyond Bowlby's Rhesus Macaques

Masako Myowa and David L. Butler

Abstract

Bowlby's theory of attachment has been hugely influential, yet his proposal and its subsequent support derives heavily from research involving rhesus macaques, the most extensively studied nonhuman primate in attachment research. Does his theory apply to other primates? A substantial amount of data concerning primate (including human) child care now challenges Bowlby's original proposal, particularly as it relates to the notion of the mother being the sole continuous care-and-contact provider: caring can be shared by various individuals, the father can serve as the primary attachment figure, and infants can form multiple attachments. This chapter focuses on the phylogenetic history of attachment among primates, identifies features of attachment that are shared or which differ between humans and nonhuman primates, and considers the possible cognitive, social, and ecological factors associated with these similarities and/or differences in attachment among primates. Current evidence suggests that the human attachment system appears to be uniquely characterized by (a) social interactions based on combined visual, tactile, and auditory modalities, (b) the use of positive cognitive empathy, and (c) certain contextual elements typically contained in human social environments.

Background

The attachment relationship between parents and their infant is an important part of the evolutionary heritage of mammals, and is particularly prominent in many primates within which a lifelong propensity for such a bond has also evolved. According to Bowlby's and Ainsworth's classic conception, common manifestations of attachment are (a) an infant's selective preference for its mother, (b) displays of agitation or distress by an infant upon maternal separation, and (c) the attenuation of such agitated or distressed states by the mother through her presence and/or soothing skills (Bowlby 1969, 1973; Ainsworth 1972). Although these characteristics are widely (and perhaps incorrectly) accepted in relation to primate development, particularly for humans, important questions remain unanswered concerning primate attachment more generally, which may have possible implications for our current understanding of human attachment: What is the phylogenetic history of attachment among the many species of primates (and not simply monkeys and/or apes) (Figure 3.1)? Are humans unique in any or all major aspects of attachment? What are the possible cognitive, social, and ecological factors associated with these similarities and/or differences in attachment among primates?

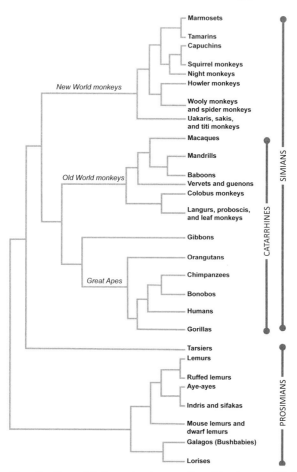

Figure 3.1 Classification of living primates mentioned in the text (Hrdy 2009). As this figure makes clear, the evolution of attachment must be considered (both generally and in more specific detail) in all primates, not simply in one or two species of monkeys or apes.

Attachment Styles and Their Variability across Primates

Contrary to the prevailing notion of "cupboard love" advocated until the 1960s (i.e., that the mother was simply a food source for her infant), Bowlby (1969) proposed that the primary function of the mother was as a source of security and tactile comfort for the infant. He conceived of the attachment system as a strong mother-infant bond that had evolved as an adaptation to ensure survival and healthy development of the infant. Importantly, however, his proposal as well as its subsequent support drew heavily from research on rhesus macaques, the nonhuman primate most extensively used to study attachment (Hinde and Spencer-Booth 1971a; Hinde 1991). From this limited scope, Bowlby and many others inferred that humans and other primates alike retained the abilities for attachment because it had originated and descended from an earlier common ancestor (i.e., homological evolution). An explosion of new data concerning primate (including human) child care has since challenged Bowlby's original homological proposal, particularly as it relates to the concept of the mother being the sole caregiver. Indeed, as described below, primate evidence now indicates that (a) among primates, caring can be shared by various individuals; (b) in some species, the father serves as the primary attachment figure; and (c) primate infants can form multiple attachments.

It is now known that many individuals among New and Old World monkeys (see Figure 3.1) serve as alloparents (Kohda 1985). Furthermore, "sharing caring" is exhibited by prosimians, such as lemur mothers (*Lemur catta* and *Varecia rubra*) who, when foraging, allow their infants (often twins) to be cared for by the father and another lactating mother, who may even feed hungry infants in the mother's absence (Pereira and Izard 1989; Vasey 2007). Similarly, galagos (*Galago sensegalansis*) and mouse lemurs (*Microcebus murinusm*) have aunts and grandmothers that can spontaneously lactate and nurse offspring (often twins) (Eberle and Kappeler 2008; Kessler and Nash 2010). Clearly, these examples show that the mother is not the exclusive attachment figure among many primates. Why is this so?

In most primate species, males are present year-round in the same social group as females with whom they have mated. Most primate species, however, have a multimale/multifemale system of mating (see Hawkes et al., this volume). This makes it difficult to identify who has fathered any given child and may contribute to why fathers in many species do not (knowingly) provide care for their young. To compensate for this lack of fatherly input, mothers may rely on other (often related) females, juveniles, and adolescents, all of whom may be eager to practice their mothering skills (see below for further discussion on the potential function of alloparenting). Callitrichid primates, marmosets, and tamarins provide exceptions to this, as they live together in large family units within which the mother, father, and older siblings all care for infants by, for example, carrying them and performing other childcare roles immediately after birth (Kostan and Snowdon 2002; Washabaugh et al. 2002; Mills et al.

2004; Mota et al. 2006). The evolutionary backdrop to this behavior includes multiparous reproduction (i.e., twins) in many marmosets (other primates are typically uniparous reproducers) and the difficulty of individually transporting and caring for multiple offspring, especially newborns (Saito et al. 2008; see Figure 3.2). For example, a newborn marmoset weighs approximately 10% of its mother's weight, and marmosets usually give birth to twins or triplets.

Although fathers may not typically play a major role in rearing their own young, it is important to acknowledge that child-rearing does occur in conjunction with the assistance of other individuals including the father (see above) and that, beyond these duties, fathers themselves can assume the role of the primary attachment figure. For instance, in titi monkeys (*Callibebus moloch*), a New World species characterized by monogamous and biparental relationships, infants seem to show an attachment bias toward their father. Indeed, despite being nutritionally dependent on the mother for the first 8–12 weeks of life, an infant spends 70–90% of its time being transported by the father (Fragaszy et al. 1982). Interestingly, young titi monkeys continue to exhibit a preference for their father when they are no longer dependent on either parent for food or transport (Mendoza and Mason 1986). This has been demonstrated in the selective approaches that 6-month-old infants take toward their fathers when simultaneously presented with their mother and father in a Y-shaped maze (Mendoza and Mason 1986). Between the age of 3–5 months, infants exhibit a greater stress response when separated from the father than from the mother (Hoffman et al. 1995). Similarly, cotton-top tamarins between the age of 9–20 weeks (and which are already independent) exhibit an attachment bias toward fathers rather than mothers, as indicated by their running to the father

Figure 3.2   Transport of offspring marmosets (*Callithrix jacchus*) by the father, with the mother nearby. Photo used with permission from Toni Ziegler.

when presented with a fearful situation (Kostan and Snowdon 2002). Quite clearly, the care of infants extends beyond the mother and can include several other individuals. In some instances, infant care and/or attachment may primarily involve the father. Does this suggest that infants are capable of forming multiple attachments rather than just one?

Let us consider this in relation to infant transfer patterns from one caretaker to the next, which usually occurs smoothly when the future caretaker takes the infant from the back of the current caretaker. Potential nonmother caretakers appear highly motivated to carry infants (Schradin and Anzenberger 2003; Zahed et al. 2008). Infants themselves may also facilitate transfers by doing so spontaneously (Tardif et al. 2002). The fact that infants experience such passive and active forms of contact with nonmother individuals relatively soon after birth suggests that infants can form multiple and varied styles of attachment with their various caretakers (Maestripieri 2003). While these examples indicate that alloparenting and multiple attachments are likely evident in many primates, whether and how such interactions involve qualitative differences in attachment between the infant and their various caregivers—particularly among great apes—remains to be investigated.

With this outline of some of the similarities and differences in attachment among primates, which contradict Bowlby's claims, we will now consider factors which might be important to explain such patterns.

Variations in Sensory Modalities Related to Attachment among Primates

It is possible that the unique characteristics of parent-infant bonds relate to their social interactions and are based on the sensory modalities used. From an early age, Old World monkey infants develop the ability to recognize a primary caretaker through various modalities. For example, within two weeks after birth, long-tailed macaques can differentiate their mother's nipples from those of other females (Negayama and Honjo 1986). In addition, at least some Old World monkey infants have shown a visual ability to recognize a primary caretaker. Surprisingly, Japanese macaque infants reared by human caretakers preferred looking at their surrogate human mothers compared to other humans after only three hours of visual experience with human faces (Yamaguchi et al. 2003). This indicates that the development of this species' ability to discriminate caregivers from noncaregivers may rely on visual as well as tactile information (as evident in their clinging-embracing behavior described below).

Recent research has also shed light on the use of visual information by apes in their early social interactions. Using the preferential looking method, one of us (MM) investigated developmental changes in infant visual face recognition in species representative of small and great apes: One gibbon (*Hylobates agilis*) and three chimpanzees (*Pan troglodytes*); (Myowa-Yamakoshi and Tomonaga

2001a). As shown in Figure 3.3, both gibbon and chimpanzee infants were able to distinguish visually each of their caregiver's faces from other faces very shortly after birth. More specifically, although the gibbon had witnessed human faces for only 2 weeks, he showed a general preference for human-type faces (particularly that of the primary human caregiver) rather than a gibbon face. In contrast, chimpanzees began to show a preference for their biological mother's face at around 4 weeks of age. It is important to emphasize that the amount of face-to-face interaction encountered by chimpanzees (i.e., mutual gaze between mothers and infants) has been found to be less than is typically experienced by humans, in terms of the duration of mutual gaze events and the amount of maternal looking (Bard et al. 2005). When reared in such human environments, chimpanzees might begin to recognize the faces of their primary caregivers earlier by accumulating more visual experience with them (the effects of human enculturation are discussed more below).

Although there are cultural differences, visual exchanges involving gazing and facial expression play a significant role in forming and enhancing parent-infant interactions for humans. For example, during social interactions, some human caregivers draw attention to themselves using several facial expressions in an "exaggerated" mode, such as raising their eyebrows and opening their mouths wider. Infants appear attracted to such changeable and attractive faces. As a possible basis for this type of interaction, human infants seem to be hardwired for visually orienting toward the faces of other individuals. For example, human newborns are sensitive to and prefer looking at stimuli-resembling faces. They also look longer at faces that exhibit direct as opposed to averted gazes (Batki et al. 2000; Farroni et al. 2002). Moreover, human newborns react to facial gestures made by others, by appearing to imitate actions

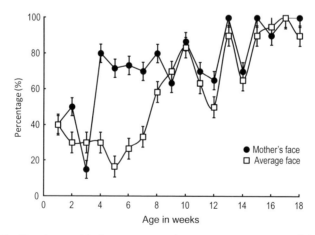

Figure 3.3 Developmental change expressed as a mean percentage of the tracking scores for a mother's face (●) versus average face (□) for each week of age, plus standard error.

such as tongue protrusion and mouth opening (Meltzoff and Moore 1977, 1983; cf. Oostenbroek et al. 2016). Such early competencies in face perception may help infants to establish face-to-face interactions with their caregivers following birth. Do other primates show such face-to-face tendencies?

During their first three months of life, chimpanzee mothers and infants, both in the wild and in captivity, also engage in face-to-face interactions through mutual gazing, although there is no evidence that a chimpanzee mother exhibits exaggerated facial expressions toward their infants as humans do (e.g., van Lawick-Goodall 1968; Bard 1994). In contrast, several Old World monkey mothers (i.e., rhesus and Japanese macaques) seldom look into the eyes of their infants; direct eye contact, in general, has negative connotations in these species and is often interpreted as a threat (e.g., Emery 2000). Nonetheless, a recent study has reported reciprocal face-to-face interactions (e.g., sustained mutual gaze, mouth-to-mouth contacts, lipsmacking) between rhesus macaque mothers and newborn pairs (Ferrari et al. 2009). Additionally, newborn apes and perhaps macaques share at least some of the features exhibited by humans in relation to face perception. Like humans, gibbons, chimpanzees, and Japanese monkeys all discriminate between face-like and nonface-like patterns, as indicated by their preferential looking shortly after birth (Myowa-Yamakoshi and Tomonaga 2001a; Kuwahata et al. 2004). Moreover, gibbons and chimpanzees are also sensitive to faces with open rather than closed or averted eyes, and they pay attention to the gaze of other agents (Myowa-Yamakoshi and Tomonaga 2001b). Finally, chimpanzee and rhesus monkey neonates appear to imitate several human gestures (Figure 3.4; Myowa 1996; Myowa-Yamakoshi et al. 2004; Ferrari et al. 2006; Bard 2007). As for how enhanced survival and/or reproduction may be enabled through this, we tentatively postulate as follows.

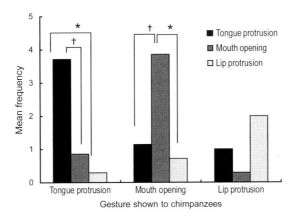

Figure 3.4 Frequencies of the three gestures (tongue protrusion, mouth opening, and lip protrusion) exhibited between one and eight weeks of age (data obtained from Pal, a chimpanzee). The x-axis represents the facial gestures shown to the chimpanzee; $*p < .05$; $^{†}p < .10$.

Hardwired competencies involving face perception that are observed in catarrhine (i.e., Old World monkey and ape) infants may help them form face-to-face interactions with caregivers after birth. In particular, the attention that an infant gives to the caregiver via direct gaze may, in turn, trigger the caregiver to provide greater attention to the infant (particularly via the sharing of positive empathy; see below). This may ultimately create increased opportunities for receiving care (Myowa-Yamakoshi et al. 2005). Still, in terms of eye gazing in humans, we acknowledge that there are strong cultural differences which may impact infant interaction; for example, some Australian Aboriginal people believe that it is disrespectful to look another person, particularly elders, in the eyes. Cultural differences involving face-to-face interactions and subsequent variations in attachment are a topic worthy of future investigation.

Moving past the role of vision (particularly face-to-face) in social interaction and attachment, what about touch? For macaque and ape infants, continuous physical contact is evident as of birth: infants cling to their mother's hair to be safe from predators, to be fed, to build a secure attachment with their mothers, and to learn socially from their mothers. Prosimian (e.g., *L. catta*) and New World monkey (e.g., *Saguinus oedipus* and *Callithrix jacchus*) infants also exhibit continual physical contact by clinging to their caregiver's hair. Yet despite this similarity, and in contrast to Old World monkeys and apes, prosimian and New World monkeys do not generally tend to embrace their infants; the infants themselves are responsible for clinging to the caregiver during feeding and transportation. The mother-infant relationship characterized by such a clinging-embracing (ventro-ventral) bond, especially observed in macaques and apes, might thus be related to the intense emotional connection between them. If so, such a characteristic may be a unique and crucial aspect of the catarrhine attachment system. As research continues, it will be important for future researchers to consider whether interspecies differences in clinging-embracing behavior among and within catarrhines are associated with potential differences in attachment. For instance, although chimpanzee mother-infant pairs are in physical contact via clinging-embracing 24 hours a day for the first three months of an infant's life (Matsuzawa 2006), and humans generally tend to engage in skin-to-skin contact, there appears to be considerable cultural differences among humans as to the amount and style of clinging-embracing which may be related to external factors such as work, on-demand nursing, and weaning (Diamond 2012). More specifically, human infants raised in WEIRD (western, educated, industrialized, rich, and democratic) societies may spend time in a playpen and sleep in rooms separate from caregivers, whereas infants raised in traditional cultures may spend considerable time being carried around by the caregiver and share the same room (if not bed) with their parents while sleeping. It is estimated, for example, that for 90% of the time during the first year of life, !Kung infants engage in skin-to-skin contact with the mother and other caregivers, an amount of time that greatly exceeds the experience of WEIRD infants (Diamond 2012). Determining the extent to which human cultures do vary in this regard, particularly among traditional hunter-gatherer societies and for children of different ages, needs to be a priority for future research.

Finally, in addition to the visual and tactile modalities, it seems likely that many humans also frequently use vocal sounds in parent-infant interactions. More specifically, human infants cry very frequently, and they gradually begin to use their cries to attract the parent's attention. In at least some (i.e., Japanese and Western) cultures, parents often reply vocally to their infants instead of actually embracing them (Takeshita et al. 2009). Whether and how such vocal interactions occur in other cultures and in other primate infant-parent dyads, and the relationship (if any) between such interactions and styles of attachment, remains to be systematically identified (for discussion on face-to-face communication between rhesus macaque mothers and their newborn infants through lip smacking, see Ferrari et al. 2009).

Primate Attachment in Relation to Alloparenting, Cognitive Autapomorphies, and (Ecological and Social) Context

Having considered the role that sensory modalities play in differentiating attachment among primates, we turn to the role of the human mind itself, as this is likely to be a major force behind the unique style of human attachment. Over the last few decades, an explosion in comparative psychological studies has uncovered not only what appears to differentiate the cognitive abilities of apes from other primates, but also what appears to be unique human capacities, such as "nested scenario building" (i.e., open-ended imagination, or stated differently, the ability for recursively reflecting on different situations) and our "urge to connect" with other minds (e.g., Suddendorf and Whiten 2001; Suddendorf 2013; Butler and Suddendorf 2014). Indeed, when combined, these unique abilities may even be responsible for transforming animal communication into human language, habitual behavior into cultural traditions, problem-solving into abstract reasoning, and empathy into morality (Suddendorf 2013). How do these (or other) human autapomorphies (i.e., traits unique to one species) contribute to attachment? As one possibility, we propose that the evolution of human alloparenting is related to the evolution of other specific aspects of social cognition that appears to be unique to humans. Before discussing this, we will set the stage by considering the presence and function of alloparenting among primates in general, with particular emphasis on one of our closest living relatives: the common chimpanzee.

Alloparenting in Chimpanzees

There is limited data on observed chimpanzee births in the wild. Field researchers engaged in long-term studies of wild chimpanzees in Africa rarely get to

witness a birth because females will quietly leave the group and disappear a
few days before going into labor. Like other nonhuman primates, chimpanzees give birth alone (Nishida et al. 2003). Although they return to the group
afterward, it is unknown where they go or what they do while they are away.
Researchers often tell of being surprised to see a female reappear suddenly
with a newborn at her breast. After returning, chimpanzee mothers raise their
children on their own. Even for a first child, a chimpanzee mother receives no
instructions or direct assistance from anyone. Nonetheless, mothers do not appear to abandon their offspring. It is possible, although as yet unconfirmed, that
a mother might return to the group alone and leave behind her newborn if, for
example, it is weak at birth or because of a lack of breast milk.

Curiously, in the early period following birth, a mother seldom wants her
infant to be touched by other group members, and thus alloparenting seldom
occurs in wild settings (van Lawick-Goodall 1968; Nishida 1983; Goodall
1986). At around one year after birth, infants develop motor skills to move
around on their own and can approach other familiar members of the group.
They increasingly spend short amounts of time being looked after by individuals other than their mothers (i.e., short-term alloparenting). Such short-
term (i.e., nonadoptive) alloparenting behavior has also been confirmed in
several Old World monkeys such as baboons (Altmann 1980), vervet monkeys (Lancaster 1971), bonnet monkeys (Silk 1980), and patas monkeys
(Muroyama 1994; Nakagawa 1995).

Many ethologists and primatologists argue on the basis of kin selection
theory that there must be some sort of (reproductive) benefit to nonmothers
who care for the young of others at a cost to themselves. The primary merits
suggested have been the building of good social relationships with the child's
mother and other members of the group (in the form of being groomed by others) and training for future child-rearing to ensure greater success with one's
own child since first-time mothers are usually associated with lower infant
survival rates. Regarding the latter possibility, Fairbanks (1990) found that
in vervet monkeys (*Cercopithecus aethiops sabaeus*), first-time mothers with
high alloparenting experience raised 100% of their first offspring to maturity,
whereas mothers with low alloparenting experience had less than a 50% survival rate for their first infants.

There are far fewer cases, however, of long-term alloparenting, i.e., adoption (operationally defined as any relationship between an adult and another
individual's child in which the adult shows maternal-like behavior, e.g., food
sharing, protection, for at least a two month period) (Boesch et al. 2010). Over
the fifty years that chimpanzees have been observed in the wild, fewer than
thirty instances have been reported (e.g., Goodall 1986; Nishida et al. 2003;
Wroblewski 2008; Boesch et al. 2010). The majority of these were orphan
situations where individuals under five years of age (i.e., not weaned) had lost
their mothers. In 23 cases, 13 represented kin relationships (e.g., grandmothers, siblings, aunts) and ten involved adoption by nonkin members. Thus,

long-term alloparenting does occur among wild chimpanzees with both kin *and* nonkin members, including males. The occurrence of such long-term alloparenting by nonkin raises an issue that requires further consideration: What possible benefits accrue to the alloparent, beyond those observed for short-term alloparenting?

It is important to note that long-term alloparenting does not seem to occur when the biological mother is still alive. Still, we have evidence of full adoption by a grandmother (the mother's mother) in a case where the mother abandoned her child (Wroblewski 2008). Whether nonkin adopt another's offspring after maternal abandonment, however, remains to be confirmed.[1]

Even though chimpanzees exhibit long-term alloparenting, and despite its potential benefits, why is it so rare? Recall that wild chimpanzee mothers seldom allow others to contact their infants within the first six months, indicating that it must be difficult for them to entrust their child to another individual. This is not to suggest that chimpanzees lack the psychological trait of trust (e.g., Engelmann and Hermann 2016). In fact, at the Primate Research Institute (Kyoto University), mothers allowed their children to be touched soon after birth by certain researchers with whom they had grown familiar over the years (Matsuzawa 2006). It is possible that captive settings might allow chimpanzee females to develop the characteristic of trusting other female individuals, as indicated by their more "bonobo-like" female coalitions that have been observed (e.g., de Waal 1998). It will be important to reveal further which factors may affect the development of trust toward other individuals in primates, and ultimately their relevance to alloparenting and attachment (see below).

Human Cognitive Autapomorphies and Alloparenting

Empathy is the ability to share and understand other people's moods through various forms of latent and unconscious transmission or high-order inferential cognitive processing. Merely being aware of another's distress or joy, for example, automatically elicits a similar response in us; this "emotional empathy" is mainly based on the mirror neuron system network (Shamay-Tsoory 2011). Further, deliberately "climbing into the shoes" of another person in an effort to imagine or feel their distress or joy, "cognitive empathy," is thought to be mainly related to mentalizing circuits (Iacoboni et al. 1999).

Humans are not the only animals capable of empathy. Along with our primate relatives, rats, dolphins, and many other animals are sensitive to the experiences of others, but mainly in regard to their feelings of distress (e.g.,

---

[1] Although we acknowledge an instance of chimpanzee adoption involving individuals known as "Gorilla" and "Roosje," documented by de Waal (1998), it is unclear whether maternal abandonment occurred.

Langford et al. 2006; Bartal et al. 2011). The phylogenetic origins of this restricted brand of empathy are thought to be directly linked to survival (i.e., an adaptation): animals may be better able to perceive quickly and take measures to avoid whatever it is that is threatening other group members. Although a small number of species (including orangutans) have shown evidence for emotional positive empathy (Davila Ross et al. 2008), we lack clear evidence that nonhuman animals are capable of cognitively empathizing with other individuals who show positive emotions, such as joy, pleasure, delight, or happiness (Myowa 2012). There are numerous theories as to why humans alone may have acquired this ability, but one interesting claim, relevant to our discussion on attachment, involves observed sex differences in empathizing, with more neurological activity being evident during empathy tasks in females (Baron-Cohen 2003). We assume that such a sex difference might be related to mother-infant pair bonding, especially in humans. With this in mind, it has been argued that mother-infant sharing of pleasurable emotions has been of paramount importance (either as an adaptation or a by-product) in the formation of human strategies for surviving the earliest stages of development (Myowa 2012).

Indeed, let us reconsider the evolution of human alloparenting in light of our possibly unique ability to empathize cognitively with the positive emotions of others. Humans show forms of long-term alloparenting irrespective of kin and nonkin members, as is evident in adoption and even large institutions for foster care (although we acknowledge that some instances of negative care do occur in such situations), whereas long-term alloparenting rarely occurs among chimpanzees and other catarrhine species. We assume that human alloparenting styles are unlikely to have grown solely out of the sharing of distress shown by others. In other words, it seems difficult to motivate others to participate in altruistic activities simply through the sharing of unpleasant emotions. On the other hand, a shared sense of comfort, joy, and achievement gained through the care of children serves as a positive mental reward to others who participate in the process. What at first appears to be an altruistic act, therefore, may be the result of pleasurable emotions derived from caring for children. It seems difficult to otherwise explain why only humans would make such an effort to raise children other than their own. Could it be that reinforcement through the evocation and sharing of pleasurable feelings played a role in the evolution of alloparenting? This is nothing more than guesswork at present (and extensive cross-cultural research is required), but one of the keys for understanding both attachment bonds and the establishment of alloparenting in humans is likely to be the cognitive function(s) of empathy. Similarly, further insights may be obtained by considering whether and how other human cognitive autapomorphies could influence attachment (e.g., mental time travel, theory of mind, nested scenario building, and our fundamental urge to connect with other minds).

Contextual Factors Related to Attachment
and Alloparenting among Primates

While we believe that human cognitive autapomorphies, such as cognitive empathy for positive emotional states, are crucial factors related to the evolution of the human attachment system, attention must also be given to the role of context. What role does one's (ecological and/or social) environment play? A few observations are offered for consideration on this matter.

Recall that wild chimpanzees rarely appear to abandon their young. This stands in stark contrast to over 100 cases of birth with chimpanzees in captivity, where one out of every two mothers that give birth do not care for their infants (Matsuzawa 2006). What causes this abnormal attachment behavior? Over half of the chimpanzees kept in captivity in Japan live in groups of five or fewer members. This is quite different from the wild, where groups typically range from 20–100 members. Although caretakers and researchers are making efforts to increase captive group numbers, small groups mean that most captive chimpanzees do not have the chance to observe or interact directly with their own or other's offspring. This suggests that a lack of social learning opportunities during an early period in life may be one of the primary reasons why some captive mothers end up abandoning their offspring. Clearly, this indicates that a chimpanzee mother's attachment to her infant and her infant-rearing behavior are largely affected by habitat (for a discussion of other great apes and the effects of learning and experience in the development of good maternal skills observed within zoos, see Bard 2002; Abello and Colell 2006). What other evidence is there for enculturation influences?

Tomasello et al. (1993) reported that enculturated chimpanzees developed more imitative abilities than mother-reared chimpanzees. Also, when young nursery-reared chimpanzees were exposed to a novel object, they exhibited gaze alternation between this object and the face of their primary caregiver, a phenomenon called human social referencing (Russell et al. 1997). Indeed, in day-to-day interactions between some human caregivers and infants, social turn-taking behaviors may play an important role in the formation of attachment. Recall our proposal that many human caregivers (in at least some cultures) attract attention to themselves by introducing infants to several "exaggerated" facial expressions, such as raising their eyebrows, opening their mouths wider, smiling, and often imitating the responses of the infants. In turn, infants may be attracted to the caregiver's changeable and attractive gestures and respond to them. Such socially responsive, turn-taking interactions (perhaps based on positive emotional empathy) may play a crucial role in attachment formation and may even increasingly enhance infants' cognitive abilities for things such as imitation (for further discussion on how caregiving practice may have influenced attachment, see van IJzendoorn et al. 2006). In any event, findings involving enculturated primates indicate that cognitive abilities related to attachment develop flexibly, depending on extended exposure to varying rearing environments after birth. This point has been further reinforced by Bard et al. (2005), who found that mother and infant chimpanzees at a Japanese center (Primate Research Institute, Kyoto University) exhibited a higher rate of mutual gaze than those at an American center (Yerkes) (Figure 3.5). Compared to the latter, Japanese chimpanzees have established long-term relationships with human researchers in everyday life and are much more familiar with the human environment and the social ritualizations involved, such as active mutual gaze and joint attention. In other words, Japanese chimpanzees may have higher levels of human enculturation. These results suggest that there is flexibility in chimpanzees' development of mutual gaze and that infants learn group-specific patterns as observed in humans (Keller et al. 2004b). Finally, the possibility that enculturation can influence (and result in) alloparenting has actually been reported, albeit to our knowledge only once: alloparenting among captive lowland gorillas (*Gorilla gorilla gorilla*) while the mother is alive has been observed by Nakamichi et al. (2007), including when nonkin gorilla mothers mutually exchanged and reared their children. It is unknown whether this instance of alloparenting involved the form of 100% child care or was more restricted on a needs basis for the child.

Further longitudinal developmental, cross-cultural, and comparative studies are clearly needed to reveal the effects of socioecological experience and its relationship with species-specific biological foundations in the establishment of attachment between parent and infant pairs among captive and wild primates, including humans. For example, given that mothers in polyandrous-structured primate species show a tendency for having twins (e.g., titi monkeys), and that this seems to result in alloparenting involving the father, what possible relationships exist between attachment and other primate social structures:

Figure 3.5   Mutual gaze between mother and infant chimpanzees (*Pan troglodytes*) in the Primate Research Center of Kyoto University, Japan. Photo used with permission from T. Matsuzawa.

single mother and child, monogamous, polygynous, multimale/multifemale, and fission-fusion (for discussion of these types, see Kappeler 1997). We may similarly ask this question for humans, as they are the only primate to exhibit all of these structures (across rather than within cultures). A related issue is whether and how socioecological factors driving these differences in group structure also impact on attachment (e.g., food availability, group size, genetic relatedness of group members, dominance hierarchies, threat of infanticide). For example, does the threat of infanticide in chimpanzees contribute to the reluctance of wild chimpanzee mothers to disallow contact from other females during their infant's first six months? Is this exhibited by mothers of other primates affected by infanticide (i.e., gorillas and orangutans) or by those which are not, most specifically, gibbons and bonobos?

Human Attachment Summarized: What Is Shared and What Is Unique

Clearly, our understanding of primate attachment has progressed considerably since it was first outlined by Bowlby. Rather than being a uniform phenomenon among primates based on a homology, we know now that attachment varies across primates. Humans appear to be unique in relation to their combination of modalities used in forming attachments (i.e., large amounts of looking and the additional use of tactile and vocal cues), the use of positive cognitive empathy, and possibly certain contextual elements typically associated with human socioecological environments. More research is required to substantiate and potentially extend these claims. Nonetheless, it is time for researchers to adapt attachment theory to the extensive primate literature that has accrued since Bowlby's passing.

Acknowledgments

This study was supported by funding from the Grants-in-Aid for Scientific Research from the Japan Society for the Promotion of Science and the Ministry of Education Culture, Sports, Science and Technology (24119005, 24300103 to MM), the Center of Innovation Program from Japan Science and Technology Agency, JST to MM, and the Mayekawa Houonkai Foundation to MM (2015–2016).

4

Primate Infancies

Causes and Consequences of Varying Care

Kristen Hawkes, James S. Chisholm, Lynn A. Fairbanks,
Johannes Johow, Elfriede Kalcher-Sommersguter,
Katja Liebal, Masako Myowa, Volker Sommer,
Bernard Thierry, and Barbara L. Finlay

Abstract

Bowlby recognized that studying other primates could help identify the needs of human infants; his evolutionary perspective has had a wide impact on understanding of human development. Much more is now known about evolutionary processes and variation, within and between species. This chapter reviews aspects of evolutionary theory and primatology relevant to Bowlby's theory of attachment. Beginning with primate phylogeny, ecological and social forces that contribute to the varieties of primate sociality are considered and some reasons canvassed that explain why primatologists do not all agree on the choice of words to describe the relationships between animals, including use of the term "attachment." To appreciate primate variation, interactions between infants, mothers, and others are characterized in a range of species. Variations and commonalities are identified and used to explore how development in human infants can be understood in terms of social relationships and maturational state at birth and weaning compared to other primates. Infant experience has long-term effects in primates other than humans. Some of that evidence is summarized and special attention is given to interactions between particular chimpanzee mothers and infants in an unusual setting, where trusting relationships between mothers and human researchers reveal variations in mothering style that appear to result from early life events, recent experience, and social context.

An Evolutionary Perspective

Since the theory of natural selection is about "descent with modification," phylogeny (descent) and fitness-related effects (modification) are both paramount. The language of evolutionary biology draws attention to theory missing in descriptions of human attachment, bonding, and welfare, as natural selection does not maximize kindness, happiness, health, or serenity—only inclusive fitness. More of one thing usually means less of something else. Allocation problems are everywhere for all living things. Interactions among individuals are complicated by pervasive conflicts of interest, situated in varying ecologies. For our topic here, the necessary conflicts of interest that pervade the most intimate relationships (i.e., between mothers and fathers, among closest kin, even between mothers and their developing fetuses) constitute the central features of biological theory about the evolution of physiology, morphology, and behavior. Evolutionary expectations about tradeoffs and conflicts of interest distinguish our perspectives from those of attachment specialists, who are concerned primarily with the well-being of mother and child, a difference we will not resolve but underline.

Because there is no starting over in evolution, phylogeny is crucial. Selection can only drive adjustments in the capacities and tendencies of individuals from features already there. Over time, that process has produced the astonishing variation in living things, from the diversity in primates to all of the organisms in our microbiome. For questions about our own species it matters that we are hominids (i.e., our closest living evolutionary cousins are the great apes), that hominids are primates, that primates are mammals, and that mammals are vertebrates. We share most of our physiology with them by descent from our recent common ancestors.

Our charge is to examine the diversity in primate infant-caregiver interactions, but the value of a wider taxonomic perspective should not be ignored. The dazzling variation in vertebrate social systems and parental behavior shows how features of socioecology alter selection. In fact, the diversity that we will describe in primates alone underlines how little close phylogenetic relatedness constrains key features of social organization and parental roles. Conversely, across vertebrates, an impressive degree of conservation of fundamental brain structure (Yopak et al. 2010), neurotransmitter, neuromodulator and hormonal systems (Hofmann et al. 2014), and basic physiological mechanisms (Gerhart and Kirschner 1997) often produces a startling similarity in the mechanistic solutions that adapt brains and behaviors to similar socioecological contexts, even when housed in bodies as diverse as a 30 g house sparrow compared to a 1000 kg bison.

Parental care has evolved independently, hundreds of times in nonmammalian vertebrates as well as in birds and mammals. The nature of care—from single parent to biparental to multiple alloparental systems—shows predictable transitions when viewed from both phylogenetic and ecological perspectives

(Reynolds et al. 2002). This range of parental protection and provisioning illuminates the central requirements. Parental care is nested within a wide variety of social structures (from solitary, to familial, to flocks of millions) as well as within territorial preferences (from territory-independent, to transitorily defined, to multigenerational defense of a huge resource). Only in looking widely at taxonomic interactions of the variations of parental care, social structure, and territoriality can the dependence and independence of the mechanisms of affiliation and care be understood (Goodson et al. 2005). In mammals, the size of the brain, developmental duration, and the possibility of extended parental care covary to a remarkably high degree (Charvet and Finlay 2012). As large brains per se are associated with particular behavioral capacities (MacLean et al. 2014; Stevens 2014), the similarities we discern when comparing our caregiver-infant interactions with those of other large-brained mammals (e.g., dolphins and elephants) may be much more than anthropomorphic projections. When investigating mechanisms of attachment, the insight gained from examining the differences in the neural circuitry associated with the neuro-peptide oxytocin in monogamous prairie voles (*Microtus orchrogaster*) versus promiscuous montane voles (*M. montanu*) (Young et al. 2001) not only overshadowed previous work in primates but initiated an explosion of research in new psychiatric treatments for human developmental and social disorders (Young 2002). A full review of diversity and conservation in social structures and their mechanisms across taxa is far beyond the scope of this report. These few examples should, however, demonstrate the benefits of a wide taxonomic perspective from which we return to our primate focus.

Primates and Primatology

Primates are specialized to be unspecialized; that is, monkeys and apes have the capacity to act in different ways depending on circumstances. Laypersons have historically believed that there is such a thing as a "monkey mother" (or "the chimpanzee," "the gorilla," "the macaque") who behaves in stereotypical, universal ways. Nothing could be further from the truth, because behavioral plasticity is the very hallmark of primates, including our own species (Strier 1994). Understanding the state of affairs is even more difficult, since scientists studying primates will often disagree about "the facts." Like any scientific discipline, primatology, apart from being loosely grounded in evolutionary theory, does not embody a unified method or philosophical approach, generating conflict by its very scientific nature.

In what follows, we introduce a basic portfolio of terms and models, including snags, caveats, and some of the disagreements within primatology. (General introductions and overviews to primate biology, ecology, sexuality, sociality, and cognition that track the discipline's development can be found in Altmann 1980; Smuts et al. 1987; Dunbar 1988; Martin 1990; Hrdy 1999;

Laland and Brown 2011; Campbell et al. 2012; Dixson 2012; Mitani et al. 2012; Strier 2016.) This brief review prepares readers for the concrete examples of primate-infant socialization presented herein. This multiauthored paper includes observations by primatologists from various schools of thought. Thus, while all examples are informed by firsthand research expertise, information about "mother-infant attachment" in primates is colored by the context in which a primatologist learned the ropes. Some will say that "attachment" as a universal, unique, and unchangeable bond is pure fiction, while others find the term useful. We have not homogenized the language that different academics employ as contributors to this chapter. As semantics indicate implicit methodological assumptions (see section on "Intellectual Fault Lines in Primatology"), the attentive reader may take that as a challenge to identify schools of thought that underlie descriptions and assertions.

Primate Taxonomy

Humans, along with several hundred other species, belong to the mammalian order of primates (Boyd and Silk 2015). This taxonomic unit was originally classified *phenetically* (based on appearance; see Figure 4.1, left panels) or more currently *cladistically* (based on ancestry, principally genomic analyses; Figure 4.1 right panels). The phenetic classification divides the group into prosimians and anthropoids (or simians), with the position of the tarsiers constituting the major difference to the cladistic approach. Most primatologists prefer the latter, as it reflects the currently accepted phylogenetic tree, although the term prosimian is still widely used.

The cladistic approach also identifies two major branches of the primate tree. The first is strepsirrhines, the "wet-nosed" primates. These are mostly nocturnal and benefit from a good sense of smell enabled by a mucous membrane around the nostrils. They include the prosimian lemurs and lorises, now typically small creatures found in Africa and Asia. All other primates belong to a second branch: the haplorrhines or "dry-nosed" primates, which generally rely more on vision than olfaction. These include the small-bodied and nocturnal tarsiers of Southeast Asia as the remaining prosimians. Almost all other haplorrhines are diurnal, representing the "true" monkeys. These are again divided into two kinds. Species native to South and Central America are called New World monkeys (platyrrhines), encompassing the small callitrichids (marmosets, tamarins) as well as capuchin, howler, and spider monkeys. Species living in Africa and Asia are called Old World monkeys (catarrhines) consisting of two clades, the preferentially folivorous (leaf-eating) colobines (e.g., langurs, colobus, snub-nosed monkeys) and the more omnivorous cercopithecines (e.g., macaques, guenons including vervets, drills, baboons).

The ape radiation (hominoids) is divided into two branches. The *small apes* (hylobatids) are confined to South Asia and comprise the siamang and various gibbons. All hylobatids are specialized fruit eaters that swing through the

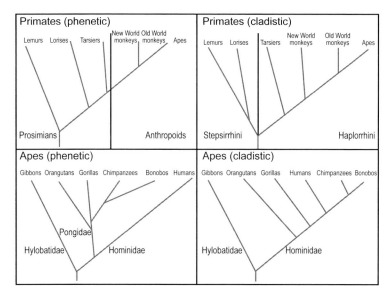

Figure 4.1   Primate taxonomy, with groupings according to appearance (phenetic, left panels) and ancestry (cladistic, right panels). Note that the graph separating the other great apes from humans (lower left panel) implies they are closer to each other, with our lineage at the far right supposedly progressing toward higher forms. This anthropocentrism is mitigated in the cladistic taxonomy (lower right panel) where actual ancestry nests humans among other great apes. Reproduced with permission from Volker Sommer.

canopy using brachiation as their characteristic mode of locomotion. The *great apes* (hominids) include orangutans (genus *Pongo*), the "red apes" of Sumatra and Borneo, as well as the African gorilla (genus *Gorilla*) and the two sister species, chimpanzee and bonobo (genus *Pan*). Humans (genus *Homo*) belong to this clade. We also originated in Africa, but have since populated the globe. Even though members of genus *Pan* and genus *Gorilla* are hairy knuckle-walkers with moderate to good climbing abilities that make them closer to each other than to humans on phenetic grounds (lower left panel), the cladistic approach takes genetic data and the fossil record into account (lower right panel). Those lines of evidence show that genus *Pan* shares a more recent common ancestor with humans than with gorillas, whereas gorillas share a more recent common ancestor with humans than with orangutans. Taxonomically, *Homo* and *Pan*, as well as extinct genera like *Australopithecus* (known only from fossils), are united in a tribus, the *hominini*.

Primate Sociality

Like many other animals, primates need to engage with conspecifics to survive and breed. John Bowlby's understanding of primate sociality was influenced by Robert Hinde (1976). According to this pioneering animal ethologist, individual acts of *behavior* can be distinguished: a behavior directed toward somebody is social, thus constituting an *interaction*; repeated interactions between the same individuals amount to a *relationship*. These technical terms circumvent questions about the actors' internal states (see below), while still allowing reconstruction of social dynamics.

A further set of terms distinguishes basic arrangements of sociality (cf. Sommer and Reichard 2000; Dixson 2012). In numerous species, males and females are more or less solitary and only meet occasionally. The cardinal types of *permanent male-female associations* are:

- one-male/one-female (*monogamy*),
- one-male/multifemale (*polygyny*),
- multimale/multifemale (*polygynandry*), and
- multimale/one-female (*polyandry*).

The neutral term "association" is consciously chosen because it may refer to patterns of grouping, mating, or breeding, and these different dimensions of social organization are not necessarily congruent (Kappeler and van Schaik 2002).

Thus, we can ask how individuals group ("Who resides with whom?"), mate ("Who has sex with whom?"), and reproduce ("Who generates offspring with whom?"). For example, female A may live with male A, the father of her first child (grouping monogamously) while engaging in occasional extra pair copulations with neighboring male B who lives with childless female B. The extra pair copulations lead to the conception of female A's second offspring. This means that female A is monogamously grouping but polyandrously mating and breeding; male A is monogamously grouping, mating, and breeding; and male B is monogamously grouping and breeding, but polygynously mating. Within the constraints and opportunities of these associations, females engaging in sexual encounters may be fertile (i.e., periovulatory) or infertile (i.e., in non-ovulatory cycle stages, pregnant, in lactational amenorrhea, postmenopause). Although species tend to express typical grouping, mating, or breeding patterns, a great deal of intraspecific variability has been observed. For example, populations of gorillas or langur monkeys may live in polygyny or polygynandry, while hylobatids or callitrichids may live in monogamy, polyandry or polygyny. Moreover, in some species such as geladas, hamadryas, or Guinea baboons, and humans, social systems are modular or multilevel, with smaller monogamous, polygynous, or polygynandrous units clearly distinct within larger communities that travel, forage or sleep in proximity.

Another way to look at primate societies is in terms of natal *dispersal*; that is, to record who remains in their birth unit (philopatry) and who leaves to join another group. In many species, one sex may be philopatric whereas the other sex will emigrate. If females stay, *matrilineal* groups result (typical for, e.g., macaques, olive baboons, and at least some orangutan populations). With relationships among close female kin, such groups have been called *female bonded*, although some primatologists stick to the more neutral term *female philopatry*. *Patrilineal* groups result if males remain in the area where they were born. This is typical of chimpanzees and long thought typical of human hunter-gatherers, but careful appraisal has now falsified that assumption. Instead, among hunter-gatherers, either or both sexes may stay or leave, local group composition changes frequently, and bi- or multilocality is most common (Alvarez 2004; Marlowe 2004; Hill et al. 2011). Gorillas of both sexes disperse, with males going further in at least some populations (Roy et al. 2014). Both males and females also emigrate in some hylobatids and callitrichids as well as in many mercantile and industrial human societies.

Thus, the order of primates is characterized by a wide array of social features that differ not only between but also within species. With this considerable *interspecific* and *intraspecific* variability in which individuals are physically together, the frequency and modality of infant interactions with other group members, including its mother, must depend on numerous factors. These include social systems, kin relations, probability of paternity, parity of mothers, number and age of co-residing siblings, and dispersal pattern. We should not expect, therefore, to find a single pattern of "mother-infant attachment" (cf. Keller and Chaudhary, this volume).

Primate Socioecology

The sketch of primate sociality above was purely descriptive and did not address *why* these animals interact in particular ways and not in others. However, potential selection pressures were identified that may cause or at least influence different modes of gregariousness. The problem is complicated by the fact that while natural selection is expected to design the anatomical and psychological features of living things as adaptations to local environments, conditions change. This leaves organisms with phylogenetic legacies that do not indicate current selection pressures, but selection in the past (Grafen 1988; Chapman and Rothman 2006; Clutton-Brock and Janson 2012), producing misleading correlations, or lack thereof, that must be laboriously teased apart.

Multiple environmental pressures might affect social behavior. To illustrate, we present several examples and then describe one well-studied factor, food distribution, at greater length. One pressure that affects social behavior is *infectious disease* (Nunn and Altizer 2006). High frequency of infectious disease may favor smaller group sizes or, in the case of sexually transmitted diseases, may bias mating effort away from promiscuity and toward monogamy. In another scenario, increased *predation pressure* (Miller 2002) may select for groups where multiple males protect vulnerable females and offspring. Alternatively, females may benefit from mating with multiple males to distribute the possibility of paternity and reduce the risk of male-committed *infanticide* which may likewise create multimale/multifemale groups (van Schaik and Janson 2000). In yet another framework, trade-offs between cooperation and competition are proposed to select for *social intelligence* (Dunbar 1992), a cognitive capacity presumed to enable greater numbers of conspecifics to reside together.

Efforts to link the quality, quantity, and distribution of food to sociality have taken particular prominence (Wrangham 1980; van Schaik 1989). The starting point for this socioecological model is the different investment of the sexes in reproduction, beginning from the level of ova and sperm (Fisher 1930; Trivers 1972; Queller 1997; Kokko and Jennions 2003; Lehtonen et al. 2016). Moreover, the internal fertilization, gestation, and lactation of mammals further specializes primate sexes in reproduction. Whereas female reproductive success is limited by food and safety, reproductive success for males is limited by access to fertile females. These dynamics are encapsulated by the epigram that females go where the food is, while males go where the females are or, more technically, that males map themselves onto the distribution of females.

Other things equal, a female might avoid competition for food by living alone. Hence, the fact that females nevertheless are found in groups needs explanation. As in many other animals, females may reduce the risk of predation by staying together. Two other main reasons related to food distribution likely influence grouping. On one hand, costs of gregariousness can be low when females subsist on spatially dispersed, low-quality foodstuffs, such as grass or leaves, that no single individual can monopolize. On the other hand, when females depend on spatially clumped, high-quality food such as ripe fruit, competition with other group members may be high. Females in larger groups, however, may have greater success defending food patches against competing groups.

These conditions influence female social relationships via different competitive regimes. In this model, low-quality, dispersed food is correlated with mild *scramble competition*. Severe aggression is rare and dominance hierarchies are all but absent so the society is more or less egalitarian. By contrast, high-quality clumped food is correlated with *contest competition* where defending food gives winners greater shares. This behavior leads to steep and despotic dominance hierarchies. However, that steepness can be lessened if competition between groups results in benefits to higher ranked members while allowing subordinates enough access to resources to make staying in the group worthwhile.

The spatial distribution of females, whether they are dispersed or clumped, influences whether males are able to defend multiple females, as does the temporal distribution of female fertile periods. Hence, if fertile periods are synchronized, a single male, however strong, cannot monopolize the whole batch, because while copulating with one female, he cannot prevent competitors from mating with a second or third. Synchrony is expected in pronounced seasonal environments where pregnancy, lactation, and/or weaning are unlikely to be successful during certain months when food availability is insufficient. If groups are very large, even without seasonal breeding, multiple females will be fertile on any given day. As a result, multimale groups develop. Conversely, if female fertility is not synchronized and groups are not too large, single males may succeed in defending them.

Thus, we can relate the principal patterns of male-female associations described above to selective forces that are brought about by the spatiotemporal distribution of food and its consequences for the distribution of females:

- If females live solitarily, males cannot monopolize more than one of them (*monogamy*). A textbook example is found in gibbons.
- If females form groups, but their fertile periods are synchronized, multimale/multifemale societies will develop (*polygynandry*). Some macaque species provide a textbook example.
- If females form small groups, but their fertile periods are not synchronized, one-male/multifemale groups or "harems" can form (*polygyny*). Langurs are a textbook example.
- If females give birth to more than one infant at a time, multiple males may need to assist the female in carrying, grooming, and protecting the offspring to reduce mortality, thus resulting in one-female/multimale societies (*polyandry*). Tamarin species provide a textbook example.

Of course, real-life situations are far more complex, and this intentionally limited set of factors does not capture all possible variations. Notable exceptions and intraspecific variations exist, as we will describe. The socioecological model has its critics (e.g., Thierry 2008). Nevertheless, the model has produced valuable predictions about the willingness of females to let other females engage with their infants (Hrdy 1976; McKenna 1979). In fact, we can distinguish between permissive mothers who may spend considerable time away from their infant while it is carried around by troop mates, and restrictive mothers who will fiercely resist attempts by others to take hold of her youngster. As a rule, infant sharing characterizes societies with scramble competition (e.g., langurs). Here, a mother can easily retrieve her infant as the dominance hierarchy is more relaxed. Moreover, infants will not grow up to be future food competitors, and thus troop mates have little reason to be aggressive toward them. The situation is different under contest competition (e.g., some macaques, some baboons). Here a low-ranking mother, in particular, would be unable to retrieve her offspring if a high-ranking troop mate resisted her efforts. Numerous cases of infant "kidnapping" with subsequent "aunting to death" have been observed. Thus, in species with contest competitions, infant sharing is typically restricted to close kin (older sisters, grandmothers), who have an overlapping genetic interest with the mother and her newborn. Examples from nonhuman primates, like those from different cultural settings among humans, thus warrant "a radical change from a dyadic perspective to a network approach" (Keller and Chaudhary, this volume).

Intellectual Fault Lines in Primatology

Primatology is conducted by human primates who have been educated as scholars and academics within an attachment network of mothers and others situated in a cultural context. As primatologists, therefore, we have our perspectives and senses of belonging. We identify with certain modes of "we"-ness and "other"-ness. In other words, we are not blank slates (Chisholm, this volume). Some of the more subtle factors that shape our research approaches follow. Being primates, we are, in fact, subject to intraspecific variation ourselves. Whether a primatologist prefers or detests certain phrases often reflects their school or academic circle.

Let us begin with what initially seems to be a semantic point regarding the relationships in which infants engage with others. Some primatologists will happily employ the term "infant caregiving." Others point out that such words render the infants as passive recipients whereas the infants might well manipulate others through signals of distress or "cuteness" into attending to them (Hrdy 1999). Similarly the term "allomothering" indicates that individuals other than mothers engage with the youngest group members (Hrdy 1977). This includes other females as well as juvenile or adult males who may show affiliative interest in babies or be rough or abusive. The relatively neutral expression "infant handling" can apply to mothers, fathers, siblings, aunts, uncles, nonrelatives or peers alike, whether they tend to the baby in a friendly mode or mistreat it.

Most primatologists agree that comparing wild populations with captive animals adds useful information to our understanding of the mechanisms and functions of primate sociality (Setchell and Curtis 2003). It is difficult to control the multitude of factors that influence behavior observed in the field, such as seasonality and food availability or pressures related to disease, predation, or intergroup competition. In the wild it is also more difficult (but not impossible) to obtain biological samples (urine, feces, blood, saliva, hairs) that can be analyzed to determine endocrine state, nutrition, pathogenesis, or genetics. Finally, it has been notoriously difficult (again not impossible) to conduct experiments in the wild that explore, for example, behavioral or cognitive suites. Without studies in captivity, we would not know that some orangutans can dive, that a nonhuman ape can operate a joystick to play the "Pac-Man" computer game, or that bonobos and gorillas readily use tools, because in their natural habitats, they have practically never been seen to do this. Thus, captive studies help us to understand a species' breadth of potential behavioral responses.

The community of primatologists also tends to be split between those who emphasize traits of human uniqueness (nicknamed "exceptionalists") and those who stress the evolutionary continuum ("gradualists") (cf. McGrew 2004; Finlay and Workman 2013). We might simply distinguish between those who are more interested in differences between humans and other primates and those who are more interested in similarities. Current research focus is on cognitive abilities (e.g., shared intentionality, mental time travel, language; Tomasello and Herrmann 2010; Hawkes 2014), which may or may not be specific to humans. Some primatologists identify the major cognitive rift not between humans and (other) animals, but rather between great apes and other primates (Russon et al. 1996). These opposing views generate lively debates, often to the puzzlement of a lay audience that expects clear answers, for example, as to whether chimpanzees have language abilities or not (Hurley and Nudds 2006).

The wider arena of cognition (Tomasello and Call 1997) brings us to the contentious topic of *internal* states, which has immediate relevance to the study of attachment (Bekoff 2007; Cheney and Seyfarth 2007; Stamp-Dawkins 2012). This time-honored conundrum is dominated by grand and often poorly defined vocabulary. Thus, we may ask if nonhuman animals possess "emotions," "feelings," "empathy," or "consciousness" (de Waal 2016), or if, in relation to our core question, they experience an inner "bond" of "attachment." Traditional Cartesian dualism denied that animals, which were seen as machines, can experience qualia (i.e., have an experience of internal states and sensations). One would be hard pressed to find current animal behaviorists who ascribe to this historical orthodoxy. Still, many students of behavior will avoid terms that imply private feelings, not because they deny that other animals can have emotions or feelings. On the contrary, modern-day ethologists and comparative psychologists tend to be at least moderate gradualists. Thus, they think it unlikely that internal experience emerged only in the more immediate ancestors of our own lineage (e.g., Panksepp 1998; Toda and Platt 2015): they prefer descriptive and less interpretative terms (e.g., "relationship" instead of "attachment"). Still this does not prevent others from reinterpreting their data in a language that implies internal states, only that the onus of justification would be on those who add such colorful layers to their portrayal of primates. In sum, gradualists recognize that we are not only justified, but obliged to employ a certain dose of *anthropomorphism* when looking at other animals (Daston and Mitman 2006; de Waal 2016). This is in line with Darwin's original idea that as with anatomy, likewise our psyche differs from that of other animals only in degree, not in kind (Darwin 1872/1965). A gradualist will therefore assume that mental experiences are more similar between humans and other hominids than between humans and small apes, and that the latter have more similarities than those between humans and monkeys. But similarity is not only a question of phylogeny. The socioecological model predicts that similar ecological pressures generate similar social responses; parallel evolution may likewise happen with corresponding neurological solutions. Thus, highly social nonprimate mammals, such as elephants and whales, and even nonprimates, such as birds (corvids perhaps, or parrots), may be most usefully viewed from an anthropomorphic perspective.

Nevertheless many continue to avoid using emotive terms for heuristic reasons. We can simply not know if a macaque is enraged or anxious, if a chimpanzee feels guilty or ashamed, or if an orangutan mother that cradles her newborn feels happiness or love. To be precise, we are notoriously poor at introspection ourselves and cannot know for sure about the private feelings of fellow human beings. We just take their word for it as a convincing approximation. In addition, social anthropologists maintain that emotion-describing words are, even with respect to the human experience, susceptible to particular economic and political situations; that is, they are "socially constructed." The way a Nigerian baby feels when she is breastfed may differ from a Japanese infant.

Others, including hard-core Bowlbyists, might assume that such qualia are independent from the Zeitgeist and are therefore "universals." Still, prime examples of universal features have been disappearing at the same speed in which detailed knowledge about our fellow primates has been accumulating, leading us back to the elephant in the room: "attachment." Some primatologists use the word attachment, as it seems to embody how caregivers and infants engage that is otherwise hard to describe or define, while others prefer to avoid its use.

Whichever perspective one might hold, all agree about a basic flaw in Bowlby's attachment theory: a single species was recruited as the principal witness of the continuum in morphology and psyche to connect humans with other primates. Several decades of research in captivity and the wild have since made clear that one species cannot be seen as representative of another.

Some Primate Variation

The range of variation will be clear, even without a complete review of the primate order. Notably missing in our selection are prosimians, callitrichids, baboons, orangutans, gorillas, and bonobos. Comprehensive coverage is also out of the question because one population of any species might not be representative of another population of the same species, as we have learned that intraspecific variation can be extensive. In reference to humans, we speak of "cultural" variation (Keller and Chaudhary, this volume), and this term is now also commonly used to refer to behavioral diversity among nonhuman animals, in particular primates (McGrew 2004). Bowlby used one nonhuman primate as a stand-in for what happened during the course of human evolution: the rhesus macaque, not very aptly also called "the" monkey. Thus, it is there that we begin.

Rhesus Macaques

The attachment theory developed by Bowlby was directly influenced by the works of Harry Harlow, Robert Hinde, and their collaborators, who studied the effects of maternal separation in infant monkeys (Seay and Harlow 1965;

Spencer-Booth and Hinde 1967; Harlow and Suomi 1974). As all of them studied the same nonhuman primate, the rhesus macaque, characteristics of this species, like characteristics of the particular human subjects he studied, affected Bowlby's theoretical framework (see Vicedo as well as Keller and Chaudhary, this volume). We now know that the social system of rhesus macaques does not represent primates in general (Strier 1994) and is not even typical of the numerous species in their genus of Old World monkeys, *Macaca* (Matsumura 1999).

Rhesus macaques are often especially aggressive, with a temperament that drives them to threaten others at the slightest provocation; their strongly hierarchical social relations are marked by the paramount importance of dominance, submission, and kinship (Thierry 2007). Correspondingly, the rhesus macaque mother exerts close control over the social interactions of her infant, which leads to the development of an exclusive relationship between them (Figure 4.2). As a consequence, the experimental separation of mother and infant produces an intense response in the infant that typically follows two steps: (a) a "protest" stage characterized by an increase in locomotion and distress vocalizations and (b) a "despair" stage characterized by social withdrawal, inactivity, and a recognizable depressive posture (Spencer-Booth and Hinde 1967; Harlow and Suomi 1974). These stages mirror the anaclitic response reported for some human children in similar circumstances, which provided the foundation for Bowlby's views (1969).

Figure 4.2 Maternal protectiveness in rhesus macaques. A mother limits the moves of her infant. Photo used with permission from Bernard Thierry.

Broader Variation in Macaques

Comparative study across the macaque radiation provides insights about the social factors affecting the degree of exclusivity of the mother-infant relationship and the development of infant caregiving. Macaques have a flexible diet that includes a major frugivorous component. This genus is characterized by similarity in basic patterns of association by sex, and great differences in the severity and importance of dominance hierarchies. On one hand, all macaque species live in multimale/multifemale groups with overlapping generations: females form kin-bonded subgroups within their natal group, while males transfer between groups at maturation. On the other hand, they present marked interspecific contrasts in levels of social inequality (Thierry 2004, 2007). Some species, like rhesus and Japanese macaques, display an intolerant social style: severe biting is not rare and conflicts are highly unidirectional, meaning that the target of aggression generally flees or submits, producing steep dominance hierarchies.

Other species, like crested and Tonkean macaques, have more tolerant relationships: weaker group members often protest or counterattack when threatened by higher-ranking individuals, biting is neither frequent nor injurious, and quarrels often end with mutual appeasement between previous opponents; dominance relationships are relaxed. The conditions of socialization covary with these patterns. In macaques with steep dominance hierarchies, most mothers are quite protective. Except for the highest-ranking females, mothers limit their infants' interactions mostly to close relatives. Consequently, the amount of alloparental care remains limited. By contrast, mothers from tolerant species are quite permissive, with mothers allowing most females in the group to handle and carry infants from an early age (Figure 4.3). Of particular importance for Bowlby's characterization of attachment, it has been shown that temporarily removing the mother from the group in tolerant macaque species does *not* induce the depressive state typically reported in rhesus infants because the relationship between mother and infant is less exclusive in tolerant species, and the care provided by other group members buffers maternal absence (Kaufman and Rosenblum 1969; Drago and Thierry 2000).

Several adaptive functions have been proposed to account for the occurrence of alloparental care: assistance to the mother, socialization for the infant, or a learning process for juvenile allomothers (Lancaster 1971; Hrdy 1976; Maestripieri 1994). These hypotheses do not explain why infant handling by individuals other than the mother is limited in some macaque species and frequent in others. Since female attraction to infants is vital in animals with extended periods of growth and development like primates (Quiatt 1979), selection must favor female tendencies to pay considerable attention to their own infant. For them to then show no interest in the offspring of other mothers, one would have to postulate that attachment processes only occur during brief sensitive periods, something incompatible with what we know about primate

Figure 4.3 Alloparental care in Tonkean macaques. A juvenile female carries an un-related infant. Photo used with permission from Bernard Thierry.

learning abilities. The covariation found between dominance style, maternal behavior, and patterns of infant caregiving may be best explained by the level of protection needed by infants in a given social environment (McKenna 1979; Thierry 2004). Mothers living in strict hierarchies have to be restrictive to secure their offspring. Females from more tolerant societies behave more confidently, allowing their offspring to move about unrestricted. Alloparental care could be a side effect of the expression of maternal behavior within a given social milieu and still provide advantages for nutrition and reproduction. Allomaternal care may allow the mother to devote more time to food searching (Stanford 1992) or decrease her interbirth interval by reducing the time spent with the infant at her nipple (Fairbanks 1990).

Several attempts have been made to correlate the contrasting social styles of macaques with the main ecological features of their habitats. The socioecological model, in particular, proposes that animals live in groups to reduce predator risk; that group living, in turn, induces feeding competition between individuals and groups, which varies with the character and distribution of resources (van Schaik 1989; Sterck et al. 1997). After decades of testing, however, it appears that this model fails to account for the interspecific variations observed in macaque social systems. We do not know of any ecological factor that can account for contrasts in macaque social styles (Ménard 2004; Thierry 2007; Clutton-Brock and Janson 2012). Instead of varying with the distribution of foods, social styles appear to vary predictably with phylogeny (Thierry 2007). The empirical finding that macaque social systems represent covariant sets of behavioral characteristics that travel together through evolutionary time means that no characteristic can be explained separately from others, and this includes the mother-infant bond and caregiving system.

Langurs

In contrast to the apparent lack of correlation between current ecology and dominance hierarchies in macaques, Indian langur monkeys, a species in which allomaternal care is common, fit the socioecological model outlined above (see section on "Primate Sociality"; also Hrdy 1977; McKenna 1979; Sommer 1989, 1996). In langur societies, infants interact from birth with a wide variety of group members. Typically, more than ten different nonmothers carry, groom, protect, and play with them. The only exclusive activity for mothers is nursing and closely embracing their infants through the night. Langur neonates are transferred the moment they are born, with the umbilical cord still attached (Figure 4.4).

Prospective caretakers will often quarrel among themselves as to who gets to hold an infant. As a result, during their first month of life, langurs spend on average one third (sometimes up to half) of their waking hours away from their mothers. All infants interact with multiple individuals, which sometimes includes juvenile males but is typically juvenile females and adult females of various reproductive stages (nulliparous, primiparous, multiparous), regardless of whether the female is pregnant, in menstrual cyclicity, or lactating. Limits to allomaternal investment appear in nursing: even if females are nursing their own offspring, they will not allow other infants to drink while tending to them. "Allo-nursing" is absent.

Unencumbered mothers will preferentially perform activities that are difficult with a baby on board, such as foraging in trees and interacting in friendly or agonistic ways with troop mates through mutual grooming and the occasional

Figure 4.4 Allomothering, where many females care for a newborn infant. On its second day of life, a newborn male infant (its umbilical cord still attached) is the subject of a tug-of-war between two unrelated juvenile females, while its half sister watches the event. Photo used with permission from Volker Sommer.

squabble. Hence, infant sharing in Indian langurs is functionally quite precisely described as "babysitting" (Figure 4.5).

Socioecological theory links the high frequency of allo-handling to the langur-typical food: leaves, which they can depend on thanks to a ruminant-like sacculated stomach. Due to the abundance and low nutritional quality of leaves, there is little reason to engage in fierce competition over this resource, unlike females of other species that rely on ripe fruit. These basically folivorous females can thus afford to group, which in turn often allows a single strong male to monopolize a batch of females. Consequently, all infants sired by the resident adult male are at least paternal half-siblings, while some of them are also full-sibs. Thus, over generations, a close network of kin relations develops among the permanent female residents, as they remain for life in their natal troops.

Such complex and close kin relations among females are conducive to infant sharing as every individual babysitter, to a certain degree, helps copies of its own genes carried by the infants to be transported into the next generation. Alternatively, instead of invoking kin selection, some researchers have linked babysitting to potential benefits of "learning to mother." It is the nulliparous females that take infants most often and keep them longest

Figure 4.5   Babysitting in Indian langur monkeys. Here, two unrelated juvenile females take care of a three-week-old female infant. Her unencumbered mother uses the opportunity to supplement her diet by plucking flowers from an Acacia tree, something that would be difficult to do if she were carrying the infant herself. Photo used with permission from Volker Sommer.

(Hrdy 1977, chapter 7). However, experienced mothers, too, will take care of other babies, and juvenile males may (rarely) carry them around despite the fact that as adults, males will never again hold or groom a baby. In any case, the langur example reminds us how ecology and social structure are entwined and how these dynamics will influence the interactive network of infants.

Vervets

Vervets are an African monkey species living in multimale/multifemale matrilineal social groups in woodland and forest fringe habitats. In the wild, vervets have a varied, mostly vegetarian, diet that includes fruit, pods, flowers, bark, and young shoots, supplemented by insects, lizards, eggs, and baby birds. Research on social relationships between vervet mothers, infants, and others was conducted for three decades at the UCLA-VA Vervet Research Colony, a captive facility managed to approximate the natural social conditions for this species in the wild. Research at the colony confirmed the strong interest in infants by allomothers that has been observed for vervets in the wild (Lancaster 1971) and provided the opportunity to collect detailed longitudinal data on the costs and benefits of an extended caregiving system for mothers, infants, and allomothers (Fairbanks 1990).

When a vervet infant was born in the colony, other group members typically came over to touch and inspect the new group member as early as the first day of life. All group members showed interest in young infants, including immature and adult males, but the most avid caretakers were juvenile females. If a juvenile female had an infant sibling, she was its most frequent caretaker; if she did not, she found another infant to hold and carry. There was variability in how comfortable mothers were with all of this attention, with low-ranking mothers being more protective and some high-ranking females allowing their infants to be carried by others up to 40% of the time.

Observations at the Vervet Colony are consistent with fitness benefits of caretaking for both mothers and allomothers. The mothers benefited from the time their infants spent with caretakers as it increased the time between nursing bouts, thus reducing the effects of lactational amenorrhea, increasing fertility, and shortening the next interbirth interval. Benefits were found for juvenile female caretakers when they produced their first infant several years later. First-time mothers who had above-average caretaking experience as juveniles were more likely to produce a surviving infant on their first pregnancy compared to females with less caretaking experience—a correlation that might also follow from differences in capacities and preferences that affect both juvenile caretaking and success as first-time mothers. Infant mortality in the colony was not related to the percentage of time that infants were carried by nonmothers, indicating that, at least with no predators and ample food, mothers who used caretakers were not reducing their infants' chances of survival.

Another group member with an important impact on the mother-infant relationship in vervet societies is the infant's grandmother (Fairbanks 1988). Just as the mother provides a secure base for the infant, the maternal grandmother provides a secure base for the mother-infant pair. Mothers who had their own mother in the group restrained their infants less and were more relaxed in infant care. Infants with grandmothers available were more exploratory and began their forays away from close proximity to the mother at an earlier age. Infants formed special relationships with their grandmothers and were groomed by them more often than by any adult female, other than the mother (Fairbanks 1988). Vervet grandmothers also had a significant effect on their daughters' fitness. Young mothers whose offspring had grandmothers in the group produced more surviving infants than comparable mothers without grandmothers (Fairbanks and McGuire 1986).

This example of the extended caregiving system of vervet monkeys illustrates that the mother-infant dyad is embedded in the larger social world. Mothers who can effectively take advantage of the help and security provided by others in caring for their current infant can benefit by increasing their lifetime reproductive success. Juvenile female allomothers promote their own fitness by gaining experience in infant care, and grandmothers contribute to the welfare of their descendants in the next generation.

The captive setting at the Vervet Colony also provided the opportunity to observe how vervet infants responded to a version of the Strange Situation Procedure used to measure attachment in human children (Fairbanks, unpublished). The following test was conducted when the infants were six months old, roughly equivalent to a preschool age child, and an age when they were spending most of the day away from the mother playing and interacting with other group members. The outdoor corrals at the colony each had a small, connected wooden shelter. When a six-month-old subject spontaneously entered the shelter area, the drop door was closed and it was separated from the group for five minutes. When the door was opened, every infant tested immediately ran over and contacted its mother. So, it appears that even in a species with multiple infant caretakers, the mother has a special role as the primary source of comfort.

Small Apes

The hylobatids have been traditionally described as living in small, stable groups, comprising a sexually monogamous adult breeding pair and their offspring (Leighton 1987). Thus, group size is drastically smaller than in great apes, with close social bonds between the two partners and no pronounced dominance hierarchy (Carpenter 1940; Chivers 1974). While gibbons are frugivorous and highly selective eaters (McConkey et al. 2002; Harrison and Marshall 2011), siamangs mostly feed on leaves (Gittins and Raemaekers

1980; MacKinnon and MacKinnon 1980), although there is considerable variation among populations (Chivers 1974; Mackinnon 1977). The traditional notion of mandatory nuclear families might fit a classic Bowlbyian view.

This textbook picture of a monogamous pair with dependent offspring, however, has been increasingly challenged (Fuentes 2000; Reichard 2009; Reichard et al. 2012) after groups with more than two adult individuals have been reported for several species (Malone et al. 2012). For example, two thirds of the males and almost half of the females of white-handed gibbons at Khao Yai National Park in Thailand live in at least one other type of group structure in addition to pair living, which is still the most frequent type of social organization (Reichard et al. 2012:242). Furthermore, hylobatid females are sexually polyandrous: they engage in extra pair copulations while living in pairs, which can result in fertilization (Kenyon et al. 2011; Barelli et al. 2013), or copulate with both males in multimale groups (Palombit 1994; Reichard 1995; Barelli et al. 2008). Thus, hylobatids are characterized by a considerable degree of social flexibility, with females taking an active role in pursuing their reproductive interests (Sommer and Reichard 2000; Reichard et al. 2012).

There is no strict dominance hierarchy between pair partners. However, several studies mention that partners might take different roles, with females usually leading the group while traveling and males more dominant in encounters with other groups (Chivers 1976; Reichard and Sommer 1997; Barelli et al. 2008). Intergroup interactions can account for up to 9% of the daily activities in white-handed gibbons at Khao Yai Nationalpark in Thailand (Reichard and Sommer 1997), with the majority representing chases between males, but only little contact aggression (Reichard and Sommer 1997; Bartlett 2003).

In addition to the adult pair, groups include up to four offspring of different ages, with interbirth intervals of two to four years. Infants are usually carried by the mother (Figure 4.6), with siamangs differing from gibbons because of their direct paternal care for the infant (Lappan 2008). When an infant reaches eight months of age, males do some of the carrying, although this behavior might not occur in all siamang pairs (Chivers 1974). In the rare event of twins, it has been reported that other group members, such as brothers (in addition to the father), might carry the infants (Dielentheis et al. 1991). Infants are usually weaned around 12 months of age (Chivers 1976); however, others have reported weaning to occur at 15–19 months (Fox 1977). Upon reaching maturity, original descriptions claimed that the oldest offspring becomes increasingly isolated from group activities and finally leaves its natal group (a process called "peripheralization"; Fox 1977) to establish its new group. However, recent studies challenge this generalization, showing that sexually mature individuals often remain in their group if inbreeding or delayed reproduction is not an issue (Brockelmann et al. 1998). Alternatively, they may directly immigrate into another existing group without a solitary period (Sommer and Reichard 2000).

Figure 4.6 Gibbon mothers and infants. Infant at three weeks of age (a) and at six months (b). Photos used with permission from Manuela Lembeck.

Chimpanzees

Chimpanzees are largely frugivorous and complement their diet of fruit with insects, honey, and occasionally mammals, including other primates (e.g., Goodall 1986; Stanford 1992). At some study sites, they have been observed to use a variety of tools, such as sticks to extract honey from beehives and ants from their nests, and hammers and anvils to crack nuts (Whiten et al. 1999). Predation by leopards and lions is a risk for some populations (Boesch and Boesch-Achermann 2000).

Chimpanzee multimale/multifemale communities can consist of more than 200 members (Wood et al. 2017), but individuals travel and forage in small, often changing, subgroups. Males stay in their natal community and females usually disperse to other communities at adolescence (Goodall 1986; Boesch and Boesch-Achermann 2000). Adult males as well as females exhibit a linear dominance hierarchy, with males dominant over females (Goldberg and Wrangham 1997; Wittig and Boesch 2003). One explanation for male philopatry and female emigration may be that male alliances provide protection from infanticide in preventing trespassing males from neighboring groups from entering the communal range (van Schaik 1996). Several cases of infanticide by neighboring males have been documented (Goodall 1977), and females avoid boundary areas when they have infants (Goodall 1986).

Male chimpanzees form parties that patrol the boundaries of a home range and respond highly aggressively toward male strangers (Manson et al. 1991). Adolescent females trying to enter neighboring communities face substantial costs from the aggression of females previously established there. The dispersal attempts of the young females follow development of their first anogenital swellings, which may serve as "social passports," gaining them tolerance from males and thus protection from females. As a consequence of female dispersal, male chimpanzees are more strongly associated with one another than with females, and association between males and females is more pronounced than that among females (Goodall 1986; Boesch and Boesch-Achermann 2000).

Chimpanzee mothers provide continuous care and contact (Figure 4.7) during at least the first three months of their infant's life (Plooij 1984; Goodall 1986). While infants are initially in constant ventroventral contact with their

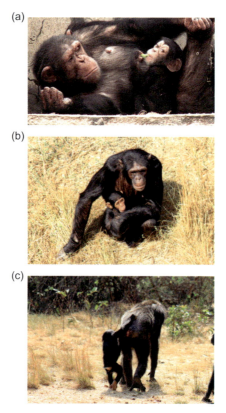

Figure 4.7 Chimpanzee mothers and infants. In (a) and (b) infants are 6 months old. In (c) the infant is about 30 months old and still remains very close to his mother. Photos used with permission from Manuela Lembeck (a) and Linda Scheider (b) and (c).

mothers, they start riding on her back by the age of five months (Bard 1995). Mothers gradually force their infants to walk and feed by themselves during their fourth and fifth year of life but continue to nurse, carry, and sleep in the night nests with them until they are five years of age. Allomothering is essentially absent, although older siblings may be allowed to carry infants (Goodall 1986). Following a mother's death, adoptions by older siblings and even unrelated adult males have been reported in chimpanzee communities at the Tai study site in Ivory Coast (Boesch et al. 2010; Myowa and Butler, this volume) Juveniles remain in association with their mothers and younger siblings, reaching adolescence at around eight to ten years. In some cases where adolescent individuals have been separated from their mother accidentally, they whimper and search for the mother even at this age. Females' age at first birth ranges from 11–14 years (Goodall 1986; Boesch and Boesch-Achermann 2000) and the mean interbirth interval ranges from 4–6 years (Sugiyama 2004; Barrickman et al. 2008).

Absence of allomaternal care may be a consequence not only of the lack of related adult females close by, mothers may also be protecting against the risk of infanticide by other females (Goodall 1986; Townsend et al. 2007). Risk of infanticide by males of their own community may be low because females mate with all the males. Widespread possibility of paternity may be protective, while low probability of paternity for particular males reduces fitness benefits for direct care, such as holding and carrying the infant.

In addition to this summary of behavioral observations in the wild, details of mother-infant interactions observed at the Primate Research Institute at Kyoto University are reported below (see section on "Long-Term Consequences of Infant Experience").

Humans

Humans live in household units within multimale/multifemale communities, clearly standing out from other primates for their abundance and ecological diversity of habitats across the globe (Brown et al. 2011; Keller and Chaudhary, this volume). Human ecological success has been linked to the evolution of human life history, which differs in several ways from that of other great apes (Hrdy 1999, 2009; Robson et al. 2006): While maturation in humans is slower—including a long childhood, late adolescence, and a remarkably increased life span—interbirth intervals are shorter and infants are weaned well before the age of independence (see below). Moreover, despite close birth spacing, the proportion of offspring that survive to adulthood is usually greater in humans, even among traditional hunter-gatherer populations (i.e., those not dependent on domesticated foods or public health care), than most corresponding estimates available from wild populations of apes (Wood et al. 2017). Obviously, humans can produce more costly offspring more efficiently in a shorter period of time, although this initially seems to contradict a fundamental life history trade-off between offspring quantity and quality.

A possible solution to this evolutionary paradox, as proposed by many researchers, is to consider offspring production in humans as "team work." In ecology, *cooperative breeding* describes a mode of reproduction in which individuals other than parents contribute to rearing dependent young (Hrdy 2009).

Of course, fathers can provide significant resources to mothers and also be engaged in directly caring for their children. However, the view of humans as characteristically living in nuclear families with fathers provisioning their wives and offspring ignores the extremely wide variation in paternal investment within and across human societies (Hrdy 2008). So the question arises: Who provides the support that allows mothers to produce new offspring while the previous ones are not yet independent (Sear and Mace 2008)?

As argued by Reiches et al. (2009), any member of a breeding group trades off direct and indirect reproduction, since contributions or withdrawals from pooled resources affect the budget available for fertility and survival. Natural selection justifies the expectation that investments are generally driven by net fitness advantages to individuals. So, researchers expect contributions to vary with genetic relatedness between donor and receiver as well as with donors' opportunity costs. For example, the opportunity cost of caring for younger siblings or providing some economic value to the family can be quite low for older children and adolescents (Kramer 2005).

In addition to older children and adolescents (not to mention childless adults), grandmothers past their childbearing years are another category of kin that constitutes a significant proportion of any human population particularly suited for providing kin support (Figure 4.8). For example, the proportion of postmenopausal women among the adult female population is about one third in the Tanzanian hunter-gatherer population of the Hadza, compared to a

Figure 4.8   A Hadza grandmother prepares her foraging tools. She is surrounded by dependent grandchildren while her daughter (the mother of two of them) watches with her new infant. Photo used with permission from James F. O'Connell.

corresponding estimate of about 3% for chimpanzees (Hawkes 2010; Blurton Jones 2016). According to the grandmother hypothesis, human reproduction takes place in a three-generation enterprise, with postmenopausal mothers assisting their offspring in reproduction (Hawkes et al. 1998).

Although this characterization has received much empirical support, there is also considerable variability among populations in the specific outcomes of grandmothering, with some studies finding contradictory effects (Hawkes and Coxworth 2013; Johow et al. 2013). One reason for the variation may be that grandparents can be related to their grandchildren either through their sons or their daughters, with different effects of (maternal) grandmothers on their daughters than of (paternal) grandmothers on their daughters-in-law (Leonetti et al. 2007). Since even parents and their offspring have conflicting fitness interests in resource allocation (Trivers 1974), it follows that the interests of genetically more distant members, such as in-laws, may be laden with much more conflict. In line with this, Voland and Beise (2002) found opposite effects of maternal and paternal grandmothers on the survival of grandchildren in a historical population of the Krummhörn region located in northwestern Germany. Here, the presence of the maternal grandmother (slightly) reduced infant mortality, whereas the presence of the paternal grandmother actually raised it.

When considering effects on maternal behavior and child development, the genetic relatedness of co-resident kin matters, and not only in the case of different genetic lineages: grandmothers' effects on grandchild survival has also been shown to vary with the probability the child inherited one of her X chromosomes (Fox et al. 2009). While most studies suggest that maternal kin are, on average, more beneficial to child survival than paternal kin, some do not fit this pattern (Sear and Mace 2008). Under some circumstances, mothers themselves opt not to invest in an offspring (Hrdy 1999, 2009). Demographic and socioecological context, along with genetic relatedness, affect whether supporting dependent young is likely to increase the supporter's own fitness, for example, by raising maternal fertility, offspring survival, or even economic productivity (Beise 2005; Kushnick 2012; Blurton Jones 2016). Varying access to resources, overall mortality risks, mating systems, and residence patterns between and within human populations provide a wide range of factors that affect rearing decisions (Lawson and Mace 2011). Depending on context, kin effects may either contribute directly to infant survival or future reproductive success (direct care) or enable parents to invest more or less in a child themselves (Kushnick 2012). Furthermore, individuals can also differ in their power to exert leverage on a mother's fertility, parenting behavior, or socioemotional development of the offspring (Houston et al. 2005). As suggested by the classic socioecological model, the fitness-maximizing consequences of behavior depend on context, and parenting in particular is known to be highly responsive to changes in environment (Royle et al. 2014). Reviewing published data on the impact of kin on offspring survival, Sear and Mace (2008) argued that investment decisions cannot easily be generalized, even if potential helpers are differenti-ated according to their age, sex, and genetic lineage. Further, they document high variability in the observed effects of different categories of kin on infant welfare (although in the studies reviewed, loss of the mother always reduced infant survival during the first two years). This variation highlights complexity associated with measuring reproductive success (Grafen 1988; Blurton Jones 2016), let alone inclusive fitness (West and Gardner 2010). Nevertheless, dis-entangling ways that direct and indirect reproduction contribute to fitness dif-ferentials within families seems feasible for the socioecologies where relevant data are available on genealogical relations, spatial proximity, and births and deaths over generations (e.g., Voland 2000; Smith and Mineau 2003).

Evolution of Human Life History

Bowlby's initial conception of the central role of early mother-infant attach-ment in later emotional health, and social and parental competence, has been revised, at least in some quarters: the concept of "mother" has been replaced with "caregiver" (inter alia), multiple definitions of "attachment" are allowed, and diverse cultural and individual forms of infant-caregiver interactions are acknowledged (Konner 2010; Quinn and Mageo 2013; Otto and Keller 2014). Still, the idea persists that early "attachment" after birth is uniquely impor-tant for humans. In support of a special role for early attachment, the period of postnatal helplessness is relatively long in humans, although some have noted similar helplessness in infant chimpanzees (reviewed in Hawkes 2006). Direct comparison of the features of human development with other primates indicates the timing of birth with respect to the infant's maturational state is somewhat early, but not remarkably so (Bard et al. 2011; see also below). It is the timing of weaning that stands out as a human distinction. Weaning is extremely early in humans compared to every other primate, occurring while the toddler still requires full provisioning and protection. The human child is thus separated relatively early from the special mother-infant bond of nursing and becomes dependent on a network of others, which can consist of multiple siblings, the mother, father, grandmothers, and a wealth of other potential alloparents, genetically related and unrelated. Below, we review information about the state of human maturation at birth and at weaning, initially with particular attention to brain maturation, so as to place early human develop-ment in a comparative context. The comparisons link brain size to the pace of development. For predictors of developmental pace, we review the coevolu-tion of weaning age, age at feeding independence, age at maturity (first con-ception), and longevity from the demographic perspective of evolutionary life history theory. These lines of evidence converge on the likely fundamental importance of uniquely early weaning on the emotional, cognitive, and social characteristics of humans.

Principles of Brain Maturation in Mammalian Development

Brain construction, maturation, and the very first behavioral capacities in placental mammals are surprisingly predictable and species-uniform (Figure 4.9) in nonhuman primates and humans, if maturational events are allometrically scaled—from conception, not from birth—with respect to eventual

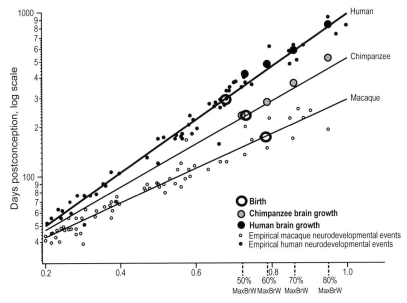

Event scale: best ordering of 271 neurodevelopmental events across 18 species

Figure 4.9 Comparison of predictable patterns and rates of maturation for the rhesus macaque, chimpanzee, and human brains around birth. The x-axis represents the "event scale," a statistical best ordering of 271 events in neural development in 18 mammalian species set to range between zero and one (data from Workman et al. 2013). These events include onsets, peaks and onsets of neurogenesis, axon extension and synaptogenesis, early physiological and behavioral events, and myelogenesis as well as specific brain volume milestones (data from Finlay and Workman 2013; Sakai et al. 2013). The y-axis is days postconception (log scale). On this representation, a steeper line for one species compared to another represents a longer time required to reach any specified maturational milestone—for example, humans take about 900 days to reach 80% of maximum brain weight, while chimpanzees take about 500 days. The linear model of the developmental "schedule" for humans, chimpanzees, and macaques are plotted with human and macaque lines generated from their empirical data (plotted) in combination with the additional data from 16 other mammalian species, while the chimpanzee line is estimated based on its brain size and gestational length (formulas in Workman et al. 2013 and http://www.translatingtime.net/). Data from the supplementary materials of Sakai et al. (2013) were used to determine the postconceptional day that percentages of maximum brain weight were reached in humans and chimpanzees. Large open circles represent the position of birth in each species on the neural maturational scale, showing considerable variation (see also Figure 4.10) compared to the high predictability of neurodevelopmental events between species.

brain volume (Workman et al. 2013). Passingham (1985) first noted that graphs of changes in brain volume versus postconception day are virtually superimposable across mammals, and that the difference between them was duration, with larger brains requiring absolutely longer to produce. The reason for the surprising uniformity of basic brain construction in mammals (not true for all vertebrate taxa) is not yet known. Perhaps, the complexity of any individual brain permits a limited range of alterations in the timing of deployment of its developmental processes, with the basic scaffolding of neurogenesis, initial tract formation and myelogenesis, and initial synaptogenesis highly conserved.

By contrast, the events of life history—birth, weaning, feeding independence, dispersal, first parenthood, and so forth—are transactional and social, defined by the competing interests and multiple goals of individuals, the trade-offs mentioned at the outset of this chapter. The timing of life history events such as birth or weaning depends not only on the maturational state of the offspring, but also on the competing and aligning interests of offspring and mother, in both individual variation and cross-species contexts (Royle et al. 2012).

Overall, long gestation requires more investment of resources from the mother than a short one. This investment involves not only the transfer of nutrients to the fetus, but also a lengthening of the interval to the next conception, and hence a decrease of maternal reproductive rate. Mothers might invest in long gestation and produce precocial offspring for many reasons, such as the need for ungulate offspring to avoid predators independently. Because adult brain mass is a power-law function of developmental duration, gestational length in combination with the degree of brain maturation at birth will also reflect adult brain mass. As Figure 4.10 shows, precocial guinea pigs have much more mature brains at birth than do altricial mice. This results from longer gestation in guinea pigs than in mice, combined with different rates of brain development during gestation. Further, brain maturation postbirth continues the unique developmental rate that is indicated by neural maturation at birth in each species and results in the adult brain mass that is predictable from total duration of development in both species.

Humans have been widely viewed as a secondarily altricial species (e.g., Martin 1990) among the primates due to the large proportion of brain growth that occurs after birth, which is frequently credited for humans' unusual cognitive abilities. Prolongation of a fetal rate of development far beyond birth is offered as an explanation for the volume increase (e.g., Coqueugniot et al. 2004), or a special adaptation for learning postbirth in humans (Sakai et al. 2013). These interpretations assume that birth occurs at the same stage in fetal development across different species, and that changes observed in development after birth are the result of evolutionary modifications of the development schedule itself. However, an evolutionary reshaping of the developmental schedule is not necessary to account for human maturation (Figure 4.10). Rather, the position of birth in humans occurs somewhat earlier on a neurodevelopmental

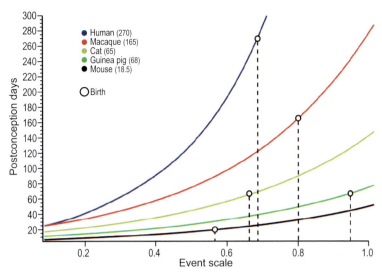

Figure 4.10   Variability in timing of birth with respect to neural maturation in five mammals. The position of birth (open circle, listed after each species name) for five placental mammals relative to the event scale (x-axis; the derivation of this scale is described in the caption for Figure 4.9); the age of each mammal in postconception days can be read for birth (or any neurodevelopmental event scale value) on the y-axis. The five placental mammals are chosen to represent close to the full range of the data set for both adult brain size and the altricial to precocial dimension of placement of birth with respect to neurodevelopment (Workman et al. 2013). The relative slope of the curves is highly correlated with adult brain mass: steeper slopes represent longer durations to reach maturity. In the most highly precocial mammal, the guinea pig, all neurogenesis and most myelination is complete at birth and brain mass is close to its adult value. For comparison, cortical neurogenesis in the altricial mouse at birth is still underway and synaptogenesis in the forebrain is only beginning. In humans and rhesus macaques at birth, cortical neurogenesis, cortical cell migration, and basic axonogenesis are entirely complete, and the succession of myelination of various tracts is in progress. The domestic cat is intermediate. From Workman et al. (2013), republished with permission of *The Journal of Neuroscience.*

schedule that is highly conserved across mammalian species. Further, the particular shape of the curve of brain growth must be calculated from the "allometrically expected" extension of human brain development required to produce a large brain (Finlay and Workman 2013). When these two factors are considered, human brain growth around birth can be seen to be very similar to that observed in other primates, appropriately scaled (Figure 4.9).

The functional state of the brain at birth is not very different across primates. In every primate studied, including humans, the basic construction of the nervous system is well over at birth, including generation of neurons, establishment of axon pathways, and initial excitatory and inhibitory synaptogenesis. The principal event underway over the range of neural maturation that primate births span is the sequential myelination of multiple brain components (Workman et al. 2013). The production of myelin, the sheathing of axons to increase their conduction velocity, is the cause of most of the perinatal increase of brain volume in primates. Each component being myelinated—spinal cord, cerebellum, optic nerve, corpus callosum and so forth—has its distinct time-table, which is experience-independent, not known to reflect ongoing learning directly. Because of a whole-brain "synaptic surge" linked to birth and independent of the conserved timetable of neural maturation and myelination, all primates appear to be relatively similarly placed in their abilities for early environmental learning (Rakic et al. 1986).

Yet, humans appear motorically incompetent compared to the great apes, perhaps more so than a somewhat earlier point of myelination would suggest. Note that the inability to walk until (typically) a full year after birth has been dissociated from neural maturation and linked instead to an insufficiency of the musculature to support body weight: infants suspended in water can produce organized stepping "reflexes" at birth (Thelen et al. 1984) with a timing characteristic of other mammals (Garwicz et al. 2009). Achievement of a critical height/weight ratio is the best predictor of walking onset (Thelen et al. 1984). The simple weight of the head may be the basic reason for the inability of human infants to support their heads at birth. By contrast, sensory and other integrative abilities can be well advanced or even sophisticated. For example, features of language structure can be learned by the human fetus *in utero* (Jusczyk et al. 1983; Mehler et al. 1988). Newborns selectively attend to contrastive stimuli and faces (Johnson 2005), with very rapid postnatal appreciation of the environment and intermodal interactions (Gibson et al. 1979; Walker et al. 1980). Statistical regularities of speech are also learned with extreme rapidity (Saffran et al. 1996). Whether delayed motoric abilities compared to general perceptual and learning abilities are "bugs" or "features" of human development remains to be determined.

The classic explanation for human altriciality at birth has been cephalo-pelvic disproportion. In this scenario, the transition to bipedal locomotion in human evolution led to a narrowing of the skeletal structure of the birth canal, conflicting with selection pressure toward larger brains that make passage through the birth canal more difficult, thus necessitating early birth so that a substantial portion of cranial expansion could take place outside the womb (Schultz 1949; Rosenberg and Trevathan 1995). A recent competing explanation is that maternal metabolic constraints rather than cephalopelvic proportion may determine the timing of birth in humans (Dunsworth et al. 2012; Huseynov et al. 2016). We might also view the shortened gestation in humans, with respect to fetal neural maturation, as part of the same forces that produce early weaning and increase the mother's overall reproductive output. All interpretations converge, however, on the idea that birth timing is a negotiation between the requirements of the fetus and the mother.

Remarkably Early Human Weaning and Infant-Caregiver Interactions

The transition from suckling to independent feeding constitutes a major shift in the behavioral and cognitive capacities required for survival (Lee 1996). Knowing the stage of brain maturity at which weaning takes place is therefore of clear benefit for understanding the relationship between development of the nervous system and life history adaptations. Across species, weaning, like birth, is an event that can be uncoupled from the highly coordinated schedule of brain development. The evidence that weaning is earlier in humans compared to other great apes is clear. In a cross-cultural survey of weaning practices, the median age of weaning across 133 nonindustrial societies was reported to be 29 months postbirth, with a standard deviation of 10 months (Sellen 2001). In traditional societies, earlier weaning is associated with shorter interbirth intervals. Early weaning should be conducive to higher reproductive output, all other things held equal. As an illustration of how late weaning can be when initiated by the child, in a large sample of U.S. mothers it has been reported that the average age of child-led weaning is 4.4 years, or 53 months (Dettwyler 2004), much later than the nonindustrial average noted above and significantly later than the median weaning age for U.S. mothers, which is around seven months (as reported by the CDC). If we compare any of these measures to reported weaning ages for other great apes (e.g., gorillas at around 3–5 years, chimpanzees at 4–6 years, and orangutans at seven or more years), we see that weaning is earlier in humans, as has been emphasized by many researchers studying human life history evolution (e.g., Kennedy 2005). These observations, however, are based on absolute duration. Allometrically corrected predictions for these species compared to brain maturation show that humans are weaned even earlier than the linear projections indicate.

So, humans have made two alterations in reproduction: (a) a relatively shortened gestation with respect to neural maturation and (b) much earlier weaning. Both are methods of reducing unique maternal investment to tractable levels, depending on biparental care, grandmothers, or other alloparents in the early postnatal period and progressively more as childhood continues (Hrdy 2009). Early weaning also serves to redirect the early learning potential of the child. From early childhood to adolescence, the brain is organized for maximum learning, but the human child is neither feeding independently nor provisioned solely by its mother. As noted above, subsidies come from a larger array of allomothers beyond the minimal nuclear family including related and nonrelated others. Humans have an extended childhood where language, custom, and allegiance are being defined by that larger social group, not by the immediate parents. Allegiance to a peer group develops even in spite of complaints from the parents (Harris 1995; Locke and Bogin 2006).

The availability of subsidies from others, which allows mothers to bear a next baby while the previous one is still dependent, may be the foundation of our extensive and unusual sociality. Because human mothers can wean early without suffering unsustainable penalties in offspring survival, they increase their own fitness while thrusting the human child into dependence on a wider network of relationships. Even though many other primates are handled by allomothers soon after birth, infants' primary dependence on nursing continues until they can feed themselves. Human infants, in contrast, are displaced from the small society of mother and child into dependence on the community of age peers, other relatives, and any number of unrelated others in the early parts of "sensitive periods" of development of any number of sensory, cognitive, motor, and social abilities. Although much evidence suggests relatively greater attunement of the human child for social interaction, imitation, and cooperation (e.g., Tomasello 1999; Bullinger et al. 2011; Haun et al. 2014; Hawkes 2014), it may be the rearing context more than motivations and preferences of the child that differ from its primate ancestors and cousins (Bard and Leavens 2014). Possessed of an exceptionally large brain constructed on a primate-typical schedule, with an allometrically predictable extended period of maturation, the human child exercises those motivations and preferences in social environments more variable in every respect than those of any primate relative.

The Evolution of Slow Human Life History

The developmental niche we inhabit is a curious mixture of a conserved neurodevelopmental schedule and a specially adapted life history. Brain size, rate of brain growth, age at maturity, and longevity all covary (e.g., Sacher 1959, 1975), but the direction of influence among these several factors remains in question. Perhaps it is larger brains that require longer development and propel increased longevity (e.g., Kaplan et al. 2000, 2003; Barrickman et al. 2008; Isler and van Schaik 2014). The causal arrow can be drawn the other way, however, using demographic life history theory (Stearns 1992), with adult mortality risk the fundamental driver of life history evolution (Charnov 1993). That demographic approach explains the range of mammalian ages at maturity and durations of offspring dependence as evolutionary consequences of variation in average adult life spans (Charnov 1993).

Among nonhuman primates, great apes have the longest life spans, oldest ages at first conception, latest ages at feeding independence, and largest brains; and compared to those hominids, longevity is much greater, first conception later, and brains larger in humans (Barrickman et al. 2008). Female fertility, however, ends at the same age in humans as it does in other great apes (Robbins et al. 2006; Robson et al. 2006), grounds for assuming this was also true of our common hominid ancestor (Figure 4.1, lower right panel). Yet unlike other great ape females, women remain productive for decades longer (Blurton Jones et al. 2002; Blurton Jones 2016), suggesting that postmenopausal longevity, not an early end to fertility, is the derived feature in our lineage (Hawkes 2003, 2010). As noted above, great apes become frail and rarely survive to menopause (Hawkes 2010), whereas even in hunter-gatherer populations (where mortality is relatively high) about one third of the women live past the childbearing years (Hawkes 2010; Blurton Jones 2016). Our postmenopausal longevity is combined with weaning ages that are remarkably early when scaled allometrically to brain maturation *and* when compared to weaning ages expected for a primate with the longevity and age at first conception of humans. The grandmother hypothesis links the evolution of our postmenopausal longevity to our early weaning (Hawkes et al. 1998).

The human lineage evolved in an ecological context where staple foods are difficult for youngsters to handle for themselves. Under those circumstances, subsidies for dependent offspring allow mothers to bear next babies sooner. The economic productivity of postmenopausal Hadza grandmothers (Figure 4.8), considered in light of demographic links among mammalian life history traits identified in Charnov's (1991) model, suggest the coevolution of these distinctive features of human life history (Hawkes et al. 1998; Hawkes and Coxworth 2013). In a two sex, agent-based mathematical model of this grandmother hypothesis (Kim et al. 2012, 2014), a life history like the other great apes evolves into a human-like one propelled by postfertile females' subsidies for weaned dependants. At the initial great ape-like equilibrium, fewer than 1% of females survive their fertility. However, when they can subsidize their dependent grandchildren, slightly longer-lived grandmothers can help more and increased longevity evolves in subsequent generations. Grandmother effects drive model populations to new equilibrium longevities with fractions of postfertile females very like those of modern hunter-gatherer populations.

These simulations do not model brain growth directly, but the links between greater longevity, longer duration of dependence, later maturity, and mammalian brain size outlined above suggest a mechanistic link. If increasing longevity did evolve as grandmothering subsidies allowed mothers to wean earlier, concurrent retardation in maturation rate would result in brains developing more slowly to larger size along an allometrically conserved primate schedule. This combination of features makes human infants more dependent on a social environment beyond their mothers earlier in development than any of our primate relatives.

Long-Term Consequences of Infant Experience

Bowlby's concerns about attachment included consequences for the infant's subsequent social behavior. Here we review observations on nonhuman primates that investigate those links. Findings on several species of monkeys—especially rhesus—have been well published. After summarizing those, we give more detailed attention to less well-known observations on chimpanzees made possible by the unusual protocol at the Primate Research Institute of Kyoto

University, which provides a window into maternal styles and the antecedents and contexts of their variation in one of our closest evolutionary cousins.

Variation in Monkeys

In rhesus macaques, the nonhuman primate central to the development of Bowlby's attachment theory, a majority of females reared without mothers are not able to provide adequate maternal care themselves: they are indifferent to their first infant or even display abusive behaviors. They can learn, however, adequate mothering skills and most turn out to be competent mothers for subsequent offspring (Suomi and Ripp 1983). Even in group-living animals, some mothers appear to be repeatedly abusive, dragging their infant by its tail or leg, pushing the infant against the ground, throwing, hitting and biting the infant, and stepping or sitting on it (Maestripieri 1998; Maestripieri and Carroll 1998). Abusive mothering occurs not only in captive populations of rhesus but also in pigtail macaques and sooty mangabeys, where up to 10% of the mothers physically abuse their infants (Maestripieri et al. 1997a, b; Maestripieri and Carroll 1998). Maternal abuse has also been observed in free-living Japanese macaques, most frequently among mothers who were orphaned after weaning and had no experience of younger siblings (Hiraiwa 1981).

Maternal style is transmitted across generations from mothers to daughters in rhesus macaques (Berman 1990). This also holds for physical abuse of infants. Cross-fostering studies in this species showed that infants who were born to, and raised by, abusive mothers as well as those who were born to non-abusive mothers but raised by abusive mothers became, in most instances, abusive mothers themselves. Contrary to that, infants born to and raised by non-abusive mothers as well as infants born to abusive mothers but raised by non-abusive mothers all became non-abusive mothers (Maestripieri 2005). This seems to echo the variation in steepness of dominance hierarchies across the macaque radiation noted above (see section on "Broader Variation in Macaques"). Tolerant species persist in their tolerance as hierarchical species persist in their maintenance of steep rank differences. The social style that infants experience sets their continuing expectations and responses in relationships.

This type of abusive mothering is noteworthy but rare in normally reared monkeys (<10% of cases observed), even in macaques. In contrast, variation in maternal style within the normal range is common and has been described along the dimensions of rejection and protectiveness for macaques and other primate species (Hinde and Spencer-Booth 1971b; Simpson and Datta 1991; reviewed in Fairbanks 1996). A relatively rejecting mother initiates fewer contacts when the infant is away, is more likely to break ventral contact, and puts limits on her infant's access to the nipple. Infants of more rejecting mothers, however, do not just passively take it. They increase their efforts to maintain contact with their mothers. They also have higher rates of contact with other group companions and begin exploring the environment at an earlier age. These differences continue into the juvenile years when daughters of relatively rejecting mothers are more likely to actively approach and spend time near others (Fairbanks and Hinde 2013). While extreme deprivation is expected to have lasting negative effects on development, overcoming challenges within the normal range of early experience can increase resilience and the ability to cope with stressors in later life (Parker et al. 2006).

Maternal protectiveness, in contrast, teaches infants that the world is a risky place. Primate mothers respond to perceived environmental risk by increasing contact with their infants, inspecting, grooming, and restraining their attempts to leave. Low-ranking mothers, mothers in groups with new adult males, and mothers who lost their last infant all have relatively high rates of maternal protectiveness (Fairbanks 1996). High levels of early maternal protectiveness delay the timing of exploring the world beyond the mother, and overprotected infants are significantly more cautious in response to novelty as juveniles (Fairbanks and McGuire 1993).

Research on early experience of the mother within the normal range supports the persistence of maternal style from mother to adult daughter described above for abusive mothering. While mothers do modify their style in response to circumstances, long-term studies of rhesus macaques and vervets confirm the significant continuity of mother-infant contact and maternal rejection across generations, even after controlling for correlated features like family dominance rank (Berman 1990; Fairbanks 1996).

Great Apes in Captivity

Incompetent or even absence of maternal care (e.g., ignoring, not nursing, or mistreating the infant) occurs not only in monkeys that lack sufficient experience but is common in captive great apes. Among chimpanzees in captivity, one out of every two mothers does not care for her infant, a stark contrast with the close attachment of mothers to infants in the wild (Matsuzawa 2006). An international survey on the breeding success of zoo-living gorillas, chimpanzees, orangutans, and bonobos, conducted by Abello and Colell (2006), revealed that the maternal skills of great ape females are associated with their own rearing histories. Incompetent mothering seems to be intergenerationally transmitted, as hand-reared mothers and/or those who lack the experience of observing maternal behavior are those who fail most often in providing appropriate care for their infants. Conversely, factors that contribute most to becoming a competent mother are being mother-reared and raised in a mixed-sex, mixed-age group where maternal and social behavior of conspecifics can be observed and/or experienced. Moreover, the presence of conspecifics shortly after parturition seems to be important in encouraging the mother's interest in her infant and stimulating caring. Intervention programs as well as the early integration of human-reared infants into a group (under rigorous supervision) may help to break the cycle.

In Japan, over half the chimpanzees kept in captivity live in groups of five or fewer. This is quite different from the wild where communities typically range from 20 to 100 members. Small groups mean that most captive chimpanzees have little chance to observe or interact directly with infants. A lack of social learning during an early period in life may be the primary reason why some captive mothers abandon their offspring (i.e., "sensitive period" for leaning to parent). In 2000, the Primate Research Institute of Kyoto University (PRI) initiated a longitudinal study on chimpanzee cognitive development (Matsuzawa et al. 2006). That year, when three chimpanzee infants were born (Figure 4.11), researchers had arranged to provide conditions of

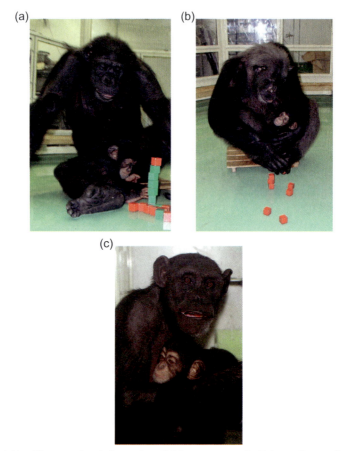

Figure 4.11 Three mother-infant pairs of chimpanzees at the Primate Research Institute, Kyoto University: (a) Ai and Ayumu (male, born in April 24, 2000), (b) Chloe and Cleo (female, born in June 19, 2000), and (c) Pan and Pal (female, born in August 9, 2000). Photo used with permission from Tetsuro Matsuzawa.

community- and mother-infant bonds to facilitate the natural development of chimpanzees, given the limitations imposed by captivity.

With the pregnancies of Ai, Chloe, and Pan, researchers sought ways to prevent them from abandoning their infants. All the females had few, if any, opportunities to observe or interact directly with the offspring of other individuals. Ai's mother was killed in the forest in Africa several months after her birth, and she was reared by multiple human caretakers. Chloe had been born at the zoo in Paris and was likely reared in a poor social environment. She came to PRI several years after birth. Pan was a daughter of Puchi, who had been born at PRI and rejected by her mother. Puchi was reared by humans from just after birth (in a human house as a human infant).

Although the mothers had already passed the likely sensitive period for learning parenting behaviors from other social group members, researchers instructed the females how to hold and take care of their infant by the following three methods. The females were shown videos of wild chimpanzees taking care of their offspring (holding, nursing, grooming) to provide observations of parenting. Second, researchers gave the mothers tactile experiences with infants by holding infant gibbons and monkeys in front of the females and encouraging them to touch them through the mesh of the cage. Third, a researcher who had reared the chimpanzee entered the mother's room and instructed her to hold a stuffed chimpanzee. Fortunately, each mother then successfully held her baby and demonstrated good maternal competence (Bard 2002).

Quantitative measures of mother-infant interactions found that in the special setting of PRI, these mothers on average protectively cradled their infants less than mothers in other captive settings (Bard et al. 2005). Of special interest here are the individual differences in protectiveness among the three mothers:

- Ai (mother) permitted only a very limited number of human caretakers who had reared her to touch Ayumu (her son) right after birth. Ai continued close proximity with Ayumu 24 hours a day for the first three months of the baby's life.
- Chloe (mother) never permitted even the human caretaker most familiar with her to touch her daughter (Cleo). Chloe also continued to keep close to Cleo 24 hours a day for over six months, and held her even after Cleo had developed her own motor competence.
- Pan (mother) easily permitted many humans to touch or hold her daughter (Pal), and often put her baby on the floor beside her just as human mothers do (Figure 4.12). Pan, who had been raised in a human home, even handed her offspring over to human caretakers.

Differences in protectiveness among the mothers may be linked to differences in their own experiences during childhood.

Before Ai, Chloe, and Pan gave birth, two others had done so at PRI: Reiko and Puchi. Reiko was born in Africa in 1966 and arrived at PRI in 1968 when she was 19 months old; she could have observed other group members

Figure 4.12   Pan with infant down. Pan put her offspring on the floor and lay with the infant from just after birth. Photo used with permission from Tetsuro Matsuzawa.

parenting in the wild. Puchi was born in West Africa in 1966. After her capture by human hunters, she was kept as a pet for 12 years in an ordinary human family. She arrived at PRI in 1979. She started social living with her conspecifics with no experience of learning or interacting with other individuals' offspring. Puchi, but not Reiko, refused to take care of her infants. Although she gave birth twice, both of Puchi's offspring were raised by human caretakers.

Following the success of rearing instructions with Ai, Chloe, and Pan, Puchi gave birth to her third offspring and again abandoned it. After artificial rearing began, researchers started instructing Puchi to hold her infant. Seven days after the birth, a researcher entered the same room with Puchi, holding her infant, and enticed Puchi to smell and touch the infant step-by-step. In the process of the instruction, Puchi gradually began to show interest in her infant. The instructions continued, and 20 days after she had given birth, Puchi successfully held her baby. These examples show that chimpanzees can learn parenting behavior by active human instruction, even when they have no experience of infant rearing early in their lives.

Concluding Remarks

An evolutionary perspective makes both phylogeny and natural selection central to explanations for variation among living things. Fitness trade-offs are everywhere, and conflicts of interest can powerfully shape social life. Primatologists do not all agree about the words used to describe social relationships, including whether "attachment" should even be used. Yet all agree that phylogeny, ecology, and social systems affect the interactions that primate infants have with their mothers and others, and that social relationships involve inevitable conflicts of interest. As shown here in a variety of primates, rearing circumstances have long-term effects on behavior later in life. In addition, infants play a role in finding the care they need and are resilient to "less than the perfect" mothering when raised in species-typical social groups with opportunities for interaction with other group members. In the detailed examples provided, later experience (at least in our closest cousins) as well as context continues to matter after infancy.

Could it be that parallel links between experience and parenting fuel the American advice industry in baby care? Among hunter-gatherers, other traditional societies, and in many non-Western cultural contexts (e.g., Keller and Chaudhary, this volume), babies are part of the daily lives of other children. Just as mothers rely on help from other adults as they bear new babies before the previous one can feed itself, young girls have continuing intimate experience with the infants before their own first conception. Even then, at least in the case of Hadza foragers, children survive better when their mothers gained experience before maturity through their own infant siblings (Blurton Jones 2016:413). In many parts of the industrial Western world, including much of the United States, contrasts with most of human experience are striking: direct exposure to babies is minimized by closely spaced births with limited family sizes, small households, and workplaces that (again unlike most of human experience) are inhospitable to babies and distant from the home.

The examples presented in this chapter show that individuals across the primate order often interact with infants they did not bear and, conversely, that the experience of primate infants often includes handling by nonmothers. By relinquishing their infants to others, mothers face risks and benefits that vary both with group composition and the mother's particular social position. Mothers, allomothers, and infants face different fitness trade-offs that include varying opportunity costs around these possibilities. While offspring might benefit from more maternal investment than is in the mother's fitness interest to give, infants can also gain from the wider interactions that shape their development. This appears nowhere more important than in our own lineage, as mammalian comparisons point to developmental correlates of the uniquely early displacement of human infants from the special mother-infant nursing relationship. Where attachment specialists and public health workers seek to implement policies that—often laudably—prioritize the welfare of infants and young children, the trade-offs and conflicts of interest faced by infants, children, mothers, and others may be relegated largely to afterthoughts. For evolutionary explanations of variation, those trade-offs and conflicts of interest are central. They make the distinctly early weaning of human infants deserving of increased attention for likely importance in expanding the capacities to adjust to varying social circumstances that we share with other members of our primate order.

5

Is the Mother Essential for Attachment?

Models of Care in Different Cultures

Heidi Keller and Nandita Chaudhary

Abstract

Attachment theory is predicated on the assumption of dyadic relationships between a child and one or a few significant others. Despite its recognition of alloparenting in some cultural environments, current attachment research is heavily biased toward the mother as the major attachment figure in the life of the developing child. This chapter presents evidence that diverse childcare arrangements exist in cultures that differ from Western norms and shows how these are equally normative in their respective cultural contexts. In these settings, alloparenting is neither chaotic nor unstable; it is the norm, not the exception. In all environments, infant care is far more than just an isolated, biopsychological phenomenon: it is an activity deeply imbued with cultural meanings, values, and practices. To account for these multiple levels, the construct of attachment must shift its emphasis away from an individual child toward the network of relationships surrounding a child. Overwhelming evidence on diverse childcare arrangements in non-Western cultures calls the putatively universal model of attachment (derived from the Bowlby-Ainsworth paradigm and still widely applied today) into question. In support of future research, this chapter proposes an inclusive reconceptualization of attachment, informed by research from non-Western cultural settings.

Introduction

Attachment theory, as formulated in the 1960s and 1970s, was based on notions of family life and relationship dynamics within a circumscribed cultural context. Does this view of "family" apply to every cultural setting? What does research from various cultures tell us about the environment in which infants are raised? How can we confirm whether infants develop best under the care of one stable adult who has specific behavioral features? What can we learn from recent work on different primate species living under varied conditions, captive or wild? How do fields of research such as primatology, anthropology, and cultural psychology contribute to expand our knowledge of infant-caregiver relationships?

Despite advancements in the study of human behavior, core tenets of attachment theory have not been altered and are still widely in place in ongoing research. If this theory is to have validity, these core constructs need to be reexamined to ensure that its basic premise reflects contemporary research and understanding. In this chapter, we revisit attachment theory to scrutinize the role it ascribes to the mother: Is the biological mother essential for attachment? What role does the mother play in diverse ecocultural contexts?

We review the core claims of attachment theory, provide background into the basic processes under discussion (child care, family life, and social relationships), and examine the emergence and expansion of attachment theory. Following this, a comprehensive critique of the theory is advanced based on examples from different childcare settings.

Centrality of the Mother

Without doubt, mothers play a special role in the lives of their children. In addition to endowing a child with genetic material, mothers make a substantial physical investment through intrauterine pregnancy, delivery, and possibly lactation. No other person in a child's life can match the maternal investment that a mother makes in offspring, and although all cultures recognize the central role that a mother plays in a child's life, the mother-child relationship is subject to varying ecological factors and cultural traditions. As Bronfenbrenner proposed (National Scientific Council on the Developing Child 2004:1):

> A child requires progressively more complex joint activity with one or more adults who have an irrational, emotional relationship with the child. Somebody's got to be crazy about that kid. That's number one. First, last, and always.

This "irrational" attitude toward a child is the foundation upon which increasingly complex activities form between children and their social partners. Stable and caring relationships are the foundation of human development the world over. For millennia, infants have survived under the caring organization of the social setting, where constraints, threats, and uncertainty are managed to enable the child to reach maturity. The solutions that get worked out in real contexts provide evidence of a wide variety of care arrangements: from the unpredictable environments of the hunter-gatherer communities to the relatively predictable lives of technologically advanced societies. Alloparenting, multiple caretaking arrangements, and distributed care (e.g., Weisner 2014) have been described in many different cultural groups in East Africa, West sub-Saharan Africa, Southeast Asia, and South America (for summaries, see Keller 2013c, 2015; Lancy 2015). These child-rearing behaviors are, however, not exclusively a non-Western phenomenon. Caregiving by siblings is common in economically challenged families in the United States. In many Western middle-class families, infants and small children are exposed to people other than just the mother and father. Social network theory (Kahn and Antonucci 1980; Lewis 1994; Takahashi 2004) argues that each individual has simultaneous close relationships with multiple significant others, from infancy to old age.

The diversity of the human condition throughout history and context must be taken seriously and accounted for as we explore children's relationships with others. The centrality of the mother, however, does not mean that investments by other caregivers are not similarly important to a child (Chaudhary 2011). Relationships between the child and one or more social others are key to growth and well-being, both during childhood as well as later in life.

Mother and Child: Monotropy in Attachment Theory

Western childcare philosophy assumes, both implicitly and explicitly, that the psychological bond between mother and child is a natural consequence of biological connectedness. As the prototype of Western childcare philosophy, attachment theory (Bowlby 1969; Ainsworth et al. 1978; Cassidy 2008) grants the mother a special and unique role in a child's development. Attachment theory is also predicated on a psychoanalytic model of the psyche. The abiding importance of the first human relationships and the early years of life for later social and emotional functioning is clearly psychoanalytic in orientation, even though Bowlby departed somewhat from the Freudian formulation of psychic energy-seeking release (Harwood et al. 1995). While investigating the nature of these interactions, observations of Ugandan mothers and babies guided Ainsworth to assume that it was not maternal warmth that accounted for different attachment qualities. Instead, the amount of caregiving and the knowledge of the baby determined the quality of care, as measured by proximity, availability, interest in and perceptiveness about, as well as promptness in responding to the baby (Mesman et al. 2016b). Despite the fact that these features were initiated through field work with Ugandan mothers and babies, we argue that each of these features assumes a culturally specific way of understanding the care of children.

Irrespective of the composition, stability, and size of a group, Bowlby claimed that there is always a special, unchangeable bond between a mother and her children (Bowlby 1969, 1988b). Yet he also acknowledged that caring for a baby or a small child can become so demanding that the mother needs to be supported by the father, a grandmother, or an older daughter (Bowlby 1980).

Ainsworth wrote that family security served as the basis from which an individual can work out other relationships later in life (Salter 1940:45). She did not mean, however, to detract from the basic premise of attachment: "there is a strong bias for attachment behavior to become directed mainly toward one particular person and for a child to become strongly possessive of that person"

(Bowlby 1973:368). Indeed, the relative mapping of others around the mother, who may play significant roles in the lives of children, received much less attention from Bowlby, Ainsworth, and their followers.

Since the original work of Bowlby and Ainsworth, the role of the mother has become more pronounced and ingrained in academic thinking. Although the possible importance of others, particularly fathers, has been recognized (e.g., Bretherton 2010), the bulk of attachment studies still focuses solely on the mother-child relationship (Cassidy and Shaver 2008, 2016).

Primary Claims of Attachment Theory

Attachment theory proposes that children develop an internal working model of relationships based on early experiences. It also holds that the mother plays a unique and central role in the attachment process as well as in subsequent developmental trajectories of a child (Main 1999; Cassidy 2008). Even when multiple caretaking is acknowledged, the central role of the mother is not disputed (Mesman et al. 2016b).

Attachment research aims at demonstrating the uniqueness of the mother-child bond, even within multiple caregiving settings. Attachment theory assumes that the mother-child bond is qualitatively different from all other relationships that the child may form: it is a specific emotional connection that the infant develops during the first year of life, embodying spatial closeness and timely extension (Ainsworth et al. 1974). Problems encountered by the child, during the initiation or maintenance of an attachment relationship with the mother, are assumed to lead to serious negative consequences and psychopathology. Attachment relationships, therefore, can express different qualities and outcomes, depending on the nature of the social experiences with the primary attachment figure. A secure attachment relationship—the golden standard for a healthy, happy, and competent developmental trajectory—can emerge when the child experiences unconditional, dyadic, and exclusive attention. Even the slightest of signals emitted from an infant must be answered responsively and sensitively (assumption of contingency) for the child to develop a sense of agency and predictability (Ainsworth et al. 1974).

Recognizing that a child may receive attention and care from multiple caregivers, Mesman et al. (2016a) propose that the sensitivity received by an infant be measured instead of the sensitivity expressed by a single caregiver. This proposal, however, does not go far enough, because it fails to account for the different modes and qualities of caregiving: "sensitivity" can be understood differently in various cultural contexts, as can the various functions that different caregivers may potentially fulfill (e.g., Yovsi et al. 2009; Lancy 2015; LeVine and LeVine 2016).

Attachment theory considers emotional expressiveness to be part of the attachment system, and social exchanges are geared toward the expression of positive emotionality. The expression of negative emotions, such as an infant's fussing or crying during duress (e.g., when separated from the mother), is thus interpreted as being indicative of a secure attachment relationship. Emotions are generally viewed as important regulatory mechanisms within attachment relationships (Cassidy 2008).

Because the bulk of attachment studies utilize between-family designs, comparisons of attachment qualities that a mother may exhibit toward her different children can easily be overlooked. This neglect is based on the assumption that the maternal state of mind regarding attachment (internal working models; George and Solomon 1989) is generally stable by adulthood; thus, mothers can be expected to interact similarly with their children and share a similar quality of attachment with each child (O'Connor et al. 2000; van IJzendoorn et al. 2000). Attachment research that addresses siblings and their mothers has concentrated on the concordance of attachment classifications (with very modest success) instead of on the variability and placement of the relationships of different children in one family within a family systems or network approach.

Monotropy Revisited

While examining the care of children across time and space, it becomes evident that the exclusiveness of dyadic relationships is predicated on several conditions. Unless these conditions are fulfilled, it is impossible for an exclusive, intimate relationship to emerge between a single adult and a single child, as assumed by attachment theory. For a mother to invest exclusive attention in a single child, she must be assured of her own safety and survival, the child's survival, a stable environment free of imminent dangers, food security, and moderate temperature. In a modern urban setting, other conditions may also be necessary to render the adult free from other life-saving or life-sustaining tasks.

Henrich et al. (2010) argue that dyadic exclusivity is a benefit of a WEIRD (white, educated, industrialized, rich, democratic) environment. In a hunter-gatherer society, by contrast, exclusive engagement with a baby at the expense of other environmental cues could endanger the lives of both the mother and the child. Scarcity of resources and ecological uncertainty are essential considerations for which there is biological and social adaptation (Morelli et al. 2014).

Like other areas of family and community life, child care is a cooperative activity that relies on the participation of kin and other group members. Childcare arrangements are thus sensitive to the environmental conditions under which families live. Dyadic exclusivity can be argued to be an adjustment to the secure surroundings of a technologically advanced and affluent society that facilitates intimate attention between mother and child. Free from subsistence activity, nuclear families in relatively affluent environments are a necessary but insufficient condition for caregiver-infant exclusivity to be expressed. Even when these conditions are met, a mother may be unable or unwilling to attend to her child exclusively.

Infant Attachment as Adaptation

Attachment theory emphasizes a strong orientation toward adaptation to the environment. Bowlby assumed that the attachment system emerged in the African savannah where our ancestors lived as hunter-gatherers. Thus, human psychology should be adapted to an "environment of evolutionary adapted-ness (EEA)."

By contrast, neo-Darwinian evolutionary theory (Wilson 1975; Alexander 1979; Tooby and Cosmides 1990; see Chisholm, this volume) places genetic variability in the center, with contextual information being crucial for defin-ing adaptation. Accordingly, in the Pleistocene EEA, it is unlikely that only one behavioral strategy—the secure attachment quality—would have been selected to be adaptive. Family life, in early human societies, was too fragile and uncertain to provide a stable context for sustaining singular and focused attention between a mother and her child. Apart from ecological and social contexts, fertility rates would have further confounded the development of a one-mother–one-child bond. Certainly it would have been more adaptable to permit as well as encourage babies to develop the capacity to bond with mul-tiple caregivers. Correspondingly, due to high infant mortality rates, a heavy investment in a single baby would have been maladaptive for the mother, given the high risk of losing her baby (Scheper-Hughes 2014). Thus un-der conditions of environmental uncertainty, high mortality, greater mobil-ity, and mortality risk, we argue that exclusive dyadic relationships would be maladaptive, both for the mother as well as for the baby (see also Morelli et al., this volume).

In 1984, Lamb et al. questioned Bowlby's premise of the adaptability of one evolutionary strategy, citing this as a misunderstanding of evolutionary prin-ciples and natural selection. They stated (Lamb et al. 1984b:146) that:

> Evolutionary biology, however, demands an evaluation not only of biologically influenced predispositions but also of the contingencies provided by the spe-cific environments or "niches" in which the individuals must manifest these predispositions.

Further, if variation and diversity in structure better predicts survival of life forms, the same argument could apply to human behavior. The more diversity we retain in the nature-culture dialectics, the greater our chance of survival will be in the future.

To many attachment researchers, attachment theory represents a univer-sally applicable account of the bond between caregivers and infants based on evolutionary and ethological considerations (Bowlby 1969; Mesman et al. 2015; Mesman et al. 2016b). This assumes that the definition of attachment and its qualities, its emergence, and consequences must be universally simi-lar. Even the discussion of the Japanese conception of *amae*[1] has not found

---

[1] Amae is a verb which Doi (2014) uses to describe the behavior of a person who attempts to

its way into attachment theory and research. This assumption has been challenged substantially from cultural, historical, and cross-cultural perspectives (for summaries, see Quinn and Mageo 2013; Vicedo 2013; Otto and Keller 2014). The basic argument is that attachment theory, like any other theory, rests on a specific, culturally bound conceptualization of an individual, with corollary assumptions about the family and relationships. In addition, family is a symbolic construction or an ideological conception with its own history and politics (Arendell 1997).

Attachment theory itself is a product of history (Vicedo 2013 and Vicedo,. this volume) and must be situated in historical time. For instance, children who grew up during the Great Depression in the United States married at a much younger age and had significantly more children than their parents' generation (Stearns 2003; Nicolas et al. 2015). German Reunification offers another example. After Reunification in 1990, the birth rate in the former German Democratic Republic dropped by almost 50% over a three-year period before stabilizing (Chevalier and Marie 2013). Major changes in family life and child-rearing attitudes accompanied the sociodemographic changes (Otto and Keller 2015). Worldwide, fertility rates are intimately linked to infant mortality rates: when child survival increases and infant mortality is minimal in a society, people tend to have fewer children. Better quality health care and literacy have been additional factors in slowing down population growth in the Global South.

Attachment Theory: A Culturally Specific View of Relationship Formation

Attachment theory rests upon the model of the Western middle-class nuclear family as it was perceived to exist around the middle of the twentieth century (LeVine and Norman 2001; Vicedo 2013; Sear 2016). In this cultural setting, parenthood would begin relatively late during the individual biography to ensure a stable economic setting. For women, first-time births would occur by the time they were in their late 20s. For men, parenthood would begin by the time they were in their late 30s, after achieving a formal education and becoming established in a profession. Despite public discourse on gender equality and shared household responsibilities at the time, mothers were viewed as the primary caretakers of infants (Georgas 2006). Depending on the availability of government support and parenthood programs, women prioritized child care over profession, while fathers remained fully engaged in their professional lives. Accordingly, the mother generally spent most of the time with the baby.

---

solicit care from an authority figure. The behavior of children toward their parents is perhaps the most common example of amae. Doi argues that child-rearing practices in the West aim to prevent this kind of dependence, whereas in Japan it continues into adulthood.

The low fertility rate of that time (between 1.3 and 1.9 in European countries) allowed child centeredness to emerge as an educational credo, expressed in exclusive attention to the baby and prompt responsiveness to all the subtle communication cues. Babies develop expectancies through contingency experiences. In this way, the environment becomes predictable and trustworthy. High formal education is associated with voluminous conversations and an inclination to mental states embodied in extensive face-to-face encounters.[2] Positive emotionality and praise stress the individuality and uniqueness of the infant while reinforcing the relationship with the mother, as the primary caregiver (Keller 2007, 2015)—the unique and unchangeable bond, which attachment theory assumes to be universal. Relationships are genuinely dyadic, and all relationships are assumed to be constructed according to the same principles, and hierarchically subordinated to the primary relationship (Main 1999; Cassidy 2008). Attachment theory also holds that adults have to be the significant relational and educational partners of children. Together, the assumptions made about adult-child relationships in attachment theory attest to it being viewed as a culturally specific phenomenon—one best adapted to white, middle-class Western families. As such, attachment theory provides a model of "neontocracy" (i.e., infants are treated as cherubs) rather than a gerontocratic model, where children are viewed as chattels (Lancy 2015).

Exporting Attachment Theory

With its emphasis on the nuclear family and the mother-child bond, attachment theory addresses a specific and narrow aspect of relationships (Takahashi 2005). It does not account for the attachment process in the environment in which children are born and raised (Chisholm 1996). Ample evidence in the anthropological and cultural/cross-cultural literature attests to the fact that children's learning environments and socialization strategies vary substantially (e.g., Lancy 2015) across cultures as well as within cultures and across time.

Despite this evidence, attachment researchers argue that cultural variability was incorporated in attachment theory, citing Ainsworth's (1967) empirical study of attachment in Uganda as evidence. However, Ainsworth adapted her Uganda experiences to fit the Euro-American middle-class context in Baltimore using the Strange Situation Procedure. The distress that she observed in the Ugandan homes when infants were separated from their mothers did not occur in the Baltimore homes. Therefore, Ainsworth relocated the observational situation into a laboratory (strange environment) and included

---

2   We note that the Western version of face-to-face routine encounters appears very different to the close physical interactions observed among Trobrianders, Himbas, and Yamonamis (The Human Ecology Archives). In these communities, babies are fondled, rocked, repeatedly kissed, and caressed during repeated contacts between the caregiver and the child. This version of the face-to-face encounter, which is accompanied by vocalizations, is very different in character from the Western model.

a stranger (strange person) to increase the stress on the infant, so that the attachment system would be activated and attachment behaviors would become visible. This U.S. middle-class adaptation was then exported by others to diverse cultural environments—from the Gusii in Kenya (Kermoian and Leiderman 1986) and Hausa in Nigeria (Marvin et al. 1977) to Western and non-Western middle-class families (for a summary, see van IJzendoorn and Sagi-Schwartz 2008; Mesman et al. 2016b)—without further questioning or verifying the cultural validity of the Strange Situation. Ainsworth was obviously not happy with this practice as she expressed disappointment "that so many attachment researchers have gone on to do research with the Strange Situation rather than looking at what happens in the home or in other natural settings....it marks a turning away from 'field work,' and I don't think it's wise" (Ainsworth 1995:12). A more field-based approach is available through the Attachment Q-sort method (Vaughn and Waters 1990), where an observer spends several hours in a family setting (with a one- to two-year-old child) before evaluating the attachment security of the child using a large number of predefined behavioral descriptions (written on cards). These descriptions "can be used as a standard vocabulary to describe the behavior of a child in the natural home setting, with special emphasis on secure-base behavior" (van IJzendoorn et al. 2004:1189). These descriptions, however, almost exclusively address the mothers' behavior and are thus bound up with issues related to sensitivity responsiveness described above.

Today, cultural differences are only acknowledged with respect to different distributions of attachment qualities—i.e., deviations from the "American Standard Distribution" as assessed in the Baltimore study (Ainsworth et al. 1978)—with 66% securely, 12% insecurely avoidant, and 22% as insecurely resistant attached children (for a discussion, see Keller 2013c). The differences were defined *ex post facto* as cultural deviations, a practice that attachment researcher Inge Bretherton (1992) found persuasive on the surface, but not based on systematic assessments of parental beliefs and culturally guided practices. Bretherton recognized the need for systematic studies of cultural differences and stressed that attachment researchers need to develop ecologically valid, theory-driven measures, tailored to specific cultures and based on a deeper knowledge of parents' and children's folk theories about family relationships. In current practice, theory and method are deeply confounded (Gaskins 2013). In a recent analysis of different cultural groups that are mainly equated with countries, Mesman et al. (2016b) argue that very little research exists to permit a systematic appraisal of attachment theory across cultures. Further, they emphasize that researchers are inadequately trained and conditions are not favorable, with researchers neglecting to consider whether methodology may be invalid or inappropriate in certain settings.

The ethnocentrism of the conclusion reached by Mesman et al. (2016b), "to support attachment theory as it stands," is a testimony to the way in which attachment theory has been applied in different cultural settings. We propose that attachment relationships need to be contextualized to reflect the diverse ecologies that impact child-rearing.

Developing an Ecological Framework

If research is to move beyond the ethnocentrism inherent in attachment theory, it must relinquish its blind commitment to the theory as it stands. Instead of working to maintain the traditional framework by adjusting interpretations of diverse cultural realities to fit the attachment model proliferated within this tradition (van IJzendoorn and Sagi-Schwartz 2008; Mesman et al. 2016b), attachment theory must be grounded in an inclusive understanding of attachment relationships that exist in diverse settings. Thus, we purposefully shift the discussion away from the centrality of the mother and focus instead on alloparenting, because this fundamental behavior is observed in diverse primate societies. Using wide-ranging examples of child-rearing, we lay the groundwork for the construction of a new ecological framework within which attachment can be better understood.

Alloparenting as the Human Condition

By assuming a central role for the mother, attachment theory has seriously disregarded other significant relationships in a child's early development. The importance of alloparenting—its meaning, patterns, structure, and impact on the emergence of attachment relationships and children's development—has received far too little attention in attachment research. Yet the involvement of others (e.g., grandmothers, older siblings, fathers, but also unrelated kin) in child care on a routine basis can be regarded as a human universal, extending back to the appearance of *Homo erectus* (Hrdy 1999, 2009; Burkart and van Schaik 2010). Cooperative breeding (Hrdy 2009; Morelli et al. 2014) allows mothers to reproduce and raise children successfully. If the task of raising children would have been the sole responsibility of the biological mother, humankind would not have survived (Hrdy 1999, 2009). Next to mothers, older siblings have the largest impact on infant survival, followed by maternal and paternal grandmothers, then fathers; even grandfathers exert a 20% effect on child survival (Sear and Mace 2008).

Extensive allomaternal care can account for higher birth rates (reducing birth intervals), earlier weaning of humans (relative to the other great apes), enhanced cooperation within social units organized by cultural norms and values, and the social and cognitive capacities for social regulations. The cooperative breeding hypothesis assumes the emergence of prosocial psychology affecting social regulations. Cooperative breeding/allomaternal care requires motivational and cognitive processes that may also lead to cognitive and social capacities which are not directly related to breeding (Burkart and van Schaik

2010). It emerged early during the history of humankind, possibly with the *H. erectus*. During the course of phylogeny, it contributed to larger brain development and a large array of prosocial and cognitive competencies necessary to cope with the complexities of human life (van Schaik and Burkart 2010:484):

> Chimpanzees, and perhaps all great apes, meet many of the relevant cognitive preconditions for the evolution of human cognitive potential, [yet] lack the motivational preconditions. In humans alone, these two components have come together, the cognitive component due to common descent, and the motivational component, resulting from the selection pressures associated with cooperative breeding.

This view stands in sharp contrast to Bowlby's evolutionary understanding. He derived the monotropic conception of human attachment from the caregiving system of the rhesus macaques, in which the mother plays a unique role for the upbringing of the offspring. Bowlby took the rhesus macaque system (studied by his ethologist friend, Robert Hinde) to be representative of the entire primate world (cf. Lancaster et al. 2000; Clutton-Brock 2002). However, in over 300 primate species, parenting behavior manifests itself in very different ways (Fairbanks 2000), in terms of social systems, parenting strategies, and systems of distributed caretaking (for an example, consider cotton-top tamarins; Blum 2002). Moreover, parenting behavior in the same species varies according to their living ecology (Boesch 2012). As Suomi (2008:177) commented: "One wonders how Bowlby's attachment theory would have looked if Hinde had been studying capuchin rather than rhesus monkeys!"

Context, therefore, matters. Women who are not situated in middle-class affluence cannot afford to spend substantial parts of their day exclusively attending to a baby. Child care is thus organized mainly as a co-occurring activity (Saraswathi and Pai 1997): Carrying a baby on the hips or the back allows freedom of movement and hand use, permitting women to engage in other activities while caring for a child. Carrying also involves other channels of communication, so that interactional regulations (e.g., behavioral contingencies) are primarily proximal (Chapin 2013a). However, since mothers cannot carry infants all the time, due to the necessary balance of energy investment and domestic activities, other people's motivation to carry infants is crucial: leaving an infant alone would pose too much of a risk (e.g., from predators). Reciprocity provides such a motivational source: Women take turns in caring for babies or trade caretaking for other provisioning, like food (Crittenden and Marlowe 2008). Unrelated caregivers may also be occasionally coerced by a mother when she needs support in child minding (Hrdy 2005b).

Alloparenting, in general, benefits the mother, the child, and the alloparents. For the mother, alloparenting enhances reproductive fitness and helps her maintain domestic and economic activities (Sear 2016); it also may enhance quality of life satisfaction and maternal well-being. Allomothering facilitates caregiving as it helps first-time mothers learn to be a parent. Weisner (2005) argues that older children who are caregivers to younger children learn aspects of nurturance, dominance, and responsibility—skills that serve them well later in life.

Allomaternal care increases children's survival, growth, and lasting effects on health (Hawkes et al. 1997; Mace and Sear 2005). Multiple caregivers in larger households are able to attend to infant cries more quickly than single caregivers (Munroe and Munroe 1971). In general, alloparented children receive more physical, social, and emotional investment. The frequency of allomaternal child contact encourages the formation of strong and trusting relationships with others, and thus increases a child's sense of security (Meehan and Hawks 2013).

It is essential to understand that these practices are neither chaotic nor unpredictable. Multiple caregiving is a stable manifestation of cooperation as well as a fundamental social practice in many cultures. We make a clear separation between multiple caregiving as a normative practice and the social neglect of children in disadvantaged contexts (e.g., institutional care, war, extreme poverty, or any situation where care is constantly changing and disconnected), where children are likely to face ignorance, aggression, or abuse. Children who grow up under conditions of multiple care, as a normative practice, experience stable, shared, and sustained nurturing from several different adults and/or children because they are valued. When these practices are viewed from the perspective of dyadic exclusivity, multiple caregiving practices may appear disorderly, since children and mothers are rarely alone, and a child is passed from one person to another. Caregiving activities are performed by different people, and the infant may be frequently cared for by other children as well. We wish to stress that not all multiple caregiving settings are always child friendly: even within cooperative care, children may face difficulties as a result of individual, familial, social or collective factors. Context, again, is important.

Exactly how multiple caregiving impacts the attachment process is not well understood. Although there is plenty of ethnographic material to suggest that children in these societies are happy, playful, and curious within the prevailing range of within-group difference, limited research has been conducted on how early relationships develop in contexts of normative multiple care. Focused studies are needed on the attachment process in these settings. Importantly, methods must be derived to measure and confirm beneficial attachment behavior under different ecologies.

Examples from the Field

In general, in communities where alloparenting is encouraged, *children are actively encouraged to engage with others*. In fact, "sticking to" the mother is discouraged: others will quickly engage and playfully interact with the child, pretend to take the child away, or tease the child for "attaching" to the mother. To illustrate, consider the following example taken from a study conducted among the urban poor in Noida, a city in the national capital region of Northern India (Chaudhary 2015). A child who was reluctant to detach from the mother received the following treatment: It was midday on a hot summer's afternoon; three women were seated on the floor of a small room that comprised the home of one of them. They were chatting and grooming their babies, while some toddlers played around them. Suddenly, one of the women turned to her companion and snatched up the companion's baby (around 10 months of age), allowing the baby in her arms to crawl away. She said aloud, playfully, that the baby was "sticking too much" to the mother. She briskly lifted up her shirt and made an attempt to breastfeed the child, laughing heartily and teasing the baby when she turned her face away, fussing. The exchange was followed by a chorus of teasing of the infant by all of the women, much to the distress of the baby.

Such playful actions by women, the exchanging of babies, and the teasing of young children who are viewed as too close to the biological mother all seem to indicate that the exclusive mother-child bonding is being discouraged in this cultural setting, putatively for the well-being of the mother and the baby, in case something goes wrong. Correspondingly, babies are encouraged, playfully, to engage in several social games of exchange and interaction with others, among family members as well as neighbors. Fictive kinship terms are always used for such relationships. In India, there is ample evidence of similar exchanges, which hark back an enduring tradition of distributed care and multiple mothering practices, in other cultural settings. For example, in Tamil Nadu (Southern India), the sharing of children among women is an ancient custom, as evidenced in Tamil Sangam poetry from almost two millennia ago (Trawick 1990:155).

Multiple caregiving may involve varying arrangements between caretakers and responsibilities. The mother may play a special role among other caregivers, be equal to others, or may not be a special caretaker at all. Moreover, these arrangements can vary over time. An example of this can be seen in the cultures of the Aka and Efe hunter-gatherers. In an assessment made with four-month-old children, these infants were passed on to different people seven to eight times per hour and held by 7–14 different individuals during the eight-hour observation periods. Overall, Aka children showed attachment behaviors to about six people out of 20 who they encountered daily (Meehan and Hawks 2013). Scheidecker (2017) offers another example of the socialization experiences of village children in southern Madagascar. Here, mothers play a special role in an infant's life during the first two years but it disappears thereafter. From then onward, children are not exposed to adults but rather develop their psychology in the context of peer groups (Scheidecker 2017). The abrupt change of the caregiving environment has also been described by Du Bois (1944) for the Alorese community in Indonesia. In this case, the nurturing relationship between mother and child during the first year of life declines suddenly to complete inattention by the mother, even to the point of potential food deprivation—a condition that attachment theory would consider as a major precursor of psychopathology (Cassidy 2008). The opposite may also occur. Hewlett (1991) describes a dramatic decline in allomothering over the first year of age for the Aka. By eight months of age, Aka infants receive substantially less care from others and relatively more care from the mother.

The primary attachment figure may not be the biological mother at all. In the Nigerian Hausa, mothers live together and share childcare responsibilities. Hausa infants seem to become attached to the person who interacted most with them, which in 8 of 14 observed cases was not the mother (Marvin et al. 1977).

From the moment of birth, infants are passed on to other caretakers in many cultural environments. In the Efe, the mother may not be the first to nurse an infant and others participate in nursing during early infancy (Tronick et al. 1987). Aka mothers are not the first to touch and hold an infant. An older female-in-law cleans the infant and takes it to the hut until the mother arrives, since Aka women give birth outside their camps (Hewlett 1991). Multiple attachment relationships may develop simultaneously that are similar in importance and significance (Morelli 2015). Multiple caregiving may thus be the dominant mode under certain conditions (e.g., when Aka hunter-gatherers are in a camp) whereas in others (e.g., during foraging activities such as net hunting), the mother may be the dominant caretaker (Hewlett 1991).

Alloparenting cultures utilize fictive kinship. Kin terms, including "mother," may be awarded to different people and go beyond blood ties. The designation of "mother," for example, can be accorded as a term of affection to an aunt, a grandmother, or any other female relative. It is also used as a term of respect to nonrelated elderly women, such as a senior researcher. This phenomenon has been recorded in several cultures, especially where the ideology of relationality prevails. For instance, in conversations with young children, Indian mothers most often use kin terms to refer to other people, even when they are not related to the child (Chaudhary 2004). In a particular example in Hindi (a northern Indian language), the kin term for a mother's sister is "Masi," or "like-mother." Among many Hindu communities, the relationship between a man and his wife's sister is a "joking relationship," one that generates social tension, usually released by socially accepted forms of teasing (Radcliffe-Brown 1940). This can be explained by the fact that the relationship has potential, practical, and maternal implications for the offspring, such as in the case of a sororate marriage (i.e., marriage to a wife's sister, usually upon her death).

Different caregivers may all perform the same responsibilities. With the exception of maternal breastfeeding, this phenomenon has been observed in the Nso farming community in Cameroon. *Alternatively, caregiver roles may be differentiated*, as has been observed by Scheidecker (2017) while observing the roles of mother and siblings in South Madagascan villages. Both would certainly impact the formation of internal working models. In contrast to the model of primary attachment to a single mother, which is subsequently conveyed to other social partners, these experiences may promote the simultaneous formation of different styles of attachment and eventually conceptions of the self (Scheidecker, pers. comm.).

Let us continue by looking at the most common caregivers across cultures: grandparents (both paternal and maternal), siblings, and fathers. Although the influence and significance of day care is not denied, discussion is focused on home-based multiple caregiving.

Grandparents as Caregivers

Grandparenting is probably the most common mode of alloparenting. Grandparents may adopt divergent roles in infant care, sometimes substituting, sometimes supplementing the mother. Grandmaternal involvement is usually higher than grandpaternal involvement because the maternal grandmother is sure that it is her genetic offspring in which she is investing, whereas grandfathers can never be certain, due to paternity insecurity (Voland et al. 2005). Grandmothering is most plausibly an adaptation through which aging females achieve better fitness returns after they have produced and reared their own children (Hawkes et al. 1998). The extended lifespan beyond the reproductive years is assumed to allow older women to assist in the effective rearing of their grandchildren (Lancy 2015).

Although evolutionary considerations apply to all grandmothers, this does not imply that there are no contextual variations. The involvement of grandmothers is obviously bound to availability, which is closely related to life expectancy. Moreover, birth order of the grandchildren may play a role. Grandmaternal involvement is dependent on settlement patterns, family structure, family relationships, and cultural dimensions. In the Apiao (southern Chile), for instance, grandmothers raise the children of their daughters, while the daughters are expected to learn and collect experiences in faraway cities (Bacchiddu 2012). In many Chinese families the involvement of grandparents in caregiving is highly valued (Mjelde-Mossey 2007), especially with respect to child-feeding practices (Xie and Xia 2011). Due to labor migration, Chinese couples live and work for most of the year in big cities or abroad, while their children are raised in the grandparent's household (e.g., Xie and Xia 2011). Grandmothers also play a special role in the upbringing of Nso farmer children in northwestern Cameroon. Grandmothers are among the most preferred caretakers during a child's first three years, while the mother is not. Eight out of nine children who changed households at about two years of age, changed to their grandparental home (Lamm and Keller, in preparation). Direct involvement is also common in rural Turkish farming families, where grandmothers raise children alongside mothers. By contrast, grandmothers in Western middle-class families understand their role as being fun partners for their grandchildren to spend leisurely time with them. They do not consider themselves as educational authorities (Lamm and Teiser 2013).

Sibling Caregivers: Playful Partners and Powerful Protectors

Sibling or polymatric caregiving may contribute over 90% of the infant care that is not provided by the mother. Siblings offer care to infants that are older than two months of age. In Nigerian families, small children were observed to interact with other children 48% of the time, compared to 10–15% interaction with the mother, leading to strong and enduring attachment bonds between siblings (Weisner 1997). In a study on alloparenting in the Hadza hunter-gatherers in Tanzania, Crittenden and Marlowe (2008) found that children between 1,5 and 17,9 years of age[3] spent up to 20% of their time holding related as well as unrelated kin, and that this was beneficial to both the caregivers as well as the babies.

In a study involving 58 families living in and around Delhi (Northern India) across social class and ecological settings, it was found that older children were expected to be caring and nurturing toward younger siblings and cousins (Chaudhary 2015). Based on long sessions of play among children in the absence of adults, several mutual benefits of these interactions were observed that look very different from adult-child interactions:

- Older siblings were caring, but not always so. Unlike adults, they would place demands on younger children, even with a bit of bossing, sometimes allowing other children to tease the younger one playfully when they were in a group.
- Younger children seemed to learn quickly that older siblings would be on their side, caring for their needs, but not always.
- Older siblings would also extract compliance in play from the younger siblings.

In one rural joint[4] family, two sisters (5,2 and 3,1 years of age) were playing by themselves in a courtyard on a pile of gunny sacks filled with grain. The girls were always together, and the older one took care of the younger one's every need when the mother and other adults were not available. During one play sequence, the older child acted out a session at school: she ensured that the younger one would comply with her every demand to act, to run about creating the perfect scene, opening imaginary gates, sitting quietly like a student, answering questions when demanded. A gentle rap on the arm was also delivered to the younger sister when something was not in order. The sheer awe that the younger child displayed toward her older sibling was dramatic. Sometime

---

[3]   Notation refers to year, month: 1 year, 5 months.

[4]   A joint family constitutes multiple generations of kin members residing together: a couple, their married sons along with their families, and unmarried daughters would constitute a joint family. This should not be confused with an "extended" family, which refers to additional kin members who may reside with a couple and their children on a temporary or long-term basis (e.g., an unmarried aunt).

later, the younger sister spilt some buttermilk while opening the refrigerator, just out of view of the several women who were chatting in the courtyard, weaving baskets or caring for their babies. Both sisters were in the adjoining room with the spilt buttermilk; when the mother looked up and realized what had happened, she scolded and spanked the older daughter. Quietly, the older sister took the punishment, not even once declaring her noninvolvement in the accident. She took the rap for the act, protecting the younger one in a silent commitment. This pattern of complementarity could be evaluated as adaptive for both partners. The younger ones were assisted, and they also accessed a world about which even the adults may have little knowledge, the world of the street, school, or playmates. For this privilege they had to adapt to the demands of a sometimes dominating older sibling during play. The older siblings for their part were protective and nurturing toward younger ones, but not always; not when, for instance, it interfered with their activities with friends or their own desires during play. The older ones seemed to learn quite effectively how to take care of a younger child in the absence of an adult without abandoning their playful endeavors.

Sibling care has been found to be nuanced and ubiquitous, providing both partners with essential ingredients of social life and mature participation (Weisner 2014). The dismissal of these interactions as being only playful and not amounting to emotional attachments is an underestimation of the possible bonding that is likely to develop. Any theory or paradigm that attempts to represent infant relationships cannot ignore these relationships: they persist from childhood to adulthood and are found in almost all communities, in Western and non-Western families. To assume that the built-in mechanism of an infant searches only for one specific and major figure for security and protection grossly underestimates the fluidity and flexibility inherent in human relationships.

Fathers

Attachment theory has increasingly considered the role of fathers as attachment figures for infants (e.g., Bretherton 2010; Palm 2014). Yet, the father-infant attachment bond is assumed to serve purposes different to the mother-infant attachment relationship: play and excitement versus nurturance and consoling. Father-infant attachment is assumed to develop according to the same principles as the infant-mother relationship, based on the quality of dyadic social encounters in everyday interactions. An important assumption in attachment theory is that the fathering role is consistent across cultures (Sear 2016).

From an evolutionary perspective, women and men pursue different reproductive interests (Møller and Thornhill 1998). Genuine paternal investment has been found to be much lower than that of mothers (female gametes are big, rare, and valuable whereas paternal sperm cells are small and abundant). Also, fathers cannot be certain that they are the biological genitors due to paternal insecurity. It is in the interest of women to select reproductive partners who demonstrate the readiness and competence to engage in enduring investments. Therefore, paternal child care is not only part of paternal investment but also of mating effort. Except for breastfeeding, fathers have the same evolutionary predispositions for infant care, including the formation of attachment relationships (for overviews, see Shwalb et al. 2013; Roopnarine 2015). Paternal involvement in child care, however, varies considerably across cultural environments, depending on a variety of contextual factors. For example, Gusii farmers in Kenya are almost never in close vicinity to their infants (LeVine et al. 1994). Fulani herder fathers in southwestern Africa keep an emotional distance from their own offspring; fathers are supposed to be the primary authority figures in the family, and emotional distance is the means used to build respect and obedience, maybe even anxiety in the children (LeVine et al. 1994; Lamm and Keller 2012). Beng fathers at the Ivory Coast and Kipsigis fathers in Kenya do not participate in infant care at all since they believe that the power of the paternal eye could harm the infant. Kipsigis also believe that the "dirtiness" of the baby could compromise the masculinity and reproductive capacity of the father. Kipsigis fathers as well as Cameroonian Nso farmers and many other sub-Saharan fathers view their role as supplying material goods. Caring for food, clothing, potential medical treatment, and school fees later in life are seen as the primary responsibilities of fathers. In addition to economic responsibility, the induction and maintenance of obedience and respect exacted by the paternal figure are viewed as core values in these hierarchically organized social systems (Lamm and Keller 2012).

By contrast, Western middle-class fathers are understood to be emotional partners of their infants, equal to mothers, and providers of cognitive stimulation and play experiences. Their participation in infant care has increased since the middle of the twentieth century, although much less than public discussions of the "new father" would suggest. In Germany, for instance, time budget studies reveal that the amount of time that fathers participate in the care of their children has increased an hour per week, with daily contact time averaging 83 minutes (Friedrich Ebert Stiftung 2009). At the same time, professional working hours per week increased by one hour after the birth of a child!

On a global scale, paternal investment is higher when there is low accumulation of material resources, absence of wars, low population density, monogamous family organization, and regular cooperation of men and women in domestic and economic activities. These characteristics are typical for hunter-gatherer communities, where paternal investment is generally higher than in herder groups or farming families (Hewlett 2004). A particularly impressive paternal investment has been demonstrated in the Aka Pygmies: babies spend 47% of their waking time with their fathers, who entertain affectionate relationships with their infants, even permitting them to suck on their fathers' nipples (Hewlett 1991). The Aka hunter-gatherers practice complete, but not permanent, role reversal with women and men doing the same activities.

Household structure substantially influences paternal participation in child care. In extended families, less than 1% of children's interactional efforts are directed toward the father, whereas the father-child interaction increases in nuclear families (Whiting and Whiting 1975). Fathers participate more in child care when marital relationships are egalitarian and cooperative, and when the child is developing normally and healthily. They participate less when post-menopausal relatives are available in patrilocal households (Fouts 2005). In India, this trend has also been observed in urban, middle-class, educated, nu-clear families when fathers are actively involved in all tasks of caregiving and no other family members reside in the household (Roopnarine et al. 1992).

Among Northern Indian families, men (fathers, grandfathers, uncles) usu-ally play with children differently: they may take them in turns for piggy-back rides, encourage them in play, or walk around with babies in their arms while the women work (e.g., tending cattle or completing household chores). Although men are frequently seen carrying babies, especially during peak working hours in the household, their interaction decreases when a researcher enters a home. It seems that matters related to children are to be discussed solely with the women: some men might stand around and listen, but only a few engage with the researcher (Shwalb et al. 2013). This might explain why the participation of men in child care in general, and fathers in particular, is frequently underrepresented and underestimated (Chaudhary 2012).

Fathers are certainly important for the development of their children, even if they are not present in the daily family life, as a result of extensive and long hunting trips (Hill and Hurtado 1996). Their role as attachment figures varies considerably according to context. Understanding the role of the father in at-tachment theory is equal to that of the mother based on nuclear family life. Any attempt to represent infant relationships must incorporate the wide variety of roles that fathers can manifest in different cultural settings.

Cradles of Care: A 2 × 2 Paradigm

To analyze the diverse contexts of care that research has evinced, a paradigm is needed to account for the different possibilities in child-rearing conditions that may exist in any geographical place and for any age group. We propose the use of the "cradles of care" model (Figure 5.1). Best imagined as a dynamic system, the model can account for a single child moving through different care settings and does not prescribe definitive categorization. It can, however, indi-cate a dominant mode of care for a particular family setting.

The model emerged out of a 2010–2012 study of 58 families who live in and around Delhi, India (Chaudhary 2013, 2015, 2018). This study focused on child-rearing in diverse family settings that were illustrative of the different

	One child	Many children
One adult	One adult, one child	One adult, many children
Many adults	Many adults, one child	Many adults, many children

Figure 5.1 Cradles of care: a 2 × 2 model that can account for diverse child-rearing conditions in any culture and for any age group. All family settings that involve two or more adult caregivers and two or more children are termed "many."

ecological contexts of the region: rural areas, a small town community, the urban middle class, and the urban poor. As the study progressed, it became evident that only a handful of the families matched the textbook version of a one-mother–one-child exclusive dyadic relationship. While examining the various settings, patterns in family relationships became apparent that were based on numbers of children and adults in the home who came into direct contact with the target child on a daily basis. Other research studies also report these same patterns (Trawick 1990; Seymour 1999). After thorough review of the video data, it became evident that the observed diversity of settings could not be fully accounted for by simply separating caregivers (single mother vs. multiple caregivers). Some settings were made up of many children with one adult, and the corresponding interactions appeared substantially different.

This model emerged to demonstrate four independent possibilities. Assessing the observed interactions according to this model provided an analysis framework for other dimensions of caregiving, such as attention and focus. It also showed that the care received by a child and the attention given by a caregiver did not always coincide. In the setting of multiple caregivers, for example, the attention received by the child proved different in quality and quantity from the setting of a single adult. Thus, dimensions of caregiving (e.g., attention, focus, contingency, trust) must be assessed according to the setting in which they occur. Using a one-mother–one-child format is only adequate and appropriate in one of four possible settings:

- *One child, one adult*: This setting matches the textbook template for the care of children: the child spends most time with a single adult caregiver, usually the mother, and forms an exclusive, close bond with this person. Urban educated middle-class families in the study had such an arrangement with their first child, although it should be noted that the child was exposed to many other adults, who always interacted with the child while in the home. This additional exposure deviates from the traditional experience of a Western home (see Figure 5.2).
- *One child, many adults*: This setting was observed in homes where other adult caregivers (e.g., a grandparent, aunt or uncle) lived with the child's parents in an extended household. In this setting, firstborn children were cared for under the constant gaze of several adults who regularly interacted with the child, substituting and supplementing each

Figure 5.2 A mother holds her firstborn son at a temple, where she has gone to express her gratitude to the gods.

other's care (see Figure 5.3). In fact, the assignation of child-minding tasks to others, especially the elderly, is considered important, as it strategically includes older family members in child care and reinforces the family unit (Tuli and Chaudhary 2010). Many nuanced details of caregiving arrangements came to light once the cradles of care model was used to analyze the families (see Figure 5.3).

- *Many children, one adult*: The setting of a single adult caregiver with many children in a home was infrequent. It occurred, for example, when a mother had more than one child to care for and the father or mother were the only adults available to the family for most of the day (see Figure 5.4).
- *Many children, many adults*: The most commonly occurring setting among the families studied was one in which many adults cared for many children. In this dynamic caregiving environment, children experienced a variety of input from adult caregivers (may also include care by siblings), who assumed different tasks and shared responsibilities. In general, supervision was also less rigid, and caregiving was supplemental in nature: people just took over tasks if they were around, with

Figure 5.3 An infant enjoys care by many: mother, aunt, grandmother, and siblings.

Figure 5.4   Soaking in the warm winter sunshine, a father looks after the family's children.

> very little (if any) specific assignment of responsibility. Much of the interaction with the child seemed to move seamlessly from one adult to another. Another feature of the many children context (both with one adult as well as with many adult caregivers) was the observed caregiver-like behavior among the children: older children often assumed responsibilities for their siblings and cousins. The greater the age difference among the children, the more alloparenting among siblings/ cousins was observed, with adults actively encouraging grouping and mutuality in their interactions (see Figure 5.5).

After settings are delineated, it becomes possible to disentangle two different dimensions in the caregiving process: the attention that a child receives and the focus of the adult. These two features coincide in the one-child–one-adult setting, but they do not in other settings. For instance, when a mother in a one-child–one-adult setting exhibits co-occurring or concurrent care (e.g., did other things around the home while caring for the child), the child receives distributed attention from the adult because the adult is focusing on the child as well as on other things (Saraswathi and Pai 1997). Under the two settings where "many adults" are involved, a child could receive concentrated attention from another adult (e.g., a grandmother) during the time in which the primary caregiver focuses on other activities.

Figure 5.5   Yours, mine, ours, and a few others: A group of mothers and children at a village health camp.

	One child	Many children	Total
One adult	7	7	14
Many adults	10	20	30
Total	17	27	44

Figure 5.6 Distribution of caregivers and children in families, based on the cradle of care model for contexts of care.

A follow-up study (Chaudhary 2015) tracked the distribution of caregivers and children (Figure 5.6). Results show that it is far more commonplace for "many" adults to care for children (30 cases) than for a single adult to do so (14 cases). A similar pattern was found in the number of children per family: homes with more than one child (27) were more frequently encountered than homes with single children (17). These results further highlight the serious inadequacy of using the one-child–one-adult template as normative and universal. It also questions a simple dichotomy of single versus multiple caregiving, since the setting with one adult and many children was distinguishable from the setting of many adults and one child or many adults and many children.

Attachment Relationships in Different Child-Rearing Settings

Attachment has been defined by Bowlby and his followers as the emotional bond between an infant and caregiver, a psychological construct expressed in mentalistic terms of cognitions and emotions. This definition is rooted in the conception of the self as a separate individual and a mental agent who "owns" cognitions and emotions that are distinct from those of others. This conception of self has been found to characterize Western middle-class individuals (for a discussion, see Keller and Kärtner 2013).

Mind-mindedness, defined as a measure of the caregiver's proclivity to treat the young child as an individual with a mind, has become a major dimension of parenting quality. It is considered to be more closely related to attachment security than sensitivity and has become more of an umbrella term. In interactions with babies, Western caregivers are expected to verbalize the infant's inner world of intentions, cognitions, emotions, and preferences. "Mind-mindedness focuses on the caregiver's willingness or ability to read the child's behaviour with reference to the likely internal states that might be governing it" (Meins and Fernyhough 2006:2).

However, there are other conceptions of self, mind, and relationships. For example, the "opacity doctrine" offers a different perspective since it defines the human psyche as a "private place" (Duranti 2008:485), which includes an indifference toward others' mental states (see also Mead 1934; Ochs 1988). Also, Everett's principle of the immediacy of experiences represents a different conception of the mind. The Pirahã Indians in the southwestern area of the Brazilian State of Amazonas value talk of concrete immediate experiences instead of abstract, unwitnessed, non-immediate topics (Everett 2009, 2014).

Mind-mindedness is a recent phenomenon in the Western world. It is related to the concept of "inward turn," which is seen as a consequence of the decline of fixed traditions and the loss of power of societal institutions. Thus, as a consequence of the "disembedding" of society's ways of life, identities can no longer be defined to the same extent by social group membership (Taylor 1989).

What is defined as an attachment relationship in a particular cultural environment needs to be based on the prevalent conceptions of self and mind. If we take the development of security and trust as the essence of forming attachment relationships, it certainly makes a difference whether these developmental processes are co-constructed in an exclusive dyadic relationship or embodied in a relational network. We would also need to qualify how the network operates. The development of trust to a larger social network extends the range of trusting relationships and thus promotes security as a contextual/environmental dimension and not a personality characteristic or an interactive process between two individuals. The construct of attachment then moves from within the individual child and into the relationships. It is the social network that invokes a secure foundation and not only a person, the mother, which is an instance of a specific context. Attention in those caregiving networks is wide-angled and abiding with the diffusion of a single focus, but it does not imply disorganization and neglect. Perhaps in socially dense settings, a singular focus would be maladaptive, for contingencies (mother absence, work outside the home) as well as inclusion (active participation of older family members in the next generation). Such multiplicity is likely to have consequences for the developing relationships. The wide attention distribution on the part of the caregivers is mirrored in infants' learning of being attentively monitored so that they do not need to seek for attention actively and explicitly (Gaskins 2013). At the same time there is the production of the diffusion of affect, so that the psychological balance and well-being is not concentrated and dependent on one single person (Gaskins 2015). This condition may imply more stability and equilibrium than fragile emotional bonds that need permanently to be negotiated (Keller 2013c). This view definitely departs from attachment theory's implicit understanding that diffusion of affect may compromise the one important relationship (Bowlby 1973). Almost the reverse of this ideology is apparent in folk wisdom regarding the care of young children in Indian homes. More specifically, a child who is capable of getting along with, and seeking out, several adults is regularly applauded and rewarded by others as well as by the mother. A mother who reserves exclusive rights over her baby is also likely to receive much criticism (Chaudhary 2004).

Obviously the socialization agenda with multiple caregivers emphasizes a different model of personhood and paints a picture of a child and childhood that is different to the exclusive dyadic caregiver (mother)-child relationship. With multiple caregivers, individual uniqueness and self-enhancement are not fostered but rather harmony and proper demeanor to fit in with the social surround. Children are believed to belong to the wider social network, and anyone can approach them to engage with them, even if it is a fleeting interaction.

During their field work on the subject of public health, Nichter and Nichter (2010) were traveling through Southern India with their young child, Simeon, collecting data. Apart from the research on pregnancy and childbirth, the Nichters kept extensive notes about the way in which their baby was received by the local community. They provide a rich account of the constant presence, social games, and active interaction that people had with Simeon (Nichter and Nichter 2010:75):

> Adults subjected Simeon to constant teasing, offering him something to play with and then, moments later, asking for it back, citing a kinship term: "I'm your mothers' brother, mava, can't I have it now?….We came to understand that teasing a child and then observing the response was a way villagers could evaluate a child's character and personality.

In such situations, the concept of the self that is promoted centers on hierarchical relatedness: the positioning of the self in the family hierarchy with corresponding obligations and responsibilities. Here, conversations do not revolve around the child's wishes and intentions but on clear instructions, moral obligations, and social roles and responsibilities. It is not mental-state talk addressing the future and the past, but the behavior in the here and now (for an example from the Amazonian Pirahã Indians, see Everett 2009). Toys or play objects are not in evidence; instead, real-world utensils, including sharp knives or machetes, can be found in children's hands (Lancy 2016). Nevertheless, autonomy and individual agency is highly valued: not in terms of the mental way of being, but rather of independent functioning (i.e., early motor independence) and action competence. Self-perception is thus mediated through social relationships.

When clearly differentiated roles are linked with multiple relationships during the early years of development, such that each of these relationships serves distinct functions, it is possible that a child develops simultaneous modes of multiple attachments. When a child's care is systematically broken up into multiple, differentiated forms and aspects of care, such an outcome is easy to imagine. Perhaps the consequences of multiple relationships seamlessly performing the same roles (e.g., grandmother, mother, aunt) would be different. Further research is thus necessary to investigate such nuanced investigations.

Future Directions in Researching Attachment Relationships

Researchers have recognized for a while the need to utilize a relational approach in the study of attachment relationships. For example, van IJzendoorn and Sagi-Schwartz (2008) stressed the need to expand the study of attachment to include multiple relationships as well as to incorporate conceptions and assessments of the child's and caretaker's modes of relationships. They also acknowledged contextual variations found in their review of non-Western attachment studies. Similarly, Heinicke (1995:307) stated "that the study of attachment needs to be expanded...to include multiple relationships." These positions align with our fundamental position: attachment research needs to move radically away from a dyadic perspective toward a network approach.

Still, many researchers appear reluctant to relinquish a commitment to attachment theory, even after its deficits have long been recognized. This may reflect, in part, a fundamental lack of understanding regarding the role that culture plays in children's development, including attachment relationships. It may also reflect the fact that cultural/cross-cultural attachment studies lack a clear conception of culture, which is often equated with country or ethnic groups, without specifying the contextual differences that exist within and between these groups. Contextual conditions provide frameworks for the development of norms, values, and beliefs as well as behavioral conventions; that is, culture (Keller 2015). Therefore, context and culture are fundamentally interconnected. Thus, assessing middle-class families in different countries cannot represent a test of cultural differences or universality (e.g., Posada et al. 1995). Mesman et al. (2015) have assessed important sociodemographic information; however, they do not use it systematically to define cultural groups. They include samples with multiple caregiving arrangements but assess only conceptions of maternal sensitivity from mothers using the Q-sort method. This standardized instrument, if valid at all, can only produce a partial picture, at best. Other dimensions of sensitivity may well exist that are not listed in the Q-sort cards.

Universality and cultural specificity are profoundly intertwined. What universal predispositions exist to acquire contextual information to solve equally universal developmental tasks? The development of attachment relationships is certainly a universal developmental task (Keller 2013c, 2015). No child would survive infancy without a caring environment and the development of trust in others as well as itself. This is important for the development of competence in all environments. Therefore, the core assumptions of attachment theory can certainly claim universality:

1.	Universality: When given an opportunity, all infants will become attached to one or more specific caregivers. However, the definition of attachment and the definition of caregiver need to be culturally defined.
2.	Normativity: The majority of infants are securely attached, yet the definition of security varies across cultural contexts (see Chapter 8, this volume).

3. Sensitivity: Attachment security is dependent on child-rearing antecedents, particularly sensitive and prompt responses to infants' attachment
   signals. However, responses can come from distributed sources, and
   with varying content.
4. Child-rearing patterns vary tremendously across cultures with respect to
   structure and content. Sensitivity may mean completely different things
   in different environments and highly valued practices in one cultural
   context may be regarded as abusive or pathological in another culture.
5. Competency: Secure attachment leads to positive child outcomes in a
   variety of developmental domains. Yes, but what constitutes a "positive child outcome" is largely culture specific. Moreover, the same developmental achievement may be predicted by different precursors in
   different cultures.

Culture is all about meaning. Assessing cultural meaning systems for each of
these core assumptions is of utmost importance, before claims of universality
or cultural specificity can be studied. Field work and multi-method approaches
are crucial. To apply the same method with the same coding and analysis systems in different cultural environments with different shared beliefs and practices distorts different realities.

Thus far, attachment researchers seem to take universality and cultural
specificity for two distinct dimensions, which are geared to confirm and recon-
firm universality. However, universality is not interpreted in terms of predispositions but as fixed phenotypes. This practice contradicts evolutionary as well
as cultural and cross-cultural theories.

In the face of such strong evidence of diversity and plurality in care arrangements, better designs for culturally informed research on attachment processes
need to be developed. For this to happen, we first need to broaden our conceptual and theoretical framework: we need designs that are able to capture what is
critical for security and trust in young children under diverse situations. It will
be necessary to use integrated mixed methods (Hay 2015; Chapters 8 and 13,
this volume) that can incorporate and isolate context. Further, we also need to
identify outcome measures that reflect the adaptive contexts in which children
are developing. One place to start would be with the suite of methods necessary to truly understand attachment and trust within the family and ecocultural
context. Such tools include ethnography, naturalistic observation, qualitative
interviews, locally developed scales to assess constructs such as "trust," "security," "sensitivity" of care, "emotionally appropriate" child behaviors at different ages, and so forth.

By including culture, family context, beliefs, and experiences of parents
and others in the designs and methods of attachment studies, better science will
result. This does not require those in the attachment field now to change their
core assumptions and identities (whether or not they should consider doing so).
It simply asks them to do better science.

Conclusion

To understand any developmental process, it is important to adopt a culturally informed perspective that accounts for the history and diversity of humankind. As researchers, our cultural affiliations predispose us to advance our own "normative framework" as a standard by which to evaluate differences (Harwood et al. 1995). This is similar to thinking locally yet acting globally (Gergen et al. 1996). As Cole (1996) wrote about cultural experiences: like fish in water that fail to perceive their surroundings, attachment theorists have failed to notice the cultural specificity of the single-child–single-adult template, even when multiple caregiving has been recognized.

Researchers can no longer ignore the resounding evidence of diversity in the care of children. Evidence that clearly shows sharply divergent cultural settings in the care of children must be incorporated into the academic mainstream. Any theoretical proposal about children's development must be culturally informed. Such a framework needs to provide for the study of any dimension of children's care throughout history, across cultures, and species. The cradle of care model proposed in this chapter offers one possible framework: it is inclusive and thorough, allowing for different care settings (i.e., from a single child with a single mother to many adults with many children). In addition, results from multiple methods (e.g., ethnography, interviews, observations, and assessments) need to be consolidated as this will enable a more detailed examination of early attachment relationships in different settings. Utilizing an expanded historical and cultural perspective, it can be argued that attachment theory, as it is understood to this day, represents a folk theory from an anthropological perspective (Bretherton 1991), or a model of virtue with strong normative assumptions and implications (LeVine and Norman 2001).

Evaluating one culture based on the normative framework of another is not only invalid, it is unethical. The implications of this are far reaching. For instance, in a recent study, Gernhardt et al. (2016) evaluated children's drawings of their family with attachment-based coding systems deemed to be valid in different cultures and found that the majority of middle-class children in Berlin, Germany, would be classified as securely attached, whereas the majority of West Cameroonian farmer children would be classified as insecurely attached. Because attachment theory is widely applied in clinical and educational work—in particular for children and families who come from cultural backgrounds other than the Western middle class—the use of inappropriate frameworks can result in discrimination and exclusion, instead of the intended facilitation and integration.

Psychology's original sin is to look for a single idealized developmental trajectory (Levinson and Gray 2012). Although the original reference was made to cognition, it applies to all psychological domains.

Is the mother essential for attachment?

Our position is that the mother fulfills a biological necessity, but that "mothering" (or caring for a child) is an attitude and activity that is not necessarily bound to biological function. Other nonbiological roles can be and are fulfilled by others: mothering can be distributed, supplemented, and substituted by one or several other individuals. This understanding necessitates a major revision of attachment theory and requires better science.

Acknowledgments

We would like to thank Gilda Morelli and Tom Weisner for their extensive and most helpful reviews. We greatly appreciate the comments from Robert LeVine, Akemi Tomoda, Abraham Sagi-Schwartz, Kim Bard, Elfriede Kalcher-Sommersguter, David Butler, Alma Gottlieb, and Gabriel Scheidecker. Thanks are also due to several participants of this Forum, who helped us refine our arguments and develop this chapter.

6

Taking Culture Seriously

A Pluralistic Approach to Attachment

Gilda A. Morelli, Nandita Chaudhary, Alma Gottlieb,
Heidi Keller, Marjorie Murray, Naomi Quinn,
Mariano Rosabal-Coto, Gabriel Scheidecker,
Akira Takada, and Marga Vicedo

Abstract

This chapter presents an alternative view to classic attachment theory and research, arguing for systematic, ethnographically informed, approaches to the study of child development. It begins with the observation that the attachment relationships children develop are locally determined and insists that these features of attachment can only be captured through observing, talking with, and listening to local people as they go about living their lives, including caring for children. It reviews the profound ways in which child care around the world differs from the Western model, upon which attachment theory was founded and myriad recommendations have been derived. This worldwide account perspective of child care is profusely illustrated with ethnographic examples. Network theory is then discussed: from the full range of social networks to relational ones (i.e., smaller sets of individuals to whom children may become attached). The chapter considers attachment theorists' resistance to the idea of multiple attachments, historically and still today. Discussion closes with a summary of the implications of our theoretical rethinking and the questions that remain.

Introduction

The lives of infants, young children, and their families differ in many ways around the world for a great number of reasons. Communities differ in the

Group photos (top left to bottom right) Gilda Morelli, Nandita Chaudhary, Alma Gottlieb, Mariano Rosabal-Coto, Heidi Keller, Naomi Quinn, Marjorie Murray, Akira Takada, Marga Vicedo, Heidi Keller, Gabriel Scheidecker, Naomi Quinn, Mariano Rosabal-Coto, Marjorie Murray, Nandita Chaudhary, Gilda Morelli, Gabriel Scheidecker, Marga Vicedo, Akira Takada, Nandita Chaudhary, Alma Gottlieb

ecosystems of which they are a part, as well as in the availability and predictability of physical and social resources. The real-life problems that people
must solve in their communities also differ, as do the ways in which they are
approached and managed. Communities vary in how they are organized and in
how community members relate to one another. Within communities, families
differ in ways that often change with status, privilege, social class, wealth,
ethnicity/race, and religion. Within and across generations, people react to
community expectations and create new ones. Despite this heterogeneity, all
communities have in common variation and change and thus must be viewed
as dynamic systems, responsive to both the opportunities and the constraints
that people encounter in their everyday lives.

Given the many ways that families and communities diverge, is it reasonable to expect that there is just one "best" way to care for infants and young
children that will promote their ability to survive and thrive in the communities
of which they are a part? We pose this question because attachment theorists
propose a view of care for all children, worldwide, and describe the role that
this care plays in attachment relationships and later development (Mesman et
al. 2015). These theorists posit that typically developing children form attachment relationships in the same way and for the same reasons. They identify
patterns of care and attachment relationships, the relation between the two, and
the implications of both for healthy development.

Attachment theory has been widely accepted by scholars as well as the
general educated public in many nations since Bowlby's original formulation.
This most likely transpired because theorists claimed support from a multidisciplinary platform that drew from evolutionary biology, animal behavior,
psychiatry, neuroscience, and psychological research on non-Western communities (van IJzendoorn and Sagi-Schwartz 1999, 2008; Mikulincer and Shaver
2014). Today, attachment theory is psychology's most influential theory of relatedness, setting standards for what constitutes healthy relationships for all
people. The reach of attachment theory to real-life situations is impressive.
Pediatricians are trained in the principles of attachment to identify problematic parent-child interactions in an effort to promote healthy ones. Educators
use these principles to recognize and support children who are considered at
risk for poor classroom learning. There are attachment-based therapeutic approaches for children, families, and couples. In addition, courts use attachment theory to make decisions regarding parental rights, and the cornerstone
of international agencies' programming for families and children is grounded
in attachment theory.

Attachment theory has had its critics, but such views have largely been
ignored (Mead 1962; Vicedo 2013, this volume). One reason is that the theory
was safeguarded by generations of scholars who shared a deeply held philosophy of personhood, self, and human development. Another is that the theory
was disseminated in the teaching and research of generations of students educated in diverse fields of study, as well as in the practice of professionals.

Increasingly, however, attachment theory is being questioned. The break from the main tenets of attachment theory began (a) when anthropologists, cultural psychologists, historians, and scholars in related disciplines drew attention to the diverse nature of infants' and young children's care experiences, as well as the cultural and ecological processes underlying them (for summaries, see Quinn and Mageo 2013; Otto and Keller 2014) and (b) when scholars from places not well represented in the attachment research community voiced concern about the global application of culture-specific patterns of care and relationships (Chaudhary 2004; Nsamenang 2006). But these challenges were not widely recognized until recently.

Attachment theorists acknowledge the cultural and contextual nature of people's lives, and the role of these factors in close relationships. For example, certain types of otherwise problematic attachments are considered adaptive under certain conditions, such as when parents are unable or unwilling to invest emotionally in, or care for, their infants. The extent to which infants experience separation from the people who care for them in everyday life is also taken into account when their distress is interpreted in the procedure designed by these scholars to study attachment. Yet these accommodations, while enabling a more nuanced view of attachment, remain grounded in the main tenets of attachment theory (Morelli and Henry 2013). As such, we view them as deeply concerning.

In this chapter we argue that an alternative approach to classic attachment theory and research needs to be taken. We agree that the ability to develop social relationships is part of our human legacy, representing a universal need to belong to social groups and to form meaningful ties with others (Baumeister and Leary 1995; Keller 2015). We agree, as well, that children form attachments to people in relationships that are distinct in particular ways. However, our approach insists on the central role of sociocultural processes and structures, in dynamic interplay with ecological processes, in the relational opportunities available to children and in the attachments they develop. We make our case in the following way: We introduce key aspects of attachment theory to demonstrate its support of species-wide, attachment-related processes. We consider examples of children's early care and relational experiences from diverse communities that call this position into question. We present a standpoint on ethics related to personhood and self to understand systematic variation across communities in these experiences, and we present a conceptual model that situates these ethics in an ecocultural frame (Keller and Kärtner 2013). Finally, we discuss social and relational networks, including attachments, as well as features and contexts of care that are important for distinguishing relationships as attachments.

Our conclusion is that children are cared for in culturally defined and ecological responsive ways, and this care is the basis for the relationships they develop. Attachment theory and research must be nimble enough to accommodate the diversity in the realities of children's lives. As long as attachment theory does not take this imperative seriously, real-world application of attachment theory is deeply concerning. We examine these concerns separately in Chapter 14 (this volume).

Classic Attachment Research and Theory

Attachment theory is about infant survival and the evolution of child behaviors that elicited care essential for this fundamental goal (Bowlby 1958, 1969, 1982). Bowlby was interested in the physical safety function of attachment behaviors (e.g., crying). He reasoned that when a young child is afraid or distressed, it is in the child's best interest to act in ways that bring him into the protective solicitude of his primary caregiver, most likely his biological mother. With age, the child increasingly directs attachment behaviors toward his mother, as she is the person who responds most readily in appropriate ways to him; this makes the child feel safe in her presence in times of need. Based on consistent repetition of these types of experiences, the child forms an attachment to his mother.[1]

Attachment theory is also about infant psychological development. Bowlby, with his colleague Mary Ainsworth, was interested in the psychological security function of the attachment system. The attachment relationship reflects an emotional tie of a child to his mother; the child loves his mother and typically greets her with joy. The child feels sufficiently safe in his mother's presence, assured that he can return to her for comfort and protection if necessary. This "felt security" provides the child with the support needed to explore the environment on his own, at a distance from others, with confidence, and to master the physical and social world. Mothers act both as a safe haven and as a secure base from which their child can explore; in effect, mothers are the lynchpins in the link between, and the balancing of, the attachment and exploration system for the child (Ainsworth and Bowlby 1991).

It did not take long for interests in the psychological function of attachment to dominate theory and research (LeVine and Norman 2001:86; Vicedo 2017). Questions about individual differences in children's attachment relationships were studied using a laboratory procedure to assess how well children were able to organize attachment and exploration behaviors during a period of moderately escalating distress designed to activate the attachment system. In this laboratory procedure, the Strange Situation, children are repeatedly separated from their caregivers—most often their mothers—for a brief period of time, and are either left on their own or with a stranger. Children who are secure in their attachment relationships are comforted by their mother's return and are

[1] Bowlby (1982) acknowledged that the person to whom a child develops an attachment relationship depends on who cares for the child in particular ways. This caregiver did not have to be the biological mother, although references made by Bowlby to this caregiver are most always the mother.

then able to use their mothers as a secure base from which to explore. Children who are insecure in their attachment relationships are unable to do this. The difference between securely and insecurely attached children was attributed to differences in children's history of sensitive maternal care (Ainsworth et al. 1978; Cassidy and Shaver 2008).[2]

The sensitive care that fosters security in attachment relationships is care that responds to a child's explicit positive and negative signals. It is exemplified by caregivers who

- are appropriately receptive and contingent in their response to the child's signals,
- follow the child's lead to structure and support the child's endeavors in unobtrusive and fittingly challenging ways,
- respect the child as a separate person with a will of his/her own,
- rely on encouragement, praise, and reasoning to motivate the child,
- are affectionate and affectively engaging, and
- encourage expressions of positive emotionality.

This care is considered the gold standard by which all care, worldwide, is to be compared and evaluated. When children are cared for in these ways, they develop a sense of themselves as being in control, competent, and worthy of help—assured that help will be available if needed (Bowlby 1980). When children are not cared for in these ways, they feel less secure in the presence of their mothers and are unable to achieve the same confidence in themselves or in others, or mastery of the environment. They are at risk for developing behavioral problems later in life (Thompson 2006; Weinfield et al. 2008).

The way that attachment theorists conceptualize sensitive care, secure attachment relationships, and children's competencies is based largely on their understanding of well-educated, middle- to high-income, urban-dwelling families of European ancestry (living in postindustrial Western societies, sometimes referred to as "Western" families). The values of such families typically sensitize children to personal preference and choice, as well as to internal psychological qualities; these principles are important to the way that people relate to others and in the way that people experience others in relationships with them (Shweder et al. 1997; Suh 2000; Raeff 2006; Kitayama et al. 2007; Markus and Kitayama 2010; Keller and Kärtner 2013). This approach to representing "self" in relationships reflects specific cultural philosophies of personhood and self, which children develop in ways described by attachment theorists. Yet when one looks around the globe, one discovers that other philosophies of the person and the self exist, and these support other ways to care for children that

---

[2] Sensitive care only modestly predicts attachment security (de Wolff and van IJzendoorn 1997). Some researchers also consider mother's ability to verbalize and interpret the mental and psychological states of their child (e.g., wishes, preferences, and intentions), and to treat the child as an intentional agent (see Sharp and Fonagy 2008).

are consistent with the philosophies. Below, we consider a selection of these philosophies and the corresponding childcare systems.

Representing Others and Self in Relationships

Children's day-to-day social engagements provide them with the opportunities to learn what it means to be an acceptable, good, and moral person as well as how to organize, interpret, and make sense of experiences about the world (Shweder et al. 2000). These senses of personhood and self are fundamentally relational in character, cultural in origin, and significant to the attachments that children develop. Personhood is a social status granted by others to individuals who meet culturally constituted standards for legitimacy in their community. When so designated, a person is considered a social being with moral status who is obligated to act (has agency) in moral ways toward others, as others are obligated to act in moral ways toward this person. Personhood standards vary from one community to the next; in many places, a strong emphasis is placed on qualities and attributes that indicate readiness to assume one's role and responsibilities in the social world as a relational being. For the Mapuche, an indigenous people of southcentral Chile and southwestern Argentina, to be *che* (a "person") means being capable of productive sociality: to be a giver and receiver (Course 2011). The idea of giving as a condition of personhood is observed in many other communities, including the Baining of Melanesia (Harris 1989). Baining infants and elderly are not considered persons because they are not able to give food. For them, personhood is transient: it is something acquired and then withdrawn.

Alongside relatedness-based notions of personhood are notions that emphasize autonomy (choice and volition). The individuality of Mapuche children, for instance, is recognized at an early age (Sadler and Obach 2006) as exemplified by the Mapuche word *püchiche*, which literally means "little person" (Quidel and Pichinao 2002; Sadler and Obach 2006; Course 2011; Williamson et al. 2012; G. Llanquinao, pers. comm.). Mapuche children are considered as being able to manifest and even (sometimes) impose their will on adults (Sadler and Obach 2006; Williamson et al. 2012). In other words, "early socialization is predominantly respectful of their *che* (personhood)" (Williamson et al. 2012:140).

Whereas *personhood*, in general terms, designates a social agent with a moral career (i.e., progressive changes in a person's moral status) as part of the social order in a particular society, *self* designates individual awareness of a unique identity as the knower (percipient) and the known (perceived) (Mauss 1985; Walker 2013). Although the two concepts are somewhat distinguishable, together they foster a person's understanding of who s/he is within the context of relationships with others. The self is an inherently socially constituted, relational construct that may only be understood in the light of an "other than the self." A child's sense of self develops by participating in the everyday life of the community. This participation is organized, in part, by the meanings and activities that community members share and convey to children, and which are differently appropriated by communities with a common cultural identity. In this way, communities adapt to local circumstances. Under most circumstances, however, communities hold onto tried-and-true practices (i.e., cultural models) because the risk of independent (individual) solutions to recurring challenges (which may fail) is too costly (Quinn 2005).

In the rich history of research and theorizing on conceptions of self in cultural contexts (Hallowell 1955; Howard 1985; White and Kirkpatrick 1985; Markus and Kitayama 1991; Holland 1992a; Quinn 2003; Kagitcibasi 2005; Wang and Chaudhary 2005), there is agreement that the "culturally constituted behavioral environment" forms the self in different ways (Hallowell 1955:87; see also White and Kirkpatrick 1985). Ideas about self put forward by Markus and Kitayama (1991) set in motion a wave of research that initially supported their thesis but later questioned it. They claimed that people constructed notions of self in two fundamentally different ways: an independent self (e.g., autonomy/agency) and an interdependent self (e.g., relatedness-heteronomous/ agentic). With time, more nuanced conceptualizations of self were advanced, but these continued to juxtapose the independent and interdependent notions of self.

There are other, quite different, views of self that acknowledge autonomy and relatedness as equally important, coexisting human needs that are part of any human action and situation. Individuals need communion as well as agency (Bakan 1966), love and belongingness as well as self-actualization (Maslow 1968), and closeness and interdependence with others as well as control over their own lives (Ryan and Deci 2000). These human needs are conceptualized in culture-specific forms and are necessary for a person's health and wellbeing. Keller and colleagues offer a way to conceptualize the coexisting needs of autonomy and relatedness that acknowledges their cultural nature (Keller 2012; Keller and Kärtner 2013; Keller 2016b). They propose that children are sensitized in different ways to the relevance, experience, and expression of each of these human needs. The ways in which this happens depend on culturally mediated, contextually based cultural regularities in children's experiences with others, which correlate closely with certain sociodemographic variables.

According to Keller and colleagues, one sociodemographic cluster represents educated (especially mothers), urban, middle- and upper-middle-class families who live in postindustrial economies. In these families, children's experiences sensitize them to value personal preferences, choice, and personal qualities such as traits, attributes, and talents. These children's autonomy is psychological in nature, based on self-reflection that centers on personal desires, wishes, and intentions. Autonomy underlies conceptions of relatedness such that relationships are defined and negotiated from the child's point of view. This self-conscious and self-contained child is the cultural ideal for families from this cluster, and this way of thinking coincides with the conception of autonomy and relatedness expressed in attachment theory.

A second cluster is made up of rural families, with little or no formal education (especially mothers), who live in subsistence economies. In these families, children's experiences sensitize them to group expectations as well as social roles and responsibilities. Relatedness is hierarchical and underlies conceptions of autonomy as self-regulated actions that meet socially constituted obligations and fulfill community responsibilities. For example, among the Mapuche, children are granted personhood status when they are able to give food (meeting social expectations). They are also expected to help with family chores (e.g., looking for firewood, taking care of animals) in the proper way, without having to be reminded.

Even though we associate particular conceptions of autonomy and relatedness with particular sociodemographic clusters, rapid social change (resulting from both global flows of information and people, as well as from internal processes such as civil war) may alter this association. For instance, after the civil war in highland Peru, adults in Quechua-speaking communities continue to expect children to contribute to the well-being of their families, animals, and communities from a young age, just as they have for generations, by observing and imitating adult work. However, a strong national emphasis on public education has greatly altered the daily routine of children. Even illiterate grandparents who do not speak Spanish encourage their grandchildren to exert their best efforts to do well at school, as this is perceived to be necessary to excel in life and get good jobs (Robins 2017). Intentionally combining components of tradition and modernity is fast becoming "the new normal" in many communities across the Global South. As people in communities adapt to accommodate such changes, we may see shifts in how autonomy and relatedness needs are met as well as concomitant shifts in children's care.

Conceptualizing Children's Care and the Diversity of Care Practices

At some fundamental level, the tasks of caring for children are the same all over the world (Benedict 1955; LeVine 1974). Most caregivers want children to survive and thrive in the communities in which these children live. This means keeping children healthy and safe as well as providing them with opportunities to learn about and from their social and physical environments in culturally organized ways. Yet the way these tasks are both defined and carried out varies, even as they are considered necessary and commonsensical in any given community (Murray 2013; Gottlieb and DeLoache 2017).

Most people care for children in the best way they know, taking into consideration (consciously or not) many factors: local ecological conditions (Keller 2007, 2016b); available economic, medical, social, and other resources

(LeVine et al. 1994); competing demands (Rogoff 2003); relationship to and history with the child (Lancy 2015); and the child's age, health status, and temperament (Scheper-Hughes 2014). The complex interplay of such factors has been represented conceptually in different ways (LeVine 1974; Whiting and Whiting 1975; Weisner 1984; Super and Harkness 1986). The *ecocultural model*, as described by Keller and Kärtner (2013), is most useful for our purposes as it considers how the dynamic co-action of the physical environment, ecosocial context, and caregiving beliefs and practices relate to children's development. The physical environment (e.g., climate, water and food supply) significantly contributes to the ecosocial context of a community; that is, to family structure (e.g., extended, nuclear) and ways of making a living (e.g., subsistence, cash-based economy). Parental education is a key characteristic of this context. The ecosocial context, in turn, significantly contributes to a community's system of care: it shapes socialization goals (what caregivers want for children), ethnotheories (ideas about how best to achieve goals), and practices (what caregivers actually do). These constituent parts, in dynamic interplay, are important to psychological processes that underlie children's developing conceptions of autonomy and relatedness.

This model clarifies that children's care and development are local phenomena that reflect the particulars of a community's physical environment and ecosocial structures in relation to the community's organized set of practices, beliefs, and traditions. To illustrate, let us consider the case of child care among the Juǀ'hoan (Takada 2005; Konner 2010; Takada 2010): a San group living in the Kalahari Desert in southern Africa—a harsh environment covered mostly with brush and grassy hills, where water is scarce for most of the year, food is unpredictable, and temperatures range from freezing to blistering hot. In the past, these people subsisted by hunting and gathering, traveling long distances on any given day. Child care among the Juǀ'hoan protects infants from the vicissitudes of this environment in a way that allows people to manage other demands. For example, Juǀ'hoan infants live their days in the arms and laps, and on the backs of people. This practice buffers infants from harm on the ground and allows them to nurse easily and stay hydrated. It also calms infants (Barr 1990; Barr et al. 1991; Esposito et al. 2015) and keeps their distress at low levels. This is important because distress is energetically demanding (Rao et al. 1993) and can compromise infant health if it persists. Moreover, it is easier for caregivers to take content infants with them on gathering and other excursions, which they often do. At the same time, caregivers encourage infant walking at an early age. The Juǀ'hoan believe that a child who is not taught to sit, stand, and walk will never perform these behaviors, and the bones of the child's back will remain "soft" unless teaching occurs (Konner 1973, 1976). Children who are able to follow caregivers on their own on foraging trips are less of a burden than are children who must be carried.

The study of hunter-gatherers elsewhere shows similarities as well as differences in child care (Konner 2010). The !Xun, for example, is a post-foraging

Namibian group of San that shows considerable sociocultural similarity to the Ju|'hoan (Takada 2005, 2010, 2015). However, in contrast to Ju|'hoan children, !Xun children are weaned earlier, and caregivers are more likely to be siblings or cousins. This observation raises questions that merit further study: Why are these changes associated with a more sedentary lifestyle in the !Xun? What other changes occurred as a result? What do these changes mean for a child's social relationships?

The care of children from diverse communities around the world provides an important counterpoint to claims made by attachment theorists. The care described as sensitive by attachment theorists is different from that which is practiced by a large proportion of people in the world. Similarly, the competencies that this sensitive care fosters—sensitizing children to experience themselves as separate and distinct, with needs and desires of their own; to act based on what they think and believe; and to change their environment accordingly (Ainsworth 1976; Bretherton 1987)—are less important to many of the world's people. Instead, care that controls a child's actions, anticipates a child's needs, or dampens a child's emotional expressiveness describes the warm, responsive care practiced by most communities in the world.[3] It orients children to others and fosters children's responsiveness to them. Children learn to see themselves as others see them, and this intensifies children's social connections and strengthens actions that maintain them. These actions are acts of obedience, compliance, conformity, proper demeanor, and respect (Harwood et al. 1995; Keller 2003; Quinn and Mageo 2013; Morelli et al. 2014; Otto and Keller 2014). Care such as this supports socially oriented notions of autonomy and relatedness grounded in meeting social obligations and responsibilities in contrast to care that supports self-oriented notions of autonomy and relatedness on which attachment theory is predicated. In what follows, we describe care practices that orient children to others, drawing on previous work (Morelli 2015).

Care that controls what children do is representative of a global pattern, but it does not meet the standards of care advocated by attachment theorists: to follow the child's lead and support the child in nonintrusive, unobtrusive, and fittingly challenging ways. In their study of 12 communities in India, Japan, Kenya, Liberia, Mexico, Philippines, and the United States, Whiting and Edwards (1988) noted that, except for U.S. mothers, all mothers ranked highest or second highest in training and controlling their children. (In these communities, people in all but the U.S. and Indian communities farmed for a living.) Similar findings have been reported for caregivers in other communities in Africa (LeVine et al. 1994; Keller 2003), East and South Asia (Chao and Tseng 2002; Rudy and Grusec 2006), the Middle East (Kagitcibasi 1970; Sharifzadeh 1998), South and Central America (Posada et al. 2002; Seidl-de-Moura et al.

---

[3] The first time we reference a community, we include the community's geographic location and mode of subsistence, when known.

2012; Rogoff et al. 2015), and non-middle-class populations in the United States (Ispa et al. 2004).

In most of the world's cultures, controlling children is what good parents do. Chinese parents and parents in other Confucian cultures practice *guan* (governance): they care for and love their children by taking control, directing their behaviors, and placing demands on them (Tobin et al. 1989; Chao and Tseng 2002). Puerto Rican parents act in similar ways to teach children to be calm, attentive, and well-behaved (Carlson and Harwood 2003), as do Cameroonian Nso farming mothers, who teach children to show good manners and to share (Keller and Otto 2009; Keller et al. 2012). Balinese mothers intentionally provoke jealousy and even rage in their toddlers by breastfeeding and playing with other babies. They remain impassive to their toddlers' displays of emotion, as one way to teach these children to calm their own feelings of jealousy and promote self-restraint (Diener 2000). Bara pastoralists of Madagascar prefer calm children and see calm behavior in the presence of elders as a sign of respect (Scheidecker 2017).

In these communities, care includes expressions of warmth (Ispa et al. 2004; Schwarz et al. 2005; Halgunseth et al. 2006), which children experience in a positive way (Fracasso et al. 1994; Aviezer et al. 1999; Posada et al. 2002; Carlson and Harwood 2003; Ispa et al. 2004; Howes and Wishard Guerra 2009). Japanese and Korean adolescents perceive *guan* as a sign of parental acceptance and warmth (Chao and Tseng 2002), and Latina and Portuguese adolescents living in the United States in families with low incomes see directive care as affirming their parents' protection (Taylor 1996).

Care that anticipates children's needs is another way to care for children that does not meet attachment theorists' standards of contingent responsiveness to children's *explicit* signals. This practice of meeting a child's needs before or around the time they are expressed blurs the child's sense of self as distinct and separate from others and accentuates the group as the child's primary referent of action. Caregivers rely on situational cues, prior history, and the child's subtle signals—which are often nonverbal—to do this. This anticipatory care is observed in communities around the world: among Efe hunter-gatherers in the Democratic Republic of the Congo (Morelli et al. 2002a), the Nso (Keller and Otto 2009; Keller et al. 2012), Makassar farmers in Indonesia (Röttger-Rössler 2014), the Bara (Scheidecker 2017), Sinhala farmers and wage earners in Sri Lanka (Chapin 2013b), Yucatec Maya farmers (de León 1998), as well as across Japan (Rothbaum et al. 2006).

Care that dampens or discourages expressions of emotions (positive as well as negative) is different from care advocated by attachment theorists that responds to a child's overt positive and negative signals and encourages positive ones. Among the Chinese and other Asian peoples, intense emotions are considered immature and socially disruptive (Kitayama et al. 2004; Wang and Young 2010). Puerto Rican middle- and working-class mothers (Harwood et al. 1995) and Costa Rican mothers (Rosabal-Coto 2012) prefer calm, well-behaved children as do the Gusii farmers of Kenya (LeVine 2004) and the Nso, who believe that a calm child fits best into its social group (Keller and Otto 2009). Whereas many caregivers address negative signals by responding to them (quickly), they tend to do the opposite with positive signals. In some communities, such as the Gusii, caregivers are relatively unresponsive to positive signals (e.g., to childish babbling).

These types of social-orienting practices are important to a child's developing sense of self as connected with others, as "a part of encompassing social relationships" (Markus and Kitayama 2010), and as "defined and made meaningful in respect to such others" (Kitayama et al. 2007). There are other similarly motivated practices that complement the ones described here.

Physical closeness: Holding and carrying children in the first year of life are common practices among subsistence-economy communities as well as in many other farming communities across much of Africa, Latin America, and Asia (Lancy 2015). In a study conducted among the agricultural Beng of Côte d'Ivoire, Gottlieb (2004) calculated that young children spend two-thirds of their daily nap time in physical contact with someone (whether on a moving back or a stationary lap), and children sleep with their mothers and older siblings until they are about ten or twelve years old (for boys and girls, respectively). In postindustrial societies, co-sleeping, co-bathing, and breastfeeding are common. The Japanese and Koreans describe this form of physical intimacy with children as *sukinshippu* and *seukinsip*, respectively (both terms for kinship). There is a tactile quality to this physical closeness (Röttger-Rössler 2014) that allows for the subtle and near imperceptible exchanges on which anticipatory responsiveness depends. It enables children and caregivers to rely regularly on nonverbal ways to communicate and, as a result, to coordinate their involvement in nonexclusive and socially nondisruptive ways (Morelli, Verhoef, and Anderson, pers. comm.).

When children are not in actual contact with others, they are kept physically close to them. Balinese mothers use fake fear expressions to keep their young children close (Bretherton 1992). Gusii and Hausa (shepherds in Nigeria) caregivers prevent children from crawling away (LeVine 2014). Japanese mothers stay close to their children (Ujiie and Miyake 1985; Rothbaum et al. 2000a). In some communities, such closeness is defined broadly to include not just close family members but larger social groups. In Beng villages, toddlers as young as two years old are encouraged to roam freely around the village, in the knowledge that all adults and older children will keep a watchful eye out for their safety (Gottlieb 2004).

Social exploration: Children explore with others close by and in social ways that differ from attachment theorists' notion of healthy exploration (i.e., solitary exploration of the physical environment at a distance from others). Japanese mothers are more likely to take advantage of social opportunities to direct their child's attention to the environment and to use toys for social engagement than are Euro-American mothers (Bornstein et al. 1990). Ju|'hoan foragers (Bakeman et al. 1990), Wolof farmers (Senegal), Beng villagers (Gottlieb 2004), and Soninke and Toucouleur immigrants to Paris (Senegal, Mali, Mauritania) are all more likely to respond to children when they are engaged socially than when they are engaged with objects (Rabain-Jamin 1994). For these African immigrant children, activities are structured around people, whereas for native-born Parisian children, activities are centered on the exploration of inanimate objects.

Orienting children toward others: Some social-orienting practices may explicitly orient the child toward others. In many subsistence-economy communities, for example, children are positioned so that their gaze is directed outward when carried or sitting; this draws a child's attention to its surroundings. This practice is complemented by an expectation common to people in these communities: children learn by attending carefully to others in the absence of explicit instruction (Rogoff 2003; Rogoff et al. 2014). Efe and Mayan families (farmers and wage earners in Guatemala; Morelli et al. 2002b) as well as Costa Rican urban families (Kulks 1999) expect this of children.

Other practices orient children to social interdependencies by accentuating the social group. Teasing, shaming, and name-calling (Samoans, Samoan Islands, farmer-foragers; Mageo 2013), abrupt weaning (Pirahã, foragers, Amazon rainforest; Everett 2014), and withholding empathic attention (Bhubaneswar, India; Seymour 2013) are examples of these practices. They direct children away from individual relationships and toward the social group as a whole. In extreme cases, mothers may discourage infants from becoming too attached to them, as some Beng mothers do when they break the gaze of their infants toward them (Gottlieb 2004). Care and protection by many others instill in children a sense of dependence on them and the group, rather than on a single individual (whether the mother or anyone else), for meeting their needs (Everett 2014; Morelli et al. 2014). Beng mothers actively draw people to their babies, by applying beautiful paints and jewelry to their infants twice a day, to establish a pool of babysitters (Gottlieb 2004).

Still other social-orienting practices teach children to see the world as others do and to adjust to their reality accordingly, without reference to the child's own mental state as separate and distinct from the other (Kärtner et al. 2010a). Beng adults (Gottlieb 2014), Kaluli caregivers (farmer-foragers in Papua New Guinea; Ochs and Schieffelin 1984), and Mapuche mothers (Course 2011) speak for their children. Kaluli mothers also teach children what to say.

Social-orienting practices such as these contrast with self-orienting practices that are important to psychological autonomy and relatedness. Examples of self-orienting practices include:

• Keeping children physically separated from their caregivers (i.e., on their own and in their own space); this enables the type of exploration that attachment theorists emphasize.

- Encouraging talk as the major form of communication in face-to-face, dyadic orientation.
- Celebrating the child's accomplishments (e.g., through praise) and attributing them to the child's efforts.
- Negotiating with the child out of respect for the child's interests and needs.

Such self-orienting practices of care, and the attachments they foster, represent attachment theorists' notions of good care and healthy development. We take a very different view.

Attachments Conceptualized

In our view, the attachments that children develop are different from those proposed by attachment theorists. Our approach builds on the ecocultural model and conceptions of autonomy and relatedness as human needs: it places great importance on community conceptions of good care and good children, takes into account the ecosocial conditions that play a role in how these conceptions influence caregiving practices and thus a child's experience of care, and allows for flexibility in thinking about how attachments are expressed and the role they play for the child across both time and contexts. We begin our description of this way of thinking about attachments by calling to mind the provocative claim made by Shweder et al. (2000:219) that "the knowable world is incomplete if seen from any one point of view, incoherent if seen from all points of view at once, and empty if seen from nowhere in particular." Faced with these three alternatives, Shweder et al. opt for an incomplete or partial view of the world, and so do we. Our approach to attachments is partial in at least two ways: First, it identifies the basics of attachments but relies on the knowledge of a given community's practices and beliefs relevant to relationships, children, and their care to understand the attachments that children develop in that community. Second, it is a view in the making; both its conceptual and empirical foundations require further development.

Social Networks

Our conceptualization of attachments starts with social network theory. This approach appeals to us because it puts relationships at center stage for study. What is relevant to social network theorists is the structure of the network—the nature of a person's ties with people and the ties these people have with others—and the relation between network structure and the phenomenon of interest. Social ties are characterized by network size, the extent to which people know one another (density), the ability of the network to endure severance of ties (robustness), and more (Smith and Christakis 2008). The entire complex of ties that provides a given network its structure has properties that are not explained by or present in the parts that make up the network. This theoretical approach has a wide reach. It is used, for example, to study issues as diverse as social cooperation (Nowak and Highfield 2011), social change (Lane et al. 2009), health behavior (Smith and Christakis 2008), health-care systems (Castellani et al. 2015), schools (Daly 2010), and terrorism (Krebs 2002).

The use of social network concepts in the study of children's relationships is most welcomed because they take us beyond the mother-child dyad to the complex interdependencies that characterize many children's relational systems (e.g., Levitt 2005; Lewis and Takahashi 2005; Rubin et al. 2009). Children are seen as participating in different social systems, for different reasons, and these social systems include people who know the child, as well as one another, in different ways and to varying degrees. These people, in line with social network theory, may be involved directly with children as partners in social activities or involved indirectly, which includes children watching them. The character and contexts of children's lives are inextricably linked to their social networks. In many ways, networks determine the physical, social, and psychological resources available to children, as well as what is required of them to gain access to these resources. A child's social network, for example, provides her with opportunities to watch, interact with, and learn about and from others. Reciprocally, the people in a child's network have the opportunity to watch and interact with the child, which may affect whether and how they invest in the child, including the child's care.

In most cases, a child inherits her first social network at birth.[4] This first network may change from place to place at a given point in and over time (Smith and Christakis 2008). Many factors contribute to why a child's social network is the way it is at a particular time and place. One factor is the child's living arrangement, which partly reflects the family's residence practices (who lives near the family) and structure (who lives in the family). In an extended family structure, for example, the family unit goes beyond the nuclear family and typically includes grandparents, aunts, uncles, and cousins. This structure is common across many parts of Asia, the Middle East, Central and South America, and sub-Saharan Africa. In 2015, Child Trends (2015b) reported that more than 70% of families were extended families in South Africa; in India, Colombia, and Turkey this was true for 50%, 58%, and 58% of families, respectively (Scott et al. 2015).[5] Yet, the makeup of extended families varies considerably within and across communities.

---

[4] A child may inherit her first social network even before birth. For societies oriented around a philosophy of reincarnation, a "newborn" is seen as emerging from a previous life, where a former family has decided to allow the child to be born. In such a scenario, childcare practices revolve around the parents of "this life" paying micro-attention to the infant's momentary needs to discourage the child from "returning" to the previous family of the "afterlife" (Gottlieb 2004).

[5] This report does not include data from many African countries and may under represent the percent of extended families as a result.

The living arrangements of children in the United States and in many European countries (in particular, northwestern Europe) differ from the regions just described. Today, in many communities, for example, the typical family structure consists of mother, father, and dependent children. While this continues to be the image of what a family should be in these places, family arrangements increasingly have become diverse. In the 1960s in the United States, over 80% of children lived with two parents, whereas in 2015 about 70% did (Scott et al. 2015). Similar trends have been noted for the United Kingdom (Knipe 2015). This picture becomes more complicated when ethnicity and social class are considered (Sawhill 2013). In the United States, for example, Child Trends found that 34% of black, 60% of non-Hispanic white, and 83% of Asian children lived with two married parents.[6]

This consideration of social networks illustrates the substantial variability on a global level in children's social networks and, as a result, in children's opportunities to engage with different people and to secure resources from them. It also illustrates the importance for development of both the child's social network as a social unit and the child's individual relationships.

As increasingly more children (alone or with family) move from place to place because of conflict, persecution, economic inequality, or instability, their social networks are likely to become more fluid and dynamic. UNICEF (2016) estimates that nearly 50 million children live under transient conditions: of these, 31 million live outside their country of birth, including 11 million child refugees and asylum seekers. This unprecedented level of migration (across country borders) and displacement (within country borders) of children presses us to understand better the changes in these children's social and relational ties, and how these ties might support their positive development.

Relational Networks

The social networks in which children are involved provide them with the opportunity to develop relationships. How many relationships develop depends on whether the many or few individuals available to the child are willing and able to engage the child, and whether the child is willing and able to engage them.

At birth, children are able to take advantage of relationship opportunities, aided by basic neurobiological processes relevant for social affiliation that mature during fetal development. There is evidence, for example, that the perceptual biases of very young infants predispose them to direct their attention to social information. Newborns are more likely to attend to human speech than nonspeech sounds (Vouloumanos and Werker 2007), and to human faces and face-like stimuli than other stimuli (Slater and Quinn 2001; Farroni et

---

[6]   One reason for the low rate of married parents living with children among black families may be that single-parent families have deep roots in practices developed from the trans-Atlantic slave trade, which deliberately separated families.

al. 2005). The facial features most likely to interest newborns (e.g., upright compared to inverted faces, direct gaze compared to averted gaze) suggest that they are attracted to cues that indicate a person's readiness to engage with them (Farroni et al. 2005). These early biases and other nascent capabilities (e.g., the ability to orient gaze and body position toward people who arouse a child's interest) help the child connect socially with others (e.g., Reis et al. 2000; Lee et al. 2009). In addition, very young infants are able to sustain others' involvement and interest in them, for instance, by matching (reflexively) facial expressions and temporally coordinating nonverbal activities (e.g., body movements) with that of their social partners. Other ways develop with age, such as using "face, voice, hands, and entire body" (Shai and Fonagy 2014:187).

The nature of a child's first relationships is likely to differ from one person to the next, and this continues as the child matures. Why this is so is complicated. One explanation concerns the activities that characterize a child's involvement with others, since these activities are often related to role-based expectations. For example, among the Murik (marine foragers of Papua New Guinea), it is common for a mother's family to share completely in the care of her child during the first six months or so of a child's life (Barlow 2013). By comparison, a Makassar mother (subsistence farmers of Sulawesi, Indonesia) is usually her child's primary caregiver for the first four to six weeks of the child's life (Röttger-Rössler 2014). Yucatec Mayan adults do not usually play with children (Gaskins 1999), whereas middle-class educated U.S. adults do (Roopnarine 2011). The activities that children and partners engage in together are likely to change over time and place, as the needs of and demands on the child and her social partners change. These examples show that the socially distributive nature of children's engagements differs: in some settings, one or a few people may do most everything with and for the child; in others, many people may do most everything with and for the child; in still others, some variation in-between may exist (Quinn and Mageo 2013; Otto and Keller 2014). In addition, people may alternate with each other as the child's primary caregiver, and people's roles may change if caregiving is organized as a division of labor (Scheidecker 2017).

The incredible variation in relational opportunities for children and in the relationships they develop is striking but not surprising, given the cultural and ecological dimensions inherent in children's relationships, coupled with other features (e.g., a child's age, competencies, relationship histories). Even so, some researchers claim that there are features common to all of the relationships people develop across the life span, although some features may be more prominent than others for some relationships, for some people, at some time in their lives (e.g., Sutcliffe et al. 2012). While these features are cornerstones of all relationships, communities differ in how they are perceived and acted upon in everyday life, as earlier examples in this chapter demonstrate. Core features of relationships involve:

- Mutual influence: what social partners are doing and experiencing depends on and has consequences for one another (e.g., coordination, co-regulation, mutual responsiveness, and synchrony).
- Emotional connection between relational partners: what social partners feel for each other and how these feelings are communicated (e.g., emotional intensity, emotional regulation).
- Time: social partners have a history and anticipate a future; this allows them to act on expectations based on past experiences and future expectations.
- Holism: social partners are sensitive to patterns based on interactions that form the relationship context, and act based on the context as a whole.

The last feature, holism, is consistent with Hinde's views of relationships: "a relationship is more than its constituent interactions" (Hinde 1999:326). Thus the idea of holism extends beyond individual relationships to consider their complex interplay with the relational and social networks of which they are a part (Reis 2000; Brown and Brown 2006; Sutcliffe et al. 2012; Kuczynski et al. 2015).

What we learn from these features is that people in relationships matter to each other. They have mutual interests and feelings for one another. In addition, how people act toward each other in the present likely reflects a shared understanding of each other based on past experiences as well as a shared expectation that the relationship will continue for some time in the future. For these reasons, people in relationships are appropriately receptive and responsive to each other, and may act in other ways to sustain the relationship. Over time, people may develop a preference for each other (rendering one or both as special) and perhaps consider each other as "irreplaceable" (Brown and Brown 2006:7).

Relational attributes of preference and irreplaceability are likely to be more characteristic of relationships that people describe as "close" (e.g., Brown and Brown 2006). These relationships are likely to be more "affect laden" (Reis et al. 2000:845), and people in them are likely to show more mutual concern and caring for one another, for instance, by setting aside their own needs and interests to attend to the needs and interests of their relational partner (Brown and Brown 2006; Reis 2014). Promoting the well-being of one's relationship partner is important to trust building, and perceptions of a person as trustworthy are important to the development and maintenance of relationships distinguished as close.[7]

It is important to note that relationships, including ones considered "close," are not always positive and supportive. Multiple agendas, conflicting interests, and time constraints may alter the dynamics of a relationship. Concerning

---

[7]   The descriptor term "close," when applied to relationships, is conceptually vague (Reis et al. 2000:844), although it often implies an emotional intensity between relational partners. Given this, and the likelihood that "closeness" is differently experienced and expressed across communities, we use this term judiciously when describing relationships.

parents and infants, Trivers (1974) argued that offspring are selected to demand more than their parents are willing to give at different points in time, thus creating an inherent conflict in their relationship. Weaning is one such time. Breastfeeding is an energetically demanding activity (Dewey 1997), and at some point, mothers may decide to shift investments from their current child to the conception of another child (to increase reproductive success).[8] This period is wrought with distress for many children the world over (and their parents as well), and Trivers claims that strategies such as temper tantrums may have evolved to help children sustain their mothers' investment in them. Mothers have devised different ways to curb their infants' interest in breastfeeding. Among the Efe, for example, mothers paint their breasts with bitter-tasting substances, make their breasts inaccessible, or send children to relatives in another camp for a period of time. Similar practices have been reported among the Ju|'hoan (Shostak 1981) and the Nso (Yovsi and Keller 2003). The conflict that arises during weaning may be particularly intense if few people are available to care for the child.

Attachment Networks

Children develop relationships with people in their social networks. Over time, some or all of these relationships develop into relationships that are distinct (i.e., close) to the child in specific ways. Before we consider care that may make these relationships distinctive for children, we reflect on why people other than mothers care for children, and the psychological processes that support this care.

Caring for Children

Even though children's attachments develop in the context of relationships, most studies are paradoxically one-sided—concerned with what the child gets from the relationships (e.g., protection, security) and the child's qualities that may influence this (cf. Roisman et al. 2013; e.g., Bakermans-Kranenburg and van IJzendoorn 2015). Even research on the care of children approaches things from a child's point of view. Far fewer studies ask why people give care to children. Addressing this fundamental question could, however, broaden our understanding of attachment and help us learn more about the reasons behind attachment (e.g., integration into the social group), the distribution of attachments (e.g., many at the same time), and the commutable nature of attachments (e.g., communal, flexible). It could also help us learn more about competencies that make attachments likely (e.g., care that fosters a child's social nimbleness)

---

[8]   There are other reasons why mothers shift investments away from the child, including the child's risk of death (Scheper-Hughes 1992).

and the psychological processes that reflect and underlie them (e.g., autonomy, relatedness, theory of mind, empathy). Our grasp of these issues is partial at best, although research suggests that we are on the right track (Meehan and Hawks 2013; Morelli et al. 2014).

Our Human Legacy: What Evolutionary Accounts May Tell Us

We know from studies of extant hunter-gatherers that mothers alone are unable to meet the dietary demands of keeping themselves and their nutritionally dependent children healthy. Mothers require the help of others to do both (Hewlett and Lamb 2005; Hrdy 2005a; Crittenden and Marlowe 2008; Kramer and Ellison 2010; Meehan and Hawks 2013). A child's relatives are usually among the first to help with child care (Hamilton 1975; Briga et al. 2012), but the sharing of care goes well beyond kinship (Hamilton 1975; Briga et al. 2012), with residency playing an important role (Hill et al. 2011; Kramer and Greaves 2011).

The willingness of people to help mothers with child care is one way that mothers are able to manage the ecological and social uncertainties that threaten a child's ability to survive and thrive. The conditions that enable cooperative acts of care, as well as other cooperative acts, most likely trace back to the Paleolithic era,[9] although this is not known for certain. Nevertheless, we consider evolutionary accounts of ancestral times because they offer a window into the genesis of social networks, give support to core relational features such as mutuality (synchrony), and provide clues to children's access to diverse relational and attachment opportunities.

Evolutionary accounts posit that ecological and social uncertainties during the Paleolithic era favored a suite of biological and psychological processes which made it possible for our ancestors to live in socially complex groups and thus to cope better with the unpredictable nature of their everyday lives. Group living, for instance, made it possible for people to extend "exchange partners" beyond the immediate family, which increased their chances of smoothing over day-to-day fluctuations in food and other resources. This, most likely, affected their reproductive success favorably.[10] However, the demands of group

---

[9]   Archaeological records suggest the origins and development of early human culture during the Paleolithic era. We use "ancestral" as a shorthand notation to refer to this period.

[10] Evolutionary claims consider the expression of psychological and biological processes in terms of reproductive success. This is most commonly defined as the number of offspring an individual produces to maturity. Fitness is the metric for measuring reproductive success in evolutionary theory. It is a complex concept with a contentious history (Beatty 1992; Paul 1992). "Individual fitness" was generally identified with the intuitive concept of reproductive success. Hamilton (1964) introduced the notion of "inclusive fitness." An individual's reproductive success also included the individual's effects on the reproductive success of relatives, weighted by the degree of genetic relatedness. In the mathematical theory of natural selection, genetic fitness is the measure of the contribution to the next generation of one genotype or one allele, relative to the contributions of other genotypes or alleles.

living were many and significant (van Vugt and Kameda 2014), and mechanisms may have evolved to offset the resulting transaction costs. Evolutionary theorists posit that people were likely to stay in a group if (a) they experienced positive emotions and moods in the group and felt socially connected to others (social bonds), (b) they felt loyal to the group, and (c) they could assess and act on threats to group cohesion. Group experiences such as these were possible if people were able to socially coordinate actions, meanings, and goals. The ability for synchronous exchanges helped to make this coordination possible: as a species, we are biased in both perception and attention toward synchrony, and we respond to synchronous interactions favorably (Ravignani 2015).

Parenthetically, we find intriguing the possibility that children evolved sensitivities to cues about the precariousness of resources, as suggested by the work of Chisholm (1996). Some of us would like to suggest that children may have evolved strategies to minimize the threats to their survival that these uncertainties posed. This may include the ability to solicit care broadly from people, and to develop multiple, simultaneous attachments based on the care that is experienced.

Further empirical evidence is needed before we are in a position to support a specific account of how human sociality evolved (see Vicedo, this volume). Even though further evidence for evolutionary claims would strengthen our views, our views do not rest on them.

Care, Today, the World Over

Across the world, people beyond mothers regularly care for young children. Keller and Chaudhary (this volume) address the prevalence of shared care, and in this discussion we draw on their work to highlight several points. First, the sharing of care is present in families from different economic backgrounds (e.g., subsistence, wage-based), with different levels of formal education (e.g., none, elementary school, high school, college), and with different family structures (e.g., nuclear, extended). Nevertheless, features of child care (who provides care, in what ways, and when) vary, depending on the interplay of factors such as views about good care, economic and other resources, religious values, work constraints, and broader political structures. In some places, relatives (especially grandparents, siblings, aunts, or uncles) provide the bulk of shared care (Weisner and Gallimore 1977; Raffety 2017; Schug 2017). In others, people unrelated to the child—neighboring children (Gottlieb 2004) or professionals (U.S. Census Bureau 2013)—provide the majority of daytime care. Shared care may take place early in a child's life (beginning at birth) or later, and it may be all-encompassing (feeding, bathing, instructing) or limited (e.g., carrying). The essential point is that the sharing of care is practiced widely but in different ways and for different reasons (Sear 2016). Shared care offers benefits to children, their parents, and to the people providing the care (Keller and Chaudhary, this volume).

Motives for Child Care

Why are some people willing, reliably so, to set aside personal needs, sometimes at a personal cost, to care for another person's child? To explore this question, we build on our consideration of evolutionary accounts of human sociality, with the caveat that only some of us see this as a fruitful endeavor to help understand contemporary childcare practices.

Costly investment in other people's children, with whom there is little or no genetic relatedness, posed at one time an evolutionary conundrum. This enigma was partly addressed when theorists considered the reproductive needs (e.g., protection, food) that people have in common. They reasoned that when people, alone, are not able to meet these needs—but people together are able to do so in the moment or sometime in the future—cooperative care makes evolutionary sense. How, exactly, people cooperate in the care of each other's children has a lot to do with culturally defined rules (Bogin et al. 2014). These rules are likely to adapt to changes in local conditions, with implications for people's investment in children (for similar arguments, see Bentley and Mace 2009; Quinn and Mageo 2013; Otto and Keller 2014).

Caring for the children of others may be in the reproductive interest of many, and this interest may continue as long as the relationships are beneficial to all. We presume that relationship features, such as mutual influence and emotional connection (noted earlier), are part of what makes these relationships (marked by costly investment) beneficial. If such features cease to exist, these relationships may continue, but differently (Brown and Brown 2006); perhaps as "a pool of recruits for more intense relationships" (Sutcliffe et al. 2012:159).

The benefits of cooperative care networks extend beyond relational partners to the group as a whole. Dunbar and Schultz (2007) suggest that a person's reproductive success depends more on long-term considerations made possible by the group: "relationships provide the key to fitness benefits at the group level" and "trickle-down benefits are reaped by the individual" (Dunbar and Schultz 2007:1346).

The Attachments Children Develop

Children may best learn about themselves, others, and the world around them by relying on people whom they trust to meet their needs. These people provide the child with resources to survive and thrive in the community of which the child is a part, and they do so in a way that fosters the child's attachments to them. Here we provide examples of the types of care, and reasons for care, that contribute to a child's sense of a person as trustworthy, and thus to the child's attachments. When we speak about reasons for care, we tap into two interest areas of attachment theorists: (a) threats to the child that emotionally overwhelm the child in negative ways and (b) the function of care providers to help the child manage these threats and thus regulate emotions (e.g., by providing the child with a secure base or haven of safety) (e.g., van Rosmalen et al. 2014). We believe it is necessary to address both the care that a child solicits and the care that is given without apparent solicitation. Especially at a young age, children are unlikely to intuit all threats. Threats to social group function are an example of this, and social-orienting care practices described earlier may help children learn about them. As our thesis assumes that multiple attachments are common for children, we conclude with thoughts about why the idea of multiple attachments remains a knotty issue for attachment theorists.

Care and Attachments

We consider relational features (along the lines we described earlier) that many claim to be important for both individual relationships and social group living. Given the centrality of these features for human sociality, we propose that they may figure importantly in the attachments children develop. These features are unlikely to be sufficient, however, and may vary in importance across communities and contexts (e.g., McElwain and Booth-LaForce 2006). In addition to proposing features of care, we give examples of their expression in one or several communities.

One feature that stands out is mutual influence characterized by behavioral synchrony. Behavioral synchrony makes it possible for interacting partners to coordinate their behaviors in time, intentionally or not. In turn, this makes it possible for people to coordinate actions, meanings, and goals, and thus to benefit from the advantages of group living. Synchrony, however, is a construct for which there are many meanings: reciprocity, adaptation, shared affect, turn-taking, and more (Leclère et al. 2014). These depictions are not equally relevant globally across all communities. However, we are drawn to what these depictions of synchrony have in common: temporal concordance. Feldman (2012c, 2014) defines synchrony in this way for processes (e.g., non-verbal behaviors, arousal) that occur at the *same time* or *close in time*; in other words, as temporally matched interactions (Feldman 2007a:329). The very young child is sensitive to the temporal organization and rhythmic qualities of stimuli (Gratier 2003) in the first days of life (Shai and Fonagy 2014), and care providers take advantage of this sensitivity by coordinating what they do with the infant's state of arousal, thereby providing the infant with her first experience with social contingencies.

Synchrony extends beyond dyadic interactions. Gordon and Feldman (2008) studied triadic synchrony among educated, urban, middle-class mothers, fathers, and their five-month-old children during play episodes. They found that infants were able to detect changes in the support each parent provided the other and changed their behavior (social focus) accordingly. Synchrony extends beyond temporal concordance between behavioral events. It takes place within and across behavioral *and* physiological systems, for each partner *and* among partners. What this suggests is that synchronous processes help create biobehavioral connections among people, which may be important to social group living (Feldman 2014).

Repeated experiences of synchrony are important to a child's development in many ways, including social and emotional regulation (e.g., Shai and Fonagy 2014). In addition, the familiarity these synchronous processes make possible with the style, manner, affective state, rhythms, pace, and so on of others (Feldman 2014:150); the positive feelings they engender (Watson 1985; Spoor and Kelly 2004); and the sense of "we" that they create (Baimel et al. 2015) may help distinguish relationships marked by repeated synchronous involvements over time in ways important for attachments.

Synchrony includes overlapping and sequential temporal concordance of events, which we believe are conceptually distinct and, for this reason, are differently important to attachments. To illustrate this point, we consider research on the temporal structure of events that sensitizes children to experience autonomy and relatedness in particular ways. Contingency responsiveness is an example of sequential concordance, thought to be particularly salient to infants in the early months of life because of their limited memory span. It takes place very quickly, within a second of the young infant's signal, and is not done consciously or deliberately. The overall contingent responsiveness of mothers across communities appears remarkably similar (Keller et al. 1999; Kärtner et al. 2008, 2010b). This is not surprising, given the intuitive nature of this reaction to infant signals.

The ways in which mothers are contingently responsive diverge with a child's age. Kärtner et al. (2010) examined mothers' auditory, proximal, and visual contingent responses to infant signals at age 4, 6, 8, 10, and 12 weeks in 20 families in Münster, Germany, and 24 rural Nso families in Kumbo, Cameroon. The extent to which mothers relied on each of these communicative modes was similar up to their infants' second month of life. After that, community differences were observed. Compared to Nso mothers, for example, the use of visual contingent responses by German mothers in face-to-face episodes increased with child age, while the use of proximal responses decreased (in line with experiences that foster psychological autonomy and relatedness). This trend was related, in part, to the growing reliance on this distal form of communication by these German mothers. In contrast Nso mothers relied on visual and proximal responses, which were relatively consistent over child age: low for visual responses (and significantly lower than the German sample) and high for proximal responses (in line with experiences that foster hierarchical relatedness and autonomy as self-regulated action that meet socially constituted obligations).

The temporal concordance of events can also be overlapping or simultaneous; Keller et al. (2008) refer to this as synchrony. This temporal structure, for example, can make it difficult for the child to take center stage, instead sensitizing the child to unity with others. Gratier (2003) studied vocal interactional patterns between mothers (middle-class, urban living, with at least one year of university-level education) and their two- to five-month-old infants in France, the United States, and India. Indian mothers and infants were more likely to participate in simultaneous vocalizations than were U.S. or French mothers, and this was seen in mothers of infants as young as two months of age. A similar pattern of simultaneous speech has been observed among Japanese compared to U.S. mother-infant dyads (Kajikawa et al. 2004).

There are additional features of care important to the attachments children develop that we would like to mention. One is the emotional connection *warmth* that people express to the child (and vice versa). Warmth is associated with positive feelings and it engenders affiliative experiences important to relationships and social group living (MacDonald 1992). It is not, however, inextricably linked to the care of children by others. For example, MacDonald (1992:762) notes that research on the Gusii strongly suggests that responsive parenting can take place in the absence of warmth and affection. Keller et al. (1999) confirmed this view by showing that contingent responsiveness and warmth are two separate dimensions of parenting. There are different ways to define warmth (e.g., positive affective tone, giving and expressing affection, pleasurable affective response) and to express warmth (e.g., affectionate contact, facial expressions, empathic affect, and mutual sharing of affective displays) (MacDonald 1992; Keller 2013b). This relates, in part, to the way that children's needs for autonomy and relatedness are understood by their caregivers and others. Warmth is often characterized by (a) physical contact and closeness, when relatedness is hierarchical and autonomy consists of self-regulated actions that meet socially constituted obligations, and by (b) facial expressions, when autonomy and relatedness are psychologically based.

Children must spend enough time with people to learn about them and the resources they are able and willing to provide. This is best learned when a child's experiences with and of others (directly or indirectly) are *reliably consistent, predictable, and repeated* over time. As children become familiar with the ways of others, they are better able to anticipate what others are likely to do and to plan for it (Sroufe 1979). This allows children to better manage their social involvements. There are different ways that people are able to create reliably consistent and predictable experiences for children. People can act in consistent ways toward a child, and this is the consistency often referenced in attachment research. However, people can also create consistency for the child by organizing their physical and social worlds in particular ways. The Efe are interesting in this regard. Efe infants are cared for by many people on any given day, yet the people who care for them often vary (Morelli et al. 2014). Here, consistency is not about who cares for the child, but rather a child's experience of care. For example, when young Efe children (7–15 months of age) fuss, they do so only for about 25 seconds. This is because many people are willing to comfort the infant quickly (as quickly as the infant's mother), and within seconds of a response, infants quiet down (Morelli et al. 2002a). There is a commonsense reason for similarities in the Efe child's experiences of care. Efe live their lives in full view of one another most days for the entire day. People witness the care that others give and comment on it. Public displays of "good" care may also be one way that people maintain their relationships with others and their good standing in their social group. The following example supports this view: A ten-year-old girl was punished for dropping her infant brother while in her care. The girl was ostracized from the group, made to stand at the edge of the camp, and not allowed to return for several hours. All the while, she wailed.

This point about consistency is similar to Quinn's view on the constancy of children's experiences (Quinn 2005). She notes that for the child to experience constancy, the experience should be repeated with regularity and be undiluted by contradictory experiences that create confusion and ambiguity. This is possible through "habitual, embodied practices" which pattern "the child's experience deliberately, vigilantly, and persistently" (Quinn 2005:481). Thus there are consistencies in the cultural world of children in matters of great importance created by a child's wider community.

Threats and Reasons for Soliciting and Eliciting Care

Attachment theory claims that children need to feel safe in contexts of threat (Labile et al. 2015:37). Threats to a child's solitary exploration of the physical world figure importantly in attachment theory, in part, because this type of exploration is tied intimately to the development of psychological forms of autonomy and relatedness. There are other threats that may undermine a child's ability to learn about and participate fully in her community, and we consider some of them.

A young child's state of arousal is an important mediator of her experiences with the social world. Feldman (2007b) notes, for example, that mothers use contingent responsiveness to heighten their infants' alertness and are more likely to provide visual and tactile stimulation when their infants are alert. When the young child's alert states are disrupted by crying and irritability, the child disengages socially from others and is less able to benefit from the opportunities that engagement provides. Furthermore, the distress indicated by these signals is metabolically costly (Rao et al. 1993). If either persists, the health of the child may be compromised.

Hunger, illness, fatigue, and pain are some of the reasons why young children cry and fuss. There are many ways to calm a child who is fussing or crying that go beyond attending to her physiological needs. Infants, for example, exhibit a calming response to being carried, and carrying may go beyond quieting a crying child: its calming effect decreases a child's heart rate and voluntary movements (Esposito et al. 2015). We mention this because, in many communities, in the first years of life, children are carried for a good part of the day, often by older children (Weisner and Gallimore 1977; Gottlieb 2004).

Threats that elicit a child's distress are likely to evoke intense negative arousal and inhibit cortical and subcortical processes important, for example, to attention and memory (Fonagy et al. 2014). Fonagy et al. (2014) claim that attachments as sources of reassurance may preempt (or abate) the threat and, as a result, reduce the extent to which important neurobiological processes are disrupted. They posit that attachment evolved to change internal (e.g., stress) and external conditions (e.g., threats) associated with threats to infant survival (Fonagy et al. 2014:35).

Threats to a child's health pose great concern to the majority of communities worldwide yet, surprisingly, the relation between child care and health, and the attachments that children develop as a result, has been of little interest to attachment researchers. This in itself is puzzling since protection from malnutrition has been cited to be "the most dramatic demonstration of the adaptive value of attachment security" (van IJzendoorn and Sagi-Schwartz 2008:900). Perhaps the relative disinterest in health—specifically nutritional health—on the part of attachment theorists can be traced to Bowlby's claim that "food plays only a marginal part in the development and maintenance of attachment behavior" (Bowlby 1969:224).

Finally, *children have a need to belong*; that is, a need for relatedness. This need can be threatened as well, but differently for children who are sensitized to notions of autonomy and relatedness as psychological, or to notions of relatedness as hierarchical and autonomy as self-regulated actions that meet socially constituted obligations. Rothbaum et al. (2011) suggest that threats perceived by children who are sensitized in the first way relate to exploration and self-esteem; here, one purpose of attachments is to make exploration at a distance from others possible for children and to affirm their self-esteem and efficacy. For children who are sensitized in the second way, perceived threats relate to social rules and responsibilities; here, one purpose of attachments is to reassure children of their social place in the group or that it can be regained by correcting problematic behaviors. To elaborate on threats to social group living, we provide two examples.

The first concerns the discipline of Murik children (Barlow 2013) and we have chosen this example precisely because discipline lies outside of attachment theorists' thinking about attachments (Kuczynski et al. 2015). Murik children are disciplined by their mothers when they do not share food, and mothers are extremely consistent in enforcing this behavior. Others shame children into sharing by teasing them. Food sharing is something a good Murik person does; it is the "quintessential expression of relatedness, caring, and belonging" (Barlow 2013:177). Murik children are also disciplined for misbehaving (e.g., nursing when a child is considered too old). Mothers, however, do not typically discipline their children for this reason; rather they invite senior figures to do so. Barlow (2013) claims that these seniors express their authority in this way, and, as a result, stress the age-grade system of control observed among these people. Children are often distressed when disciplined. Nonetheless, they are able to return quickly to their good standing in the social group (e.g., by being given the opportunity to share), and are often quickly comforted and reassured by the people who disciplined them. For Barlow (2013:174), ordinary discipline "guides attachment emotions and behavior along cultural norms" and many people have disciplinary functions.

The second example illustrates distancing practices used by Samoans (Mageo 2013). The purpose of these practices is to direct children—beginning around the age of three or four—away from affiliative relationships with individual people toward an affiliative relationship with the group as a whole. Samoans sensitize children to attend to others, to serve elders obediently, and to assume their proper status in the group. Children who demand attention, act in self-centered ways, or are aggressive toward others for individual gain threaten their tie to the group and are shamed, teased, or hit by adults and other children alike.

Children's Trust of Others and Attachments

Children are likely to consider people trustworthy when (a) their experiences with people are appropriately contingent (synchronous), (b) people help children meet the threats they experience in a manner that engenders feelings of safety and security, and (c) these and other experiences take place in the steadfast ways we have described. Feelings of trust are important for attachments to develop.

Children's Multiple Attachments

Even though much of early attachment theory was informed by the care practices of U.S. middle-class, college-educated families, attachment theorists do not ignore the possibility that children may form attachments to others who care for them besides their mothers: "a child can also get attached to other caregivers who are in regular contact with the child and make it feel secure in times of need" (van Rosmalen et al. 2014:12). Views such as these, however, often safeguard the role of mother as the primary attachment figure: "Human babies, however, do not have an instinct that causes them to become attached to the first living thing they encounter. They get attached to the person *who cares for them the most* during the first few months of life. *In most cases, that is the mother*" (van Rosmalen et al. 2014:13, italics added for emphasis). Attachment theorists' struggle with fully integrating multiple attachments into theory and research is also visible in other ways, such as their near-exclusive selection of mothers for study. This struggle is not new and echoes Bowlby's uneasy relation with the same idea.

In the 1969 edition of *Attachment*, Bowlby acknowledged that infants may have more than one attachment figure based on who in the household cares for them (Bowlby 1969). However, he argued that infants developed attachment preferences in the first year of life based on the care they received. He reasoned that, at first, infants orient toward their social world in a nondiscriminatory way. Over time, based on differences in the ways that people "act in motherly ways" toward the infant (e.g., engage in lively social interaction and respond to signals and approaches), infants become selective in directing attachment behaviors (Bowlby 1969:306). He argued that mothers are biologically primed to behave in "motherly ways" and are likely to be better at behaving in these ways than are other people. Bowlby did not believe that plural attachments lessened the importance of each attachment, but rather that infants developed a *hierarchy of preferences*, in which the mother was normally the most preferred. Using the research of Schaffer and Emerson, he stressed that children's primary attachment to their mothers was likely to be more intense than the greater number of attachments these children developed in the first months of life (Bowlby 1969:202).

Bowlby's decision to give preference to mothers as children's first and primary attachment figures was likely based on the thinking of the day about mother-infant relationships, taken from psychoanalysis, psychiatry, and primatology (Plant 2010; Vicedo 2013, this volume). Bowlby, for example, interpreted Harry Harlow's work on rhesus monkeys as proving that maternal love and care was necessary for an infant's development (Vicedo 2011). However, Bowlby was selective in the research he used to advance his thesis. At the time of his writing in the 1960s, Bowlby knew of Harlow's work which suggested multiple, simultaneous attachments. For further discussion on Bowlby's selective use of Harlow's work, see Vicedo (2009, 2013).

Multiple Attachments, Multiple Questions, Multiple Implications

Our thesis in this chapter is that children may develop multiple attachments with different people, at the same time, because people can assume a variety of roles and responsibilities that matter for attachments. This gives children a lot of flexibility in who they are able to rely on to meet their various needs at any given time and over time, and it may allow children to seek out people who seem most able and willing to help them at a particular moment for a particular reason. The elasticity of children's experiences that are made possible by their varied social relationships raises questions for us and we offer our thoughts on some of them.

Currently, we know little about children's attachments beyond the mother, and what we do know draws primarily from procedures developed in studies of urban, educated, middle- to high-income families in postindustrial societies.

The care of children in these families typically fosters autonomy and relatedness as psychological. There are some exceptions, with field studies relying on orchestrated mother/caregiver-infant separations in which the infant is left with an unfamiliar adult (e.g., Kikuyu farmers of Kenya; Leiderman and Leiderman 1977), or on the natural comings and goings of mothers (e.g., Aka foragers of the Congo Basin Rainforest; Meehan and Hawks 2013). These studies suggest that in these particular communities, which practice multiple caregiving, infants develop multiple attachments. There is a need for more research on this issue, using people and communities not well represented in current studies, and methods adapted to community circumstances, including local values. Gaskins et al. (Chapter 13, this volume) discuss potential ways in which this can be done.

Not surprisingly, our discussions exposed more questions about children's attachments than they answered. To encourage future lines of research, these are summarized below for further consideration:

- Must the attachments children form be ranked in terms of preference (i.e., hierarchy of preferences)? We don't think so. We believe a child's relational networks, and the roles and responsibilities of people in these networks, play important roles in their lives. At one extreme is the situation where a child is cared for primarily by a single person (typically the biological mother). This child may develop a strong preference for this attachment figure. At the other extreme is a child cared for by many people who share roles and responsibilities. This child may develop a strong preference for many of these attachment figures.

- Must attachments involve dyadic regulatory systems? We don't think so. The breadth of some children's attachment networks, along with the multiparty and physical nature of their involvements (e.g., more than two people in physical contact simultaneously with the child) suggest that many people may play a role at the same time in similar or different ways in a child's regulatory processes. This may take place on a regular enough basis for it to be meaningful to the child.

- Can children's trust (and their sense of security) go beyond their caregivers to the group as a whole? We think so. If a child has trustworthy experiences with many people in her group, the child may extend these feelings of trust to others whom she knows less well but is willing to "test out." Along this line, Mesman et al. (2015:110) suggest that "the notion of secure base may be applied to a group experience....In cultural contexts where caregiving is characterized by a network of (simultaneous) caregivers, the secure base is provided by the total network, not by a single individual." We add that children with multiple attachments are unlikely to experience them as a collection of single attachments, but rather as an integrated system of relationships.

Final Reflections

The pluralistic approach to attachment proposed in this chapter is an alternative approach with substantive theoretical and empirical differences to classical attachment theory. We argue that no theories of child development and of the emotional needs of children can be developed without research from a wide range of communities. This research must rely on serious ethnographic work that investigates the role of the complex interplay of the physical environment; the ecosocial, political, and economic contexts; cultural views and practices (especially views of personhood and self) on children's care; and the relational and attachment networks children develop. For each community studied, methodological tools must be empirically sound, meaningful, and ethically respectful.

Our approach accommodates the great differences in children's living arrangements around the world without prejudice. We believe that it has the potential to revise current understanding of children's attachments for children from a diverse array of communities, including those disrupted by political and economic reasons, and to advance inquiry into the role of cumulative adversities on the well-being of children. Our approach seeks to comprehend supportive contributions of children's relational and attachment networks in the context of family and community, even as both change. Palestinian communities in the southern West Bank and Gaza provide one example of this. Many of these families share residences with relatives beyond the *'a'ila* [nuclear families]: "The atmosphere of these family-based communities is village-like, with families sharing meals if they live in the same building, and women helping each other with domestic responsibilities" (Akesson 2017:96). Within such settings, a child's network of relationships offers a source of great strength and offsets the unpredictable, often violent circumstances related to military occupation.

As we discuss in Chapter 14 (this volume), scientific methods need to engage with the diverse realities of children's lives to complement knowledge reached experimentally in psychological laboratories. Real communities are dynamic and ever-changing systems. Our science should be as well. Many of the ideas presented here may be modified by future research. We hope that is the case. Equally, however, we hope that the methodological imperative to erect general theories based on empirical research takes culture seriously.

7

Exploring the Assumptions of Attachment Theory across Cultures

The Practice of Transnational Separation among Chinese Immigrant Parents and Children

Cindy H. Liu, Stephen H. Chen,
Yvonne Bohr, Leslie Wang, and Ed Tronick

Abstract

Prolonged transnational separation between parents and children is a common occurrence for many families today. Typically motivated by the desire to create a better economic future for the entire family, parents who move abroad in search of work opportunities often face limited childcare options in their country of settlement. This causes some parents to send their infants and young children back to the parental homeland to be cared for by relatives for extended periods. In this chapter, serial attachments and separations among caregivers and children in the United States and China serve as a cultural exemplar to extend and situate the meaning of attachment. The goal is to understand how this practice might affirm and challenge various concepts within attachment theory. Attention is given to the concept of monotropy, a basic component of attachment theory that assumes children's healthy development depends on a singular attachment created by sensitive interchanges between a parent and child. In turn, new directions are proposed for its measurement and related constructs.

Introduction

Attachment theory postulates that infants form secure attachment when they receive consistent, predictable, and sensitive caregiving from a primary caregiver

(Ainsworth and Bowlby 1991). This "secure base," in turn, allows the child to explore his or her environment. A particular challenge to the underlying assumptions of attachment theory derives from the cultural practice of transnational parenting, in which parents and children live in different countries. This arrangement takes place in many Chinese immigrant families: parents send their North American-born infants to China to be cared for by relatives, such as grandparents, aunts, or uncles. These children—known as "satellite babies"— are often reunited with their parents in North America only after their family secures affordable child care, achieves a degree of financial stability, and/or arranges for the child to attend school in North America, a process that can take several years (Bohr and Tse 2009).

Because this phenomenon within the Chinese immigrant community is not well known, with the exception of some media representations (Sengupta 1999; Wang and Wu 2003; Bernstein 2009), we begin with a brief description regarding its prevalence. Reports from New York City suggest that thousands of Chinese children may be separated transnationally from their parents each year; one nonprofit agency estimates that 40% of participants in their childhood education program have undergone this type of separation (Bernstein 2009). In a separate study conducted in New York Chinatown, 57% of expectant women strongly considered sending their newborns to China; among this group, 75% intended to bring their children back to the United States after they turned 4 years old (Kwong et al. 2009). Furthermore, at one New York Chinatown Health Center, it was determined that 10–20% of the 1,500 infants born each year were sent to China (Sengupta 1999). Lastly, in a qualitative study of undocumented immigrants in New York City, 72% of Chinese undocumented mothers recruited for the study sent their infants to China by the age of 6 months (Yoshikawa 2011). In our recent data collection in Boston Chinatown, approximately 20% of our sample of Chinese immigrant parents with children from birth to 10 years of age reported being separated from their children for at least six months, or were strongly considering it. Based on in-depth interviews that we conducted with 28 of these parents, employment instability, job schedule inflexibility, and limited childcare options were identified as factors that could influence the family decision to separate. Furthermore, the practice of transnational separation also appears to occur across socioeconomic levels (Bohr and Tse 2009).

Despite its theoretical implications for attachment theory, limited scholarly work has focused on this particular phenomenon. Previous research indicates that children may form attachment relationships with different caregivers for months or years at a time during different points in their development (Leinaweaver 2014). Leinaweaver (2010) argues that the relationships formed draw upon kinship ties, thus easing the tensions produced by migration, which serves as a method to care for both the old and young. However, as repeated separation and disrupted attachments in early childhood have been associated with poorer developmental outcomes (Karen 1994; Cassidy 2008; Kobak and Madsen 2008), Chinese children who experience transnational separation may also be more likely to demonstrate problematic socioemotional outcomes. Conversely, it is possible that specific contextual and cultural factors protect children who undergo this experience or may even promote favorable outcomes. Given increased globalization, these separation experiences among Chinese transnational families may remain a cultural norm. As is true of any cultural norm, there are costs and benefits of this practice, which may have implications for our understanding of attachment formation and development. For instance, a limitation in attachment research is the lack of discussion regarding functional costs to the child (and the parents) of putatively good, or secure, attachments.

In this chapter, we provide a cultural exemplar using observations of Chinese transnational separation experiences to extend and situate the meaning of attachment. We focus specifically on parent-child separation within Chinese immigrant families that are settled in North America. We provide a brief overview of how attachment has been understood in relation to culture, followed by a description of the experience of transnational separation that underscores the structural, cultural, and individual factors necessary to consider in relation to attachment. Examples from our review of the literature as well as our direct study of this contemporary phenomenon are included. We also advocate for a framework that takes globalization into account in the development of attachment. Finally, we suggest that transnational separation among families is a cultural norm.

The commonness of transferring children between different primary caregivers, an accepted practice that is assumed to support well-being, challenges the universality and evolutionary argument of attachment theory. A major question that arises is how certain assumptions of attachment are violated through this practice, while families may still survive and thrive. In our discussion, we raise questions and propose new ideas for the measurement of attachment and for its related constructs.

Culture in Attachment Theory to Date

In their seminal paper "Attachment and Culture," Rothbaum et al. (2007) argued for greater attention to cultural differences in the core tenets of attachment theory, rather than confirming its universality. By contrast, Waters and Cummings (2000) maintained the centrality of the secure base while also emphasizing the need for boundaries of attachment constructs to be better illuminated through cross-cultural research.

In our view, both perspectives offer opportunities to expand our understanding of culture and attachment, including those that have relied on the Strange Situation paradigm in determining attachment classifications. Comparisons of attachment classification distributions across different cultural groups have generally found no major differences in the proportion of secure attachment classifications (van IJzendoorn and Kroonenberg 1988). Nonetheless, studies have shed light on cultural differences in the distribution of insecure attachments, including observations of higher proportions of insecure-avoidant types within Western cultures and higher proportions of insecure-resistant types within non-Western cultures. These classification differences suggest that cultural variations of children's exposure to strangers and parental attitudes toward child independence may be associated with attachment (Grossmann et al. 1985). To put it bluntly, it must be kept in mind that the classification system itself was derived within Western cultures from a universalistic assumption that there were only limited and particular forms of attachments, a view that excludes cultural variation. As argued by Rothbaum et al. (2007) and others (Morelli 2015; Keller and Chaudhary, this volume), most attachment researchers tend to take a Western middle-class view of development, which presumes that the mother serves as the primary caregiver and interacts with the child most exclusively. Theorists have acknowledged the need for attachment theory to move beyond the presumption of a monotropic and dyadically organized relationship (Tronick et al. 1987; van IJzendoorn and Sagi-Schwartz 2008), undergirded by research which shows that many, if not most, children in the world are embedded in a multiple caregiver system (van IJzendoorn and Sagi-Schwartz 2008; Morelli 2015). These arrangements challenge exclusive parent-child models of attachment and reveal the ways in which multiple caregiving produces functional bonds between caregivers and children.

While known cultural influences on the Strange Situation Procedure have been invoked in explaining classification differences across groups, we argue for a stronger emphasis on the broader ecological context, cultural beliefs and practices, and goals associated with self-construal and relationships (Keller 2008). The prevailing view of attachment tends to presume a parent-child dyad in which the caregiver has lived with and cared for the child since birth, and in which a shared history and shared experience in culture and language exist. As we demonstrate below, there is an opportunity for the current Western standard dyadic model of attachment theory to consider how economic and cultural practices, beliefs about transnational migration, and meanings of family, self, and relations may play a role in the formation of attachment.

The Concept and Prioritization of Attachment in Transnationally Separated Chinese Immigrant Parents and Children

Close examination of caregiving systems suggests that they are heavily influenced by the economic structures within a society (Archer et al. 2015). This underscores the link between caregiving arrangements and broader societal needs. The type of economic structure in place—whether hunter-gatherer, farming, or industrial—is tied to specific work and settlement patterns, with households and lifestyles developed to maintain these arrangements. Within these structures lie cultural ethnotheories involving family relationships and child development, as well as socialization agendas and practices that contribute to institutionalized caregiving arrangements. These arrangements, which occur in particular times and places, dictate the whereabouts of caregivers and children and expectations of bonds between the two. Attachment theory must recognize that cultural arrangements not only require additional evaluation, but that its concepts are also inherent and embedded in the way individuals and communities experience the world. That is, ethnotheories are external phenomena that are documented by observers but may also be adopted by the individual. One's own understanding of their place in the world is culturated.

Economic Migration as a Motivation for Family Separation

I thought I couldn't work if the kids were with me. And the tradition of my town is like this: everybody sends their children home. So we also decided to send our children home. We could work here [in the U.S.], and when the kids are older we will bring them here, and by then we might be better off financially.
— Fujianese mother in Boston who was separated from child from age 6 months to 5 years old (translated from Chinese)

We lived in a very small place; both of us had to work and no one could take care of him [infant son]. We didn't know where to find help for child care [in the U.S.]. All we knew was that lots of people came here for a few months, and then sent their children back. Because when he's older, he'd recognize people, he won't know people when he's four or five months old, and won't cry too much if he's back, so we thought at that time…everyone sent children back… Because my parents had nothing to do in China, and we were very busy here, so we brought the baby back and asked my parents to help…He's also our first child, we didn't have any experience in child care, didn't know what to do. And having him back [in China] will make it easier for us; I could go back to work.
— Fujianese mother in Boston who was separated from child from age 5–20 months (translated from Chinese)

In Chinese society, it is common for members of three generations to live communally (e.g., in the same household) or in close proximity (e.g., in the same village or city); family members all share the same culture and language. In the contemporary period, the three-generation childcare arrangement is maintained for many families in China even when that means that children will be separated from their parents across long distances. As the country's economy has rapidly industrialized, it has become common practice for rural parents to move to urban areas as migrant workers. Millions of these parents leave their children behind in their home city under the care of family members (Waldmeir 2015). It has been estimated that more than 58 million children are currently living apart from one or both parents, accounting for over one-quarter of all rural children in the country (Su et al. 2012).

Transnational separation may be recognized as an extension of this pattern of economic migration. With increased globalization comes the promise of financial opportunity in other countries, driving the movement of families. For Chinese immigrant parents, this encompasses a variety of work and/or educational opportunities, ranging from employment as cooks in the restaurant industry to obtaining graduate education in North America. Our research team has conducted two separate mixed-methods studies on this phenomenon, both of which used quantitative surveys and qualitative in-depth interviews. The first examined the experiences of 28 Chinese immigrant parents in Boston Chinatown who experienced (n = 25) or considered (n = 3) transnational separation from their children. The second involved retrospective recollections of 40 Chinese American adults who lived with grandparents or extended family in China, Taiwan, or Hong Kong for at least six months as young children (< age 13) while their parents were working or studying in North America.

As detailed above, the 28 parents interviewed in the first study represented approximately 20% of a larger investigation of Chinese immigrant families in Boston Chinatown. Most cited economic pressures (n = 22) and/or a lack of parenting support (n = 21) as a reason for separating or considering separation. Importantly, 15 parents noted that they felt that they had no other option, while ten endorsed low parent self-efficacy and suggested that their own parents would do a better job raising the child. Seven parents sent one child back as a strategy to manage caring for multiple children. Six participants said that their decision was motivated by their own parents' desire to spend more time with their grandchild. Cultural reasons appeared to play a much smaller role, as only one parent claimed that her decision was motivated by the desire to expose her child to Chinese culture and language (although nine mentioned that learning the language was a positive consequence for children). Of those who made the decision to separate transnationally from their child (n = 25), on average, children were 12 months old when they were initially separated from their parents, and separations lasted an average of 2.1 years.

Of the 40 adults interviewed for the second study, many reported that their parents' emigration was motivated by the search for a good job or higher income, the desire to provide the child with better educational and other opportunities (n = 25), and the opportunity to obtain an education for themselves (n = 15). Most participants were between the ages of 2 to 6 years at the time of initial separation and were separated for an average of 2.7 years. They typically reunited when parents achieved more financial stability, which tended to coincide with parents' completion of graduate studies and/or the ability to buy a house or move to an area with better schools. Thus, throughout the migration process, different family structures emerge as relatives maintain caregiving relationships across borders. For many, these arrangements adhere to cultural norms that prioritize pragmatic solutions for the entire family, resulting in parents and children living far apart from one another in different cultural and linguistic contexts.

Cultural Values and Expectations about Family and Child Care

> We didn't know much (about day care)… I only learned about day care this year that you can send very young children there. But you know she was so young, I wouldn't have felt comfortable sending her off to day care, so I decided to send her back to China.
> —Fujianese mother in Boston (quoted earlier) who was separated from her child for 4.5 years (translated from Chinese)

> In China usually people will have a babysitter and it is easy for family and everything is much cheaper. And here because babysitter fee is high, so mommy has to take the job and it is much harder.
> —Chinese mother in Toronto struggling to decide whether to send her infant back to China (Bohr and Whitfield 2011)

The lack of preferred childcare options in North America or immediate family support for parents as they pursue economic opportunity abroad is a commonly cited reason for transnational separation in Chinese culture. Bohr (2010) has argued that separation for the purpose of obtaining child care is a coping strategy that addresses the economic needs of these young families. In her work on families from Peru, Leinaweaver (2010) refers to this childcare arrangement as "the outsourcing of care," defined as "the deliberate act of drawing on social capital, particularly kinship ties…an action that allows migrants to meet both economic and social needs with a minimum of disruption." Despite the vast geographic distances between infant and parents, the salient cultural norm of grandparents and relatives providing child care in China and other countries allow this arrangement to be viewed as appropriate and viable even across borders (Parreñas 2005; Zontini and Reynolds 2007; Bohr and Tse 2009; Kwong et al. 2009). On a practical level, technological advances for long-distance communication, the use of social media, and parental travel to see their child (Bohr and Tse 2009) are perceived to facilitate greater ease for those who are separated for extended periods. Given the more limited options for caretaking as immigrant families in the United States aim to meet specific goals, these advancements allow different family structures to arise.

Accordingly, it is unknown how Chinese immigrant parents make the decision to *prioritize* these economic opportunities and childcare options versus keeping their infants in the United States or Canada. Relatedly, how do Chinese families view the attachment relationship between parents and children? The various views of parent-child relationships and child development may help us understand how Chinese migrants make these decisions to separate:

> My baby is now nine months, I'm afraid that baby will forget about [me]. Seeing her grow up, every day, I feel I can't be separated from the baby. I'm feeling that the baby and I are attached together. I would feel really bad [if the baby had to go to China], if it has to be, then it has to be, but I would feel very bad.
> —Chinese mother in Toronto considering separating from her infant (Bohr and Tse 2009)

In a study by Bohr of 12 Chinese immigrant mothers who were considering sending their children back to their home country, all mentioned preferring for their infants to stay with them. These statements reflected their views of attachment:

> Okay, so if I sent him back, let's say for two to three years and then we, we don't have a close relationship when we take him back, then I am afraid that he won't trust me and he won't listen to me and it's hard for me to discipline him. The relationship would be blocked; I would feel guilty and self-blame.
> —Chinese mother in Toronto struggling to decide whether to send her infant back to China (Bohr and Whitfield 2011)

Among the 25 parents in the Boston-based study who made the decision to separate, 13 parents stated that they would do so again under the same circumstances, while 7 indicated that they would not. Most noted that they would keep their child with them if they did not have to work or if family members were able to come to the United States to assist with child care. (Note: three interviews were incomplete due to recording problems and are thus not included).

These responses suggest that developing an attachment relationship with their child, and/or fears of losing it, are concerns to Chinese immigrant parents. On the other hand, due to structural limitations that parents face, many feel that they "had no choice"—a predicament that service providers within the Chinese community verify. One mother of four in Boston described the decision to send her first child back from the age of 3 months to 3 years as follows: "We had no other options. My husband and I needed to work and didn't have time to take care of him, so we had to send him to China." The ability to determine risk suggests that families anticipate a range of possible outcomes to separation. Bohr argued that the parents she interviewed often utilized "tolerated ambivalence." In other words, "while all acknowledged what was often a very painful ambivalence when contemplating separation from their offspring, mothers forcefully referred to the economic problems created for them by the lack of adequate childcare possibilities and the power of culture as influencing their choices" (Bohr and Whitfield 2011). Thus, parents' understanding of attachment formation and its ensuing implications may be just one consideration among many that they must weigh in making this decision.

In key ways, this situation mirrors the calculations made by urban migrant workers in China, who leave their children behind in rural hometowns to be cared for by grandparents and other relatives. In two-thirds of cases, economic pressures and the high cost of living in cities are the main motivators for splitting households (Wang and Wu 2003). Other structural factors, particularly the inability to secure urban residence permits for children that would give them access to education and health care, are primary reasons why parents choose to separate from their children (Burnette et al. 2013). Unlike the North American case, however, Chinese left-behind children are on average older at the time of first separation (9–10 years old) and also tend to be separated for longer periods of time (3–4 years) (Wen and Lin 2012). In our study in Boston Chinatown, most of the children who had separated and reunited with their parents were between 2–8 months of age at the time of separation and had lived apart for an average of 2 years. In several cases, parents even brought children back earlier than expected for different reasons, including the fear that their relationship was growing too distant or that they were being spoiled by lenient grandparents.

While such reasons together reflect the parent's own or anticipated concerns about the separation, they vary in their understandings of children's experiences of separation and reunification. Children are often infants when separated and presumed to be too young to experience strong emotional reactions to leaving their parents. The mother quoted below was separated from her daughter from age 18 months to 3 years. When asked whether she thought the child understood the separation, she stated:

> She probably didn't understand. We went to the airport with my cousin. She knew my cousin…[and] didn't understand she was going to go somewhere very far. But at the airport, she went in, and cried a little…then my cousin played with her, and she became less sad, then she got on the plane and went back…There was lots of fun stuff for her in China…and she got used to them eventually, and she didn't ask to come back to mom.

Aside from experiencing the loss at separation, after acclimating to new caregivers, children can undergo further loss when they reunite with their nuclear family, a reality that parents acknowledge. According to the mother quoted above, when her daughter first returned to the United States, "She was very sad. She was sad about leaving her grandma." Despite children's possible confusion and pain at being separated from their primary caregivers, parents tend to assume that children will adjust easily to new circumstances and to living with their nuclear family again. Hence, few of the parents we interviewed spent much time emotionally preparing their children to move back to the United States. One mother who was separated from her infant for 7 months described how much her son missed his grandmother once he had returned to Boston:

> His grandma missed him too and Skyped with him. He saw her on the screen and knew it was his grandma, so he cried. He cried and then asked for his grandma, trying to hug her…[This] lasted for two weeks.

Such responses suggest that for some children, adjustment to the second separation, this time from grandparents, was much more difficult than the initial separation from the parents.

Whether parents and caregivers prepare children for separation and unification and in what ways can provide further insight into how parents believe their children will respond to the separation or reunion. For instance, in the sample of 40 Chinese American adults who lived in China, Taiwan, or Hong Kong as small children while their parents were in the United States, most were too young at the time of initial separation to recall whether they received any emotional preparation from their parents or other family members. Of those who could remember, 18 reported that their family members provided little to no preparation for the initial separation from their parents. One 21-year-old male college student who lived in Taiwan with his grandparents from ages 4–7 while his parents pursued graduate degrees in the United States recounted his experience:

> They didn't really give me a talk. They didn't really sit me down because I figure they didn't really understand the importance of sitting kids down and then talking them through the process. I just feel like this probably isn't something that Asian parents do, whether it's transitioning through life or just, so to speak, dumping your kids in your home country.

Furthermore, 21 individuals in our sample reported receiving little to no preparation for their reunion with parents (and often, siblings) in the United States, which many viewed as a situation that made the transition even more emotionally difficult. In one case, a 28-year-old male respondent, who lived with his grandparents in China from ages 4–11, recalled the lack of information he received about his reunion:

> I really don't think I understood too much about why I was leaving China. I remember thinking that I didn't really want to leave. Like, I'm the type of person who back then really didn't like change, so any change was really stressful.

One explanation for the lack of preparation given to children about separation or reunion may be the importance that Chinese families place on kinship relationships, or blood ties. The belief that "blood is thicker than water" promotes social harmoniousness and group mindedness but also mutual dependency and relationship-centeredness (Lam 1997). This kinship norm and the underlying emphasis on maintaining a strong family network may be viewed as transcending geographic location. On one hand, biological kinship may confer the belief that attachment ought to take place indiscriminately with different members of the extended family. Another possibility is that kinship produces the expectation of automatic parent-child attachment, regardless of the time and distance apart and lack of shared experience (e.g., maternal sensitivity in the context of physical caretaking that characterize the typical dyadic attachment model). To the extent that kinship has served as a social safety net for cultures that espouse more interdependent ideals, some families may even see this as cultural preservation of relationships despite the physical separation, based on the belief that they will have a typical parent-child relationship once the child returns. Most parents from the Boston Chinatown study felt that children should not be away from their parents for more than a few years so as to maintain an intimate bond. However, one mother who sent her son to China from the age of 5 months to 4 years speculated that the ease of their reunion could be attributed to biological ties:

As soon as [my son] got back, my husband told him that I was his mom and he immediately came over to me. He's not like other kids who ignore you or cry or whatever. Maybe it's because of blood. He's not scared of us and he wanted to be with me. (Translated from Chinese)

A question that emerges from this discussion is how attachment theory considers the prioritization of this dyadic attachment within the notion of kinship, an issue that we will address in the following section.

Models for Expanding Understanding of Attachment in Transnationally Separated Families

A Bioecological Model as a Basis for Understanding Transnational Separation on Children's Development

With the implications of economic, cultural, and dyadic processes playing a role in transnational separation, Bronfenbrenner's bioecological model provides a concise theoretical framework for examining the effects of transnational separation on a child's development (Bronfenbrenner 1977). We briefly provide examples for how such processes are situated within the bioecological model's components of *person, process, context,* and *time.*

Person and Process

By definition, the experience of transnational separation is a disruption in a child's proximal processes. The child's regular interactions with his primary caregiver(s) are interrupted and replaced by interactions with others. Both the child's and caregivers' adaptation to these transitions may be influenced by a child's person-level characteristics. These characteristics, such as an infant's uninhibited temperament, may elicit sensitive behaviors from caregivers across contexts. By contrast, a fearful or inhibited temperament may exacerbate the challenges of separation and reunion with caregivers. One parent interviewed in Boston Chinatown highlighted her son's "adaptable" temperament as a key factor in his smooth reunification with his parents, even after a separation of over four years:

[It] was weird, right, because my friends sent back their children, too, and when their children came back they spent some time to get close with their children, but my child felt, I don't know how to put it into words…When I first picked him up, he was really happy. That night he hugged me and called "Mom" incessantly, and asked, "Mom, can you read me a story?"….To someone who didn't know, we looked like we had never been separated before…His adaptability is really good…he basically didn't have any problems adjusting. (Translated from Chinese)

Context

The context component of the bioecological model conceptualizes the various systems of influence in transnational separation. In addition to the microsystem-level proximal processes described above, mesosystem-level influences may play a key role in children's adjustment to transnational separation. For example, communication between a child's biological parents and his or her current caregivers may vary in both quality and quantity. Caregivers in Asia may provide a child's biological parents with regular, detailed updates on the child's development and daily activities, and may facilitate regular parent-child interactions. This has been demonstrated in other populations, such as Filipino mothers who work abroad, where media can allow for "a more complete practicing or intensive mothering at a distance" (Madianou and Miller 2012:83). At the same time, fractious relationships between parents and temporary caregivers may result in sparse parent-child engagement during the period of separation and thus provide parents only minimal insight into their child's development. In our study in Boston Chinatown, most parents described having daily online communication with their child and caregiver during the separation. This was a marked difference to our sample of adult Chinese Americans with early separation experiences, who typically had very limited contact with their parents in the pre-Internet era. Yet despite technological advances allowing for easier communication, there were also perceived limitations, as explained by one mother who sent her 9-month-old son to China for two years:

> Although the technology was very developed and we could Skype online, he knew seeing Mom on Skype was different from seeing Mom in person. It was a strange feeling, and it wasn't an intimate relationship. (Translated from Chinese)

Here the nature of the relationship may be mediated by the media rather than through "co-presence," as argued by Madianou and Miller (2012). A question is how attachment can be formed through representations of the other individual, which can vary based on the form of communication (e.g., Skype, telephone, and emails).

The broader exosystem and macrosystem levels may be seen as influencing parents' decisions to initiate, prolong, or terminate transnational separation. At the exosystem level, a parent may be working multiple low-paying jobs without provisions for health insurance or child care, or may be a full-time student in a highly demanding graduate program. Both situations may contribute to a parent's decision to send their child temporarily back to their home country. Among the 24 parents in the Boston Chinatown study who mentioned economic pressures as a motivation for separating from their children, many specifically mentioned the competing demands of parenthood and employment:

> Even if I sent [my child] to a day care…at the time I worked in a Chinese restaurant, the working time was long, sometimes from 10 in the morning to 10 at night, and on weekends, Fridays and Saturdays it's even later, sometimes until

11 pm. So, since my husband also works at a restaurant, the timing was not right. (Translated from Chinese)

Even parents in the Boston Chinatown sample who held advanced degrees found transnational separation to be the best solution for balancing work and parenting demands. One mother described how she and her husband decided to leave their daughter in China while he pursued postdoctoral training:

> [The separation] temporarily benefits my husband and me since we get time to gradually settle down and adapt to the environment. My husband can focus on his research, while we both have time to plan for our future. We can do research on schools and housing before she gets here…If we brought her over without having the time to do any of this, it would be a lot more challenging…If we brought her out with us…we would have so much stress on top of the financial, employment, environmental stress we're already facing. (Translated from Chinese)

Finally, a number of macrosystem-level influences—cultural norms regarding transitional separation, immigration policies and laws, and beliefs about the nature of the parent-child relationship—can all be salient factors when a parent weighs the benefits and consequences of transnational separation. Three parents in the Boston Chinatown sample specifically mentioned how cultural norms regarding transnational separation motivated and informed their decision to separate from their child:

> We knew that a lot of [Chinese immigrants] who had just arrived for 3 or 4 months send their children back before they get older and recognize you…so we thought that since everyone's doing it, we would do it, too.

> A lot of us who come from Fujian [separate from their child] after seeing other people do this…I think a lot of people from Fujian think this way. There are other people who want to focus on work and so they can't take care of their children. Having their children stay here wouldn't be good either. That's why they send them back.

> Where we are it's like a tradition. Everyone sends their kids back home, so we decided to do that, too and we can still work here, and bring them back when they're a bit older, and we might be doing better economically at that time.

Time

As it subsumes all other components of the bioecological model, time (the "chronosystem" of the bioecological model) may be the most relevant factor in transnational separation and its effects on a child's development. The pressures for an immigrant family to separate transnationally are often at their peak during the first few years in their new country of settlement; indeed, parents who make the decision to separate from their children often note that their primary goal in doing so is to establish a firm financial foundation for their family. One mother who sent her 5-month-old son to China for 7 months described her reasoning:

> I wanted to keep my child with me and better understand his life from all aspects. Yet I also considered many other factors. For example, at that time, only my husband worked, and expenses were high. So I thought about sending my child back for some amount of time, and when our economic status improves we could take him back and take better care of him. (Translated from Chinese)

Finally, the time course of transnational separation—namely, its duration and the developmental period in which it occurs—can also impact aspects of a child's development. For example, the son of one parent in the Boston Chinatown study was cared for by his grandfather from 3–6 years of age. The mother describes her belief that this experience, particularly the time in her child's development at which it occurred, had a lasting impact on her son's relationship with both her and the grandfather:

> My relationship with him now is…I was just thinking about this yesterday. He's actually closer with his grandpa than he is with me. He's a teenager now and he doesn't tell me a lot of things, but he tells them to his grandpa when he visits... When his grandpa visits, he still sleeps with his grandpa. He's 15 years old now and he still sleeps with his grandpa. When he can't fall asleep at night, he asks his grandpa to read him stories….He's closer with him than he is with me.

In reflecting on her experience, this mother specifically underscored the impact of the child's age and the duration of the separation:

> We sent him back at that age…Don't let them be separated from you for too long. I think those 3 years were his golden developmental time…a precious age where you start to learn how to behave, rules, language, all these different aspects. We sent him back during this time period.

Indeed, responses from a number of parents in the Boston Chinatown sample indicated thorough considerations of the role played by the child's age at separation in his or her subsequent adjustment. Some of these developmental considerations were drawn from their personal experiences, while others were observations of others' experiences with transnational separation:

> My younger sister has a son and a daughter. The older daughter came back around 4 or 5; she behaves poorly and was not close to her mom. My sister immediately got her 1 year and 8 month old son back, and the son is close to his mom.
> I don't agree that you should send your children back after they are over one and a half years old. Because after they are one and a half years old, they already have some relationship with their parents.

> I have a friend who's from Fuzhou. She sent her child back. When he came back, he was already at the age for high school. Now he doesn't even go to college and he just dates around, plays around. His parents can't do anything about it because he's old now. You weren't with him when he was young. You don't know how to

communicate with him. You have no idea what goes through his head....there's no way for you to control him now.

When he came over here, he was 5 or 6 years old. I still had the chance to guide him, communicate with him, educate him....He still needed to listen to us. When they're older and have experienced separation for that long, there are endless difficulties you need to deal with.

One mother drew a contrast between two of her children who had both been sent back to China, and concluded that the one child's "bad habits" were a result of the younger age at which he was sent back:

[My younger child] went to China when he was too little. When [my] older child was in China, he was...almost 2 years old. He already knew many things and didn't need to be taken care of too much by his parental or maternal grandmothers...but many grandparents in China pamper young children too much and just do anything for them. Thus, [my younger child] was so used to being taken care of and being pampered. He had many bad habits when he returned to the U.S.

In sum, each of these responses indicate parents' acute awareness that a child's age at the time of separation, as well as the duration of the separation itself, are critical factors in their children's subsequent adjustment. Process and context, with the time component of the bioecological model provides a theoretical framework for conceptualizing the influences of separation on children's development. With this, it is perhaps not the fact that attachment theory has focused on dyadic relationships that is the problem, but rather that attachment theory has failed to acknowledge the importance of the many systems that make up the ecologies in human development. This oversight has repercussions for our understanding of attachment across all cultures.

Caretaking Arrangements and Family Lifestyle on Attachment

With respect to the development of attachment, we build upon the bioecological model by considering the effects of serial attachments and separations that take place across a great geographic distance. Existing caregiving models that have been informative for understanding attachment relationships may be initially useful in informing this phenomenon. For instance, multiple caregiving can take the form of communal or institutional child care, although these relationships take place simultaneously, rather than sequentially. In these arrangements, there is not necessarily great geographic distance between caregivers and children. The foster care arrangement may approximate the attachment relationships that take place with caregivers sequentially, as children are removed from their parents and live with a foster parent, before possibly being reunited with their parents or being placed in a different household. While the children may live apart from their parents, as is true of transnationally separated children, in many cases foster children are removed because of some form of maltreatment and parents do not necessarily wish to separate. Transnational separation among Chinese immigrants, on the other hand, generally does not occur under such forced or adverse circumstances, and parents themselves, in large part, proactively make this decision.

However, understanding the effects of caregiving on attachment in the Mainland Chinese context may be an important starting point. In a study of Chinese mothers from Shanghai, grandparents and caregivers provided child care for most of the mothers who had to return to work within 4 months after giving birth. The study found that infants who slept with multiple caregivers and nonparental caregivers at night were associated with insecure mother-infant attachment (Ding et al. 2012). However, in a recent study of a larger sample in China, no differences were found in attachment classifications when assessed between mothers and infants where up to two-thirds of the primary caregivers were grandparents (Archer et al. 2015). Aside from the potential problems of applying attachment classifications across cultures and settings, these studies point to the need to study attachment by understanding the child's experience of having multiple caregivers, while holding the possibility that attachment with other caregivers may be culturally or developmentally adaptive.

Additionally, regional differences in cultural values in China may be associated with infant attachment. It has been argued that the Northern Chinese value independent self-construal relative to Southern Chinese, who may emphasize an interdependent self-construal (Talhelm et al. 2014). Indeed, a greater proportion of Southern Chinese infants than Northern Chinese infants were found to have the resistant insecure type (Archer et al. 2015). There is also evidence that migrant Chinese mothers moving from rural to urban settings for work emphasize a value of independence similar to Western mothers (Zheng and Shi 2004).

Despite scenarios that involve multiple nonparental caregivers, most scholarly assessments remain focused on parent-child attachment. An issue is how attachment classifications derived from parent-child dyads are ecologically valid with these variations in caregiving arrangements and cultural values, and importantly, whether these classifications are in any way predictive in diverse family communities.

Cultural Value Systems Regarding Transnational Separation and Expectations for Parent-Child Relationships

Cultural value systems of independence and interdependence, as well as their related psychological goals for autonomy and relatedness across development, have played a notable role in describing the beliefs and practices of Western and non-Western societies. The Chinese may see themselves as embedded in a web of a relational network (Bond and Hwang 1986), where individual achievements and behaviors that reflect autonomy are performed to maintain stability for the whole family.

This orientation of interdependence within Chinese culture promotes connectedness among family members and is maintained by transnational families. According to Yeoh et al. (2005:308), the transnational family "derives its lived reality not only from material bonds of collective welfare among physically dispersed members but also a shared imagery of 'belonging' which transcends particular periods and places to encompass past trajectories and future continuities."

Concepts central to attachment theory, such as maternal sensitivity and assumptions of mothers being physically present, may not be similarly prioritized in situations where the idea of "family" is maintained despite distance between members over long periods of time. This is not to say that either physical and/or emotional availability or day-to-day consistency and sensitivity are unimportant. However, the value of relationships may be maintained at a broader kinship level rather than at a dyadic level. That is, feeling secure may be "anchored" in an individual dyadic relationship or in a family or in a group. In this view, the security of attachment as equivalent to a dyadic relationship in the West is the consequence of a primary dyadic caretaking system, whereas in other caretaking systems, security and relationships do not overlap. Indeed, even in the West, children may not feel secure with individuals with whom they have wonderful relationships (e.g., grandparents, cousins). Furthermore, it could be possible that attachment needs of the adult parents may be met through the support of their kin, even within the transnational separation experience, producing greater emotional well-being for biological parents.

To the extent that a sense of security for children is important, as well as the well-being of the parents, this could be derived, developed, and maintained through a longstanding kinship between members of Chinese families even when parents are physically separated from their children for extended periods. Understanding attachment as a more general phenomenon may supersede a physically proximal and dyadic secure attachment, the development of a secure base, and the promotion of individual exploration (Madianou and Miller 2012).

Next Steps: Assessing Core Assumptions of Attachment through the Lens of Transnational Migration

Thus far, we have focused largely on macro-level factors that play a role in the separation and reunion of families, including economic and cultural influences, and culturally specific attitudes toward attachment. A major step for attachment theory and its measurement (see Chapter 13, this volume), given the understanding of the practice for transnational migration, is to determine how this practice challenges and informs core assumptions of attachment at both macro- and micro-levels.

Figure 7.1 embeds the various attachments and their formation and separation with each caregiver across two cultural contexts as characterized within a

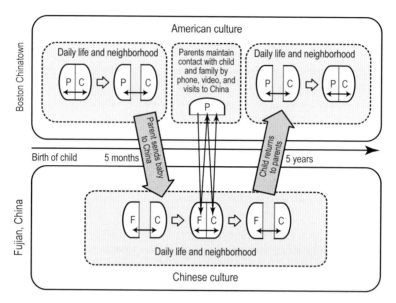

Figure 7.1 Simplified example of a transnational separation between parents (P), family member (F), and child (C) between Boston Chinatown and Fujian, China. Semicircles and gray arrows display the dyadic relationship and its formation and separation over time; proximal environments (extended family, neighborhood) are represented within dotted lines. Black arrows display the relationship processes within each dyad.

typical Chinese immigrant transnational separation. Referring back to the challenges of attachment theory articulated by Rothbaum et al. (2007), regarding the sensitivity, competence, and secure base hypotheses, we argue that each hypothesis needs to be examined at both levels among transnationally separated families, as each construct may be emphasized in different ways when considering continual shifts in practices or views of a cultural practice.

The sensitivity hypothesis assumes that the mother's ability to respond sensitively to her infant's signals leads to secure attachment. In relation to transnational separation, the practice of sending an infant far away to be cared for by others may be driven by the mother's view that the grandparents in China will be available and able to respond sensitively to the infant's signals and to provide the continuous physical caretaking that is desired for the child. Because parents must work long hours, they often feel that sending them to grandparents will allow their baby to receive one-on-one care that they themselves would not be able to provide. As one mother in Boston who sent her 9-month-old son back to China for two years reflected:

> I thought his grandparents taught him wholeheartedly, and he learned a lot. His way of thinking was easily developed. Had he been with me, I wouldn't have spent all my time with him—I would have spent half of my time at most. His

> grandma spent more time and energy taking care of him and educating him,
> which I couldn't have done myself. (Translated from Chinese)

Therefore, despite being separated for prolonged periods across a great distance, some parents find this type of one-on-one attention preferable to living in the same home with the child and placing the child in an expensive daycare setting. Furthermore, in our interviews, a number of individuals mentioned feeling low self-efficacy as parents after bearing their first child; they thus felt more at ease sending infants to live with grandparents, whom they viewed as having more childcare experience. Grandparents themselves can also exert pressure on parents to send the child to them. One mother who had her first child in her mid-20s explained why she decided to leave him with his grandparents in China for 7 months after taking him back for a visit:

> Because my child went back, and I saw my parents loved him a lot, hoping he
> could stay for some time. This was also the case with my mother-in-law. Plus,
> how to put this in words, I myself thought it would be hard to take care of the first
> child. So it was like, okay, when he grows older I would take him back.

Thus, sensitivity may be valued but not necessarily seen as a requirement for the biological mother. This may be a view that is aligned with those from multiple caretaking societies but seen as more extreme given the geographic distance and prolonged separation. However, it certainly differs greatly from the Western middle-class assumption of the mother being the primary caretaker for the child.

The secure base hypothesis refers to the concept of infants feeling safe and comforted by the presence of a caregiver, providing a secure base that allows them to feel comfortable exploring their environments. Chinese culture emphasizes close and physical caregiving; however, whether this particular experience is seen to serve as a foundation for children to develop a secure base is an important question, especially among Chinese immigrants that undergo transnational separation. First, and as is the case for other types of serial separations and reunions such as foster care, the timing and length of separation and reunion and children's experience with each of the caregivers, may impact their establishment of a secure base. Second, and in taking a macro-level perspective on the secure base hypothesis, a major question that arises is whether a secure base may extend beyond the primary caregiver to other caregivers; that is, what reference group is considered as a source of security and how is this experienced by the child? Might the importance of kinship relations reflect this secure base and reinforce expectations about a child's attachment to a caregiver? Measuring the security and how it is developed in attachment relationships as well as through other means (e.g., through the instantiation of kinship) would be an important step in determining the scope for this construct of security. Finally, there is the question of how the Chinese view the importance of a secure base and subsequent exploration. Traditionally, this exploration has been thought to characterize Western cultural values, given that it reflects the development of independence and autonomy. Nonetheless, with the growth of China's market economy and increasing globalization, Chinese parents are placing greater value on autonomy and self-sufficiency (Chen et al. 2010). It is unclear if and how possible shifts in cultural orientation affect parents' understanding of the function of a secure base.

The competence hypothesis refers to the relationship between one's attachment style and the ability to predict later outcomes, with secure attachment specifically seen as leading to competence and better coping skills later in life. However, there are several concerns with viewing competence as an outcome. At minimum, those with insecure attachments may be highly competent in various domains. In regards to cultural definitions, competence has been largely defined through a lens of Western values that prioritize independence, emphasizing exploration, autonomy, self- and emotional regulation. On one hand, individuation may be a priority for families that experience transnational migration. Indeed, many of the Chinese American adults we interviewed who had early separation experiences reported having a greater sense of independence. One 18-year-old female respondent, who experienced two different separations from her parents between ages 2–5 and 7–10, described how she was affected by these experiences:

> For me I think it definitely has made me more independent, like I don't really like relying on other people. I think the separation definitely had something to do with that…In a way I think that, I guess it's a bad thing to say, but in a way, when I was growing up I thought, "oh, I only have myself to rely on," because I was switching between so many people to live with, and I am the only thing that's constant in my life.

It is unknown whether this self-reported independence is adaptive or functional, as it could also reflect a lack of trust and what could be considered an insecure, avoidant attachment style. Thus, competence ought to be defined from within a culture. Some ways of assessing this may include examining the expectations for outcomes among families and individuals that have chosen to separate transnationally and to study the socialization practices that play a role in these outcomes. Furthermore, it is unclear whether the idea that security experienced and established in early childhood translates to affective experience holds across cultures. While Chinese caregivers often place a premium on physical safety, its relation to children's affective experience may not be considered a priority, especially given the different cultural values placed on emotion expressivity and regulation. Altogether, studying transnational families could help to identify some parameters for defining the secure base as well as its purported outcomes within attachment research.

Finally, transnational separation provides an opportunity to consider the unique experiences a child might have with different caregivers across different stages of development and, furthermore, how these experiences do or do not transfer to other current and later relationships (Tronick 2003). Of particular relevance to Tronick's theory is the "thickness" of relationships, or the variety of time-activity contexts, including feeding, diaper changing, putting to bed, and other infant-related care activities. These regulatory activities may take place with the parent in the United States prior to separation, with another caregiver such as a grandparent in China during the separation, and then transfer back to the parent after reunion. As such, regulatory patterns are developed between each caregiver and the infant, characterized with implicit knowledge within the individuals in the relationship, and co-regulated. It has been argued that "thicker" relationships are more differentiated and therefore their regulatory patterns are less likely to be transferable to other relationships. The notion of differentiation and intransferability challenges modern attachment theory to specify exactly how attachment transfers or if it is necessary that it transfers across individuals. There is an assumption that the parental/maternal primary caregiver relationship is prototypical. If relationships become increasingly unique, then serial relationships could pose a very serious problem indeed. This is an important consideration in the study of attachment processes within the context of transnational migration, since infants may be sent back to China as early as infancy, develop a (likely) thick attachment with their grandmother, and then be expected to develop a relationship with their parents after their return to the United States. As illustrated in Figure 7.1, infants may have to contend with serial, multiple ruptures to their most fundamental relationships. In our interviews with parents, many recall challenges in their children's initial adjustment after being brought back to the United States. One mother recalled her son's adjustment period when he returned at nearly 3 years of age:

> There were some language issues because he spoke Chinese in China…his spoken English wasn't good enough. Sometimes he spoke Chinese and sometimes English. After he came back to America everything changed, no matter if it was living, eating, people, the environment or other aspects, which he wasn't used to in the beginning and needed time to adjust to. (Translated from Chinese)

Despite these challenges, parents tend to believe that children are resilient and eventually adjust to the new family, cultural, and educational environment. Even so, there is a general consensus that bringing children back earlier is better for their adjustment. One full-time working mother from Boston Chinatown who was currently separated from her infant described how her thoughts on separation had changed:

> I originally planned to have the child stay there until the child was 3 years old so that the child can go to daycare centers in the school. But if possible, I want to have the child back next year, or when the child is one and a half years old. I don't want to leave her in China for too long. There are many cultural differences between China and the U.S. I don't want my child to learn some bad things, so I want to have the child back to the U.S. soon. (Translated from Chinese)

Myriad factors may play a role in children's adjustment, including new relationships with peers and teachers across development, which likely contribute to these outcomes. New research is needed to identify these moderators for later outcomes for children who were transnationally separated.

Summary

Transnational separation is a phenomenon that provides an opportunity to closely examine the assumptions of attachment theory. Important insights from transnational separation can expand the meaning and utility of attachment theory, by considering factors that are meaningful for relationships. For example, it is necessary to develop ways to assess the locus and strength of security as well as to assess other independent qualities of relationships. The bioecological model provides a framework that pulls together the multiple contexts that underlie transnational migration, and most notably emphasizes the macro-level factors involved in the decision of families to send their infants to China. Cultural value systems and the expectations family members have of their relationships, including the formation and development of attachment between caregiver and child, play a role in this decision. These cultural beliefs and attitudes may contrast greatly with assumptions of Western-oriented attachment theory, challenging the current definition of maternal sensitivity, the prioritization of establishing a secure base, and the idea that competence arises from an early secure attachment. Such systems that articulate person, process, context, and time—all which underlie the development in individuals and relationships—have yet to be fully incorporated into attachment theory. Finally, the serial attachments and separations across different caregivers and time produce unique relationship experiences for infants. This provides an opportunity to understand how the formation of a dyadic relationship with one caregiver (e.g., a grandmother in China) might transfer to the relationship with another caregiver after reunion (e.g., a mother in the U.S.), and whether other relationships and factors moderate child outcomes that produce resilience or risk. The phenomenon of transnational separation questions if and how transferability of attachment relationships takes place and also its effect on outcomes in both individuals and families. Altogether, transnational separation is a cultural phenomenon that offers a range of promising new directions for the study of attachment and child development.

8

Meaning and Methods in the Study and Assessment of Attachment

Suzanne Gaskins, Marjorie Beeghly, Kim A. Bard,
Ariane Gernhardt, Cindy H. Liu, Douglas M. Teti,
Ross A. Thompson, Thomas S. Weisner, and Relindis D. Yovsi

Abstract

As originally conceived and still practiced today, attachment theory is limited in its ability to recognize and understand cross-cultural variations in human attachment systems, and it is restrictive in its inclusion of cross-species comparisons. This chapter argues that attachment must be reconceived to account for and include cross-cultural and cross-species perspectives. To provide a foundation for rethinking attachment, two universal functions of attachment systems are proposed: they provide (a) socially organized resources for the infant's protection and psychobiological regulation and (b) a privileged entry point for social learning. Ways of understanding the nature of the cultural and ecological contexts that organize attachment systems are suggested, so that they can be recognized as culturally specific, normative behavior. Culturally valid methods for describing children's attachment systems are also discussed. In conclusion, a wide range of research strategies are proposed to facilitate the extension and contextual validity of measures of attachment across cultures and species.

Conceptualizing Attachment

Redefining the concept of attachment itself is logically prior to expanding on the ways in which researchers can engage with attachment in cultural contexts.

Group photos (top left to bottom right)   Kim Bard; Marjorie Beeghly; Suzanne Gaskins; Relindis Yovsi; Thomas Weisner; Ariane Gernhardt; Ross Thompson; Suzanne Gaskins; Douglas Teti; Cindy Liu; Relindis Yovsi; Kim Bard; Douglas Teti; Marjorie Beeghly; Ross Thompson; Relindis Yovsi; Thomas Weisner; Ariane Gernhardt; Cindy Liu; Suzanne Gaskins, Marjorie Beeghly, and Douglas Teti; Marjorie Beeghly

Thus we begin by addressing two central questions: What is an attachment system? In natural interactions, how can we recognize and characterize attachment in the daily lives of human infants and young children as well as in infants and young juveniles from other primate groups?

If we are to rethink the ways in which attachment might be studied in cultural contexts, we need to understand not only what attachment is, but also what it is not. While a particular child may well have more than one attachment figure, there is something distinctive about an attachment relationship compared to all other social relationships: not all social interlocutors have that status conferred on them just because they interact with the child, even if it is on a regular basis.

It is also necessary to consider attachment across species. Although much of our discussion in this chapter is situated within a frame of human infancy and early childhood, we more broadly refer to infancy and the early juvenile period in any primate species that may have attachment systems. In particular, we know that infant great apes (chimpanzees, orangutans, gorillas, and bonobos) develop attachment systems that function much like those of humans (e.g., Pitman and Shumaker 2009; van IJzendoorn et al. 2009). For ease of readability, however, we highlight the cross-cultural perspectives of human attachment.

Rather than beginning with a theoretical conception or attempt a decontextualized description of behavior, we begin by articulating what the functions of attachment are in an infant's life. We use "function," in an evolutionarily meaningful sense, to refer to the relation between attachment and the increased chances of an immature member of a species to reach adulthood, and thus function effectively as a member of the social group. Because all cultures (and all mammalian species, for that matter) must solve the problem of ensuring that the young survive infancy to reach and achieve reproductive maturity, the attachment system (despite its cultural variants) can be regarded as part of that species-wide adaptational challenge.

Our functional definition of attachment has two parts. The first function of attachment is to provide socially organized resources for the infant's protection and psychobiological regulation (including stress regulation when stress is present). Attachment figures serve in a privileged capacity to manage infants' safety, their behavior, and their emotional and physical well-being. Attachment thus ensures that one or more caregivers has privileged capacity to protect infants from harm and to help regulate their psychobiological systems, which are essential to their survival but which are poorly regulated at birth (e.g., systems related to feeding and nutritional intake, immunological functioning, protection). With respect to stress regulation, attachment figures have a privileged (or amplified) capacity to act as buffers of stress and thus help to regulate the child's stress reactivity and its potential consequences for emotion regulation, behavioral self-control, as well as cognition and learning (Thompson 2014). One important way that cultures differ from one another (and which has not been recognized in attachment research) is in the amount and type of stress children experience during the course of everyday activities, depending on whether care is organized to be proactive or responsive to the children's needs. In some cultures, stress or distress is thought to be detrimental to the child, so caregivers provide anticipatory care before the child expresses discomfort or desires (Keller and Otto 2009). In contrast, caregivers in Western cultures are more likely to allow their offspring to experience distress before intervening, based on a belief that doing so promotes children's self-regulatory skills and builds independence. Because there are cultural variations in strategies for regulating (and producing) children's stress, the concept of responsiveness to the child's distress signals must be considered only a provisional indicator of attachment. To reflect this range of behavior across cultures, we therefore characterize this first function of attachment in terms of psychobiological regulation rather than in terms of stress.

The second function of attachment is to provide the child a privileged entry point into social learning. Although cultures (Lancy 2015) and indeed primate groups (Whiten et al. 1999) differ widely in the specifics that infants must learn to become a well-functioning member of their particular group, infants everywhere face the task of learning to become a competent participant, and in all cultures, social learning (learning through social participation) plays a central role in this process (Lancy et al. 2010). The capacity to engage in social learning is universal but not well organized at birth; it develops over the child's first few years of life (Tomasello et al. 2005; Callaghan et al. 2011; Bard and Leavens 2014). We believe that attachment figures play a privileged role in facilitating and encouraging young children's earliest accomplishments in social learning in a number of ways (not all of which will be found in a given culture or a given primate group). These include, but are not limited to, providing motivation, facilitating nascent attempts, and providing a culturally specific road map for social expectations and practices. This social interdependence, itself a product of an attachment system (Sroufe and Waters 1977), serves as an entry point for children's more general acquisition of a cultural meaning system as the child develops a working model of the social world through daily social interactions.

We hypothesize that attachment figures become specially recognized by the developing infant because of their special roles in these diverse forms of regulation. They can be, for example, a source of salient and sometimes unique emotional experiences, the focus of infants' social expectations for responses that provide stress relief and positive arousal, and eventually the locus of motivational processes that cause the infant to seek proximity to that person and to interact with that person preferentially. Because of these cognitive and emotional processes in an attachment system, infants and their attachment figures will eventually come to share and coordinate psychobiological regulation and social learning (Tronick and Beeghly 2011). Some of the characteristics of these interactions are unique to the individuals involved, whereas others will be widely shared within a culture (or group) yet vary across cultures. These variations may be particularly significant between cultures that systematically invite the child to act independently and those that invite the child to coordinate their actions more closely with others. However, as their capacities, knowledge/expertise, and culturally organized social roles expand over developmental time, children in all cultures are likely to become more capable of comanaging all aspects of the attachment system, again in ways that are consistent with the cultural expectations for how children can and should act.

How can attachment relationships and figures be recognized? When one makes the assumption that there is a single primary attachment figure who is either "mother" or "mother-like," the answer to this question is almost trivial. However, it becomes more problematic and complex when it is recognized that there can be multiple attachment figures. It is further complicated when one acknowledges that attachment figures for a particular child may change over time, and that one attachment figure can be favored in a given context or activity but not in another. When a child's attachment system consists of multiple attachment figures, it will undoubtedly have more complexity and fluidity in its structure, making it harder to identify and describe.

In the face of this increased complexity, there is a risk of expanding membership in the attachment system to include any important social partner. We would discourage overexpansion because it threatens to dilute the significance of the privileged role of attachment figures in a young child's experiences and socialization. Thus, while we attempt in this chapter to expand the definitions and conceptualizations of attachment systems, and the behaviors that occur within those systems, we are also committed to the idea that attachment figures will be members of a closed set and that some regular social partners in a child's world will not be included. In the abstract, social partners should be considered attachment figures only if their presence in the child's world consistently serves the two functions described above:

1. Ensuring safe engagement with the environment while supporting psychobiological regulation.
2. Providing a privileged entry point for social learning.

Based on our understanding of the ethnographic and primatological literatures, we believe that while, in any infant's everyday experience, there may well be multiple social partners (Hrdy 2009) who play these roles, there will not be an extremely large number of attachment figures. When there are multiple attachment figures, responsibilities may be distributed more or less equally and interchangeably. Alternatively, there could be a hierarchy of attachment figures (e.g., when one attachment figure is not available to provide care, another consistently steps in) or a set of specialized ones (e.g., mother providing nursing, grandmother co-sleeping).

At a more practical level, one might begin to identify attachment figures by observing which figures are responsible for supporting and organizing a child's most basic daily activities (e.g., sleeping, eating, holding/carrying, bodily care, assurance of physical safety) and engagement with the world. Who regulates a child's under- and overactivity, in either a proactive or responsive manner? Who feels a responsibility to respond in a (culturally defined) "timely" manner? Who does the child turn to preferentially for such regulation (e.g., soothing when distressed), and who is responsible for organizing the child's network of care, assigning caregiving responsibility to others and supervising or evaluating them? In short, we think that attachment figures can be best identified by starting from the perspective of the children: what are their needs, and who addresses those needs?

Why is an attachment system important? We argue that it leads infants to engage safely with (and learn from) the environment through visual, manipulative, and locomotive exploration, and to coordinate or synchronize their social behavior with attachment figures and others. The attachment system will serve these purposes if infants experience social interactions with attachment figures that address the infants' needs and that are consistent and therefore predictable, so that infants not only react to behavior directed to them, but also come to anticipate, even expect, certain kinds of behavior. The particular characteristics of the caregivers' and infants' behavior may, however, vary widely across cultures.

We chose to initiate our attempt at conceptualizing attachment as a cultural activity without direct reference to the established tradition of attachment research. However, it is useful to revisit Bowlby's (1969) original work on attachment. Bowlby described pre-locomotive attachment behaviors as "goal-directed" (i.e., they draw the attention of the caregiver to the infant) and post-locomotive attachment behaviors as "goal-corrected" (i.e., infants' propensity to signal to and seek the proximity of their caregivers continually varies, depending on the caregiver's whereabouts and the infants' emotional state). We hasten to note that there is cultural variation in the degree of infant signaling, in general (Gaskins 2006; Salomo and Liszkowski 2013), as well as the degree to which infants signal distress, in particular. According to Bowlby, a distressed infant will be highly motivated to seek out the caregiver for comfort and will cease to do so once proximity is achieved. A nondistressed infant may not be motivated to seek out the caregiver and be content instead to explore the environment; in the next moment, however, this same infant may engage in proximity seeking if the caregiver moves away from the infant, a stranger approaches, etc. Infants with multiple attachment figures may not show these types of reactions in the same way (e.g., Meehan and Hawks 2013). Bowlby stated that goal-corrected attachment is characterized by continual shifts in the relative balance of the attachment and exploratory behavioral systems in response to changes in exogenous (e.g., caregiver separation or approach of a stranger) and endogenous (e.g., hunger, pain, or fatigue) conditions.

Although Bowlby (1969) identified the "set-goal" of the attachment behavioral system as maintaining proximity to the caregiver, Sroufe and Waters (1977) argued that the set-goal of the attachment system was "felt security," because the degree to which infants appear to be perturbed by separations from their caregivers, stranger approaches, etc. varies across infants. This variation is partly a function of infants' differing amounts of experience with potentially stressful events (e.g., separation from attachment figures and contact with strangers) as well as a function of individual differences in temperamental thresholds for experiencing distress.

Drawing from Bowlby's formulations about the developmental significance of early attachment (Bowlby 1969), a central tenet of attachment theory is that as goal-corrected attachments develop, infants develop "working models" of their caregivers and themselves that are shaped by the quality of care (Bowlby 1969; Ainsworth et al. 1978; Sroufe and Fleeson 1986; Bretherton and Munholland 2008). Infants who have enjoyed a history of sensitive caregiving are expected to develop a representation of their caregiver(s) as responsive to their needs, and of themselves as worthy of love and support. By contrast, infants who have experienced a history of insensitive care (e.g., unresponsive or inconsistently responsive, intrusive, and/or rejecting) are expected to develop a representation of their caregiver(s) as unresponsive to their needs and of themselves as unworthy of love. These internal working models are theorized to have a powerful organizational influence on infants' behavior toward caregivers and others, although, as the term "working" implies, they can also be further shaped by experience. Indeed, Bowlby (1973) argued that the quality of care in infancy, and the content of the internal working models that emerge from infancy onward, directly impact children's capacity to resolve subsequent psychosocial adaptations. Infants who come to trust in their caregivers' availability and responsiveness are expected to negotiate subsequent adaptations (e.g., separation-individuation and autonomy in toddlerhood, social competence with peers in later childhood) more successfully than infants who do not.

These central concepts of set-goals and internal working models sit in uneasy tension with some of the ethnographic knowledge about the variation that exists in how children's everyday environments are culturally organized (Morelli et al., this volume) and the primatological literature about the variation that exists in parental strategies (Hawkes et al., this volume). For example, whether all children will actively maintain proximity to a caregiver is not clear, even if "proximity" is defined in a more general way than physical proximity, because different normative care practices characterize different cultural systems. Konner (1976) reports that for !Kung infants raised in the Kalahari Desert in Africa, physical contact by all caregivers was observed 90% of the time at three to five months of age, and was still a high 42% of the time by 18 months; in such cultures, proximity is a given much of the time. Substituting it with the term "felt security" is equally troublesome, if not more so because of the various sources of felt security that may be derived from different practices of early care other than attachment.

Moreover, what should count as "sensitive caregiving" is also defined to a large degree by culturally organized parental goals. LeVine provided an example to the Forum from Mary Ainsworth based on her original comparisons between Ganda infants and parents, and U.S. middle-class infants and parents (Ainsworth 1967). Both communities provided sensitive care within the framework of their own culture, but the Ganda caretakers and infants did not emphasize face-to-face smiling and overt displays of affection—signs of "sensitive" responsive care in the Western context. As Ainsworth (1967:334–345) stated:

> In our American households the parents, loving relatives and interested visitors alike bend over the baby as he lies in his crib, presenting him a smiling face, and waggle their heads and talk to the baby in an effort to coax a smile. This kind of face-to-face confrontation was not observed to occur in the Ganda sample. Indeed, it was rare for an adult even to hold a baby so that there could be a face-to-face confrontation, for the baby was, at least from about eight months on, usually held in a sitting position on the adult's lap, facing outward and leaning back.
>
> By the end of the first year of life, babies in our society are able to return an embrace or kiss when it is given to them, perhaps clumsily, but in distinct response to the adult's affectionate advance. That this is largely a culture-bound pattern of response—whether learned through reinforcement or imitation—is suggested by the fact that Ganda babies very rarely manifest any behavior pattern even closely resembling European affection, and, indeed, their mothers did not try to elicit hugging or kissing in the baby, although they themselves occasionally nuzzled the baby while holding him.
>
> The fact that Ganda babies do not hug or kiss, whereas Western babies who are encouraged to hug and kiss do so, suggests that this pattern of attachment behavior is of a different order than the other patterns considered in this chapter—it is much more contingent on a specific learning process.

Since both Ganda and Western babies are receiving culturally meaningful, sensitive care and they display appropriate attachment to their mothers, the fact that Ganda babies do not experience the Western cultural patterns of engaging in face-to-face interaction and encouraging children to hug and kiss should not be considered evidence of insensitive care. To the contrary, it is a sign that the Ganda social-learning orientation is outward toward multiple others in the social setting, and less dyadic toward a single caregiver. Similar patterns are seen in many other cultures (e.g., Martini and Kirkpatrick 1981; de León 1998).

During our discussions at the Forum, we tried to respect the theoretical foundations of the traditional claims made about attachment. However, these traditional claims often make culturally specific assumptions that lead to excessively broad claims about the universal nature of attachment systems. Although we are committed to the argument that attachment is a universal process in humans and other primates, we think it is imperative to begin by looking at attachment systems across cultures with an open mind about which specific characteristics of attachment systems are universal and necessary, and which ones are culturally specific. These distinctions cannot be made by relying on the existing research, which has been formulated using Western caregiver-infant interaction as the guide for characterizing attachment (and which, in fact, has been conducted primarily in Western societies).

An interesting question arose that we are unable to answer, due to lack of adequate information: Does the development of an attachment system occur differently in contexts where there is a single primary caregiver compared to contexts involving multiple caregivers? Plural caretaking may lead to earlier social self-regulation or to children becoming more resilient. Because children have to integrate information about different attachment figures into their expectations about their social world and are dependent on more than one person, the working model developed by children with multiple attachment figures may be more complex and flexible.

Another interesting question with no immediate answer is whether proactive systems of care yield different kinds of attachment than the forms of responsive care more typical of the contexts in which attachment theory developed. How important is a caregiver's responsiveness to infant signals if the caregiver usually intervenes before such signals occur? Does infants' understanding that caregivers are continuously attentive (and therefore feeling that there is no need to attract and sustain attention of attachment figures) lead to distinct types of attachment behavior? Perhaps this pattern of proactive care changes the nature of exploration, or perhaps it changes the use of and dependence on social referencing. Answers to such interesting questions await further research.

Finally, we recognize that the concept of "psychobiological regulation" is, to some extent, culturally relative because the circumstances requiring a caregiver's support necessarily vary. How a young child responds to strangers is one example of potential stress, and this response has been assumed to be universal and to require caregiver support. There is, however, wide variation, both within and across cultures, not only in how often and under what circumstances young children are exposed to strangers, but also in how they react to them, and what their reaction should be to meet cultural expectations (Gaskins 2013; for an example of cultural differences in emotion and emotional expression, see Keller 2013a for a description of early socialization for emotional control among the Cameroonian Nso). During our discussion at the Forum, James Chisholm described an intracultural difference in fear of strangers among Navajo babies between those living in camps with extended families (who demonstrated less fear of strangers) and those living in nuclear family camps (who demonstrated more fear of strangers). Tom Weisner offered the example that young children among the Abaluyia and other communities in Kenya and elsewhere in Africa observe their parents and older children going up to a stranger and shaking hands (the culturally appropriate adult response to meeting a stranger), and come to do so themselves with a calm, solemn, respectful demeanor. In addition to the fact that strangers do not have the same cultural meaning for children in different cultures, it appears that the children's responses to strangers may be difficult to interpret without the benefit of significant cultural understanding. For instance, in Weisner's example above, is the young child's handshaking a straightforward imitation of a behavior seen many times, an enactment of a learned, socially appropriate script, or a way of managing fear? While it is worth exploring candidates for a universal stress-producing interaction, this can be done most appropriately by looking for culturally meaningful examples of such interactions in individual cultures and then evaluating whether there appear to be similarities across cultures that could be used as a point of comparison.

This culturally grounded theoretical reconceptualization of attachment will serve as the conceptual foundation for the rest of the chapter, as we consider how to study the context of attachment in a culturally meaningful way, how to assess individual differences within a particular cultural system, and what kinds of tools could be used to study attachment in cultural context.

Measuring Attachment in Different Cultural and Ecological Contexts

For some time, anthropologists, cultural psychologists, and others who study children and their development in a variety of cultures have argued that traditional and contemporary approaches to attachment have studied the phenomenon of early attachment without knowing enough about the variation that exists across cultural environments and ecologies (Harwood et al. 1995; LeVine and Norman 2001; Gaskins 2013; Quinn and Mageo 2013; Otto and Keller 2014). Practitioners looking to base their interventions on evidence have argued the same (Pence 2013). Here we describe what cultural information should be known and considered to understand how attachments form, what they look like, and what their outcomes are. A wealth of information already exists about how different cultures understand birth, infancy, and childhood (e.g., Hewlett and Lamb 2005; Keller 2007; Konner 2010; Lancy 2015). These resources can inform our evaluation of the cultural appropriateness of any claims about attachment systems. Often, they provide evidence that characteristics of children, attachment figures, social interaction practices, or everyday environments assumed to be universal do not actually exist (Gaskins 2017). This is what LeVine has called anthropologists exercising "their veto with evidence from non-Western cultures" (LeVine 2007:250).

It is, however, necessary to go beyond this reality check on assumptions about the universality of human development, and to do so we need more specific information about how attachment systems work in different settings. To pose these types of questions, we need to know more about the larger cultural system that supports attachment figures and provides the underlying rationale for the patterns of social behavior that children experience in their daily lives. Finding one or a small number of communities which do not fit a general pattern that has been thought to be "universal" serves as a starting point for the formulation of research questions, but the process does not end there. What are the patterns of variation around the world that do exist, and what might be contributing to that variation? While a number of reports about attachment systems are already available for consideration (e.g., Quinn and Mageo 2013; Otto and Keller 2014), more detailed and systematic data from a wide range of cultures, in more direct conversation with existing attachment research, is needed. Only after the range of attachment systems and behaviors across cultures is examined can we discover to what extent similar patterns of attachment behavior exist across cultures and identify potential candidates for the universal characteristics of attachment.

Likewise, studying urban populations in a number of countries and finding that they fit the proposed universal pattern does not end the research process. Most nations are not single, homogenous cultures; often, multiple well-formed cultures coexist within a single country. Thus, cross-national research is not necessarily cross-cultural (Keller and Kärtner 2013). Increasingly, urban populations, especially relatively wealthy and educated ones, are likely to demonstrate ways of thinking and behavior that are similar to Western ways. For example, Mesman et al. (2015) found some consistency across nations in urban, literate mothers' responses to what behaviors characterized an ideal mother. However, their three samples from rural populations (presumably representing distinct cultural groups with traditional belief systems intact) did not show much agreement with the other samples, nor with each other. Mesman et al. (2015:10) conclude:

> Across 26 cultural groups from across the globe, mothers' ideas about the ideal mother were found to overlap substantially with the notion of the highly sensitive mother, pointing toward a universal appreciation of the importance of contingent responsiveness in parenting young children….On the other hand, we also found a significant effect of cultural group on sensitivity beliefs that was largely, but not entirely, due to sociodemographic factors, and especially rural versus urban residence.

From their point of view, it appears that each national sample represented a "culture," and the significant difference they found across samples within a nation (samples which may in fact represent distinct cultures) was interpreted only in terms of how urban they were. In our proposals here as well as in Chapter 13 (this volume), we are not discussing cross-national research, but rather research that focuses on intact cultures, where members share a system of beliefs and practices that inform their caregiving.

There are five major categories of information that one would want to know about a cultural system or ecological context when trying to observe and interpret attachment. Listed below, we include a range of questions about specific topics that fall within each category. For some cultures, there may be enough ethnographic information to be able to find answers for many, if not all, of the questions. If there are important relevant questions for which the answers are unknown, then more ethnographic work is needed before trying to identify attachment systems or describing and interpreting the behavior of children and their attachment figure(s).

One caveat is that ethnographic research offers useful information about normative cultural beliefs and practices, but it often offers less information about intra-cultural variation. If a belief is reported, is it central to the group's cultural understanding about the world? Is there evidence of whether a belief is widely held and reliably instantiated in behavior and everyday practices? Because one of the potential outcomes of studying attachment systems is to understand the impact that systems with different qualities have on children's development, it is important to study not only the normative beliefs and practices but also individual differences. It is also important to understand whether attachment practices and beliefs are conservative and resistant to change in the face of cultural change or upheaval (e.g., immigration, war, catastrophic illness).

We believe these five categories of information support a culturally (or cross-species) informed understanding of attachment systems and behavior.[1] Equally, they would inform understanding in other areas of children's lives and development. It is a long list and thus unrealistic to imagine that any one study of attachment would be able to address every single item. Its intent, however, is to provide suggestions for topics to consider when studying attachment and security within any particular community context in a way that would ensure cultural validity. Many items, but not all, would be appropriate to the study of attachment systems in other species as well. We offer the full list of suggestions that emerged from our extensive and animated discussions to guide further enquiry:

1. Morbidity, mortality, risk of death/illness (emic/etic perspectives): The most basic factor in a culture or ecology that organizes caregiver beliefs and behavior is infant survival rate (LeVine 1980). When infants face high risk of death or impairment from threats (e.g., serious illness, physical dangers, and malnutrition), caregivers must prioritize decisions that ensure survival over other goals for their children:

    • Infant morbidity and mortality rates: How likely are children to get seriously ill or die?

    • Predictability and scarcity of resources: Are the resources relevant to infant survival regularly available?

    • Danger in environment and the risks and concerns about dangers that adults perceive: How likely are children to get hurt or have harm done to them in their daily lives? What risks are recognized as significant by caregivers?

---

[1] Our categories have a strong resemblance to the cultural learning environment model (Whiting and Edwards 1988; Edwards and Bloch 2010), which proposed three levels, of which we are discussing the first two in this section: (a) ecology, resources, risks; (b) parental cultural beliefs, practices/routines, people; and (c) child development. This tradition could be used to refine and expand this categorical framework. Also relevant are other existing ecocultural models of development (Super and Harkness 1986; Weisner 2002; Worthman 2010; Weisner 2011a) in establishing a conceptual framework for context measures.

2. Ecology, resources, and impediments (environmental, social, institutional): The everyday experiences of infants and children are strongly affected by more general aspects of the cultural environment, including how people obtain basic resources such as food and shelter, what levels of wealth or poverty exist, how society is governed, how public or shared resources are distributed, and how political events are shaping their lives:

    *   Ecology: How hard is it to get resources of various kinds?
    *   Subsistence: What are the demands and patterns of work?
    *   Political and legal resources: Are caregivers supported by their community's structure and practices (e.g., stability of leadership)?
    *   Institutional resources: Do health and social services use social models that are in harmony with the families they are designed to serve, or do they use models in conflict with those families?
    *   Pressures for cultural change in a given community: What are the current challenges to the continuity that provides safety and well-being to families (e.g., immigration, war)?

3. Parental ethnotheories (shared cultural beliefs) and other parental beliefs: All caregivers socialize with and provide care to their children informed by their worldviews about children's development and learning, appropriate cultural roles for children, and their goals for what their children should be able to do by the time they are adults. They may also hold specific ideas and expectations about individual children. Caregivers are aware of some of their beliefs and values and can articulate them clearly, but many are implicit and unarticulated, making it harder for a researcher to learn about them:

    *   What kind of person do parents want their children to become?
    *   What makes a good caregiver?
    *   What capacities does an infant have?
    *   What is a competent child?
    *   What social roles and relationships are recognized for young children?
    *   What hopes, goals, expectations do adults have for the child?
    *   What are the cultural beliefs about socialization and development and learning?
    *   What are the cultural challenges and expectations for the child (e.g., autonomy, codependence)?
    *   What are the beliefs regarding personhood and self (e.g., cultural conceptions of who children are now, and how they will change over time)?
    *   Is there a cultural model of something similar to attachment?
    *   What kind of language is used to talk about children, and what terms are used to describe early relationships?

- What social groups does the child belong to and what are adults' special understandings about those groups? Are such groups based, for example, on age, gender, social class, or caste?
- Is the child perceived as having important individual characteristics (e.g., temperament, competencies, deficiencies)?
- Do adults have ideas about how the child understands the world, and do those ideas change for children of different ages?

4. People, household (or for other species, conspecifics), and local group: Children's everyday experiences, including interactions with attachment figures, are structured not only by cultural beliefs and values, but also by the particular local environment in which they live:

- What is the social structure in the community?
- What are the norm and range for household size and family composition (and socioeconomic markers like education, class)?
- Who are the people that are physically around the child?
- Who are the people considered important to the child, even if not present (including ancestors)?
- Who is interacting with and observing the child?
- Who has preferential interactions with the child?
- What are the characteristics of children's social partners (e.g., age, gender, kinship)?
- What roles do children's social partners play, and what activities do they engage in with children?

5. Social and caregiving practices and routines, as well as consistency in everyday experience: General cultural systems, caregiver beliefs and values, and a household's local characteristics combine to produce the everyday environments of children. These everyday experiences serve as a powerful socialization tool, especially when there is structure to and repetition of events:

- Is the structure of the family's day predictable from one day to the next?
- Is the structure of the child's day predictable from one day to the next?
- What goes on in caregiver/child interaction? How does this vary across contexts?
- Who feeds, dresses, bathes, soothes, sleeps with, and plays with the child (and any other daily activities)?
- Who is responsible for the well-being of the infant?
- Are there identifiable teaching styles among caregivers?
- Who plays social games with babies, and what is the nature of those games?
- Who provides the objects used by the child (e.g., toys), and are the objects simply provided or also mediated?

Beyond the particular items, this extensive list is useful in its entirety as it indicates which issues are judged to be central by anthropologists, cultural psychologists, and other researchers for the ecological validity of studies of attachment systems. Many, perhaps most, of the items listed here have not been addressed in traditional and contemporary studies of attachment, even those which aim to study attachment from a cultural perspective (e.g., Mesman et al. 2016b). To study the cultural organization of attachment, serious attention must be given to cultural beliefs and practices as well as to the ecological context of everyday behavior. To date, however, most attachment studies which focus on cultures outside the West usually have either no, or very little, cultural or contextual evidence; they do not factor in the socioeconomic circumstances of the communities studied and often have not contextualized the measures used. We argue that conducting studies in other cultures using traditional methods of measuring attachment in isolation from studying the context is not sufficient to understand attachment from a cultural perspective. Yet studies done in other cultures are often accepted uncritically as evidence for universality and for the normativity of specific behaviors as indicators of sensitivity and competence. In the end, one needs to be able to answer "why" caregivers and infants do what they do; that is, one needs to know about caregivers' cultural beliefs and values that lie behind their motivations and actions.

Measuring Individual Differences in Attachment

In the attachment tradition, one might argue that Bowlby (1969) focused on normative attachment, describing how the system worked at the level of humans and related species. Building on this tradition, Ainsworth and her colleagues, especially with the development of and commitment to the Strange Situation Procedure, changed the focus to one of individual differences in attachment systems and how they might predict individual outcomes of well-being and mental health (Ainsworth et al. 1978). Although we think the primary focus of culturally informed attachment research, at least at the outset, should be the description of normative characteristics of different cultural systems of attachment that are shared across individuals (or at least most individuals) in a given culture, we believe that it is also important to address the issue of qualitative differences within cultural groups. Individual differences in attachment have traditionally been a large part of the research on attachment. From the perspective of cultural differences, however, they are also one of the most controversial characteristics of traditional research on attachment.

We fully expect there to be within-individual differences in any cultural group. For a given infant, there may be differences in how the child relates to different attachment figures. In studies of WEIRD (i.e., western, educated, industrialized, rich, and democratic) cultures (Henrich et al. 2010), infants have been found to form different qualities of attachment with different social partners (e.g., fathers versus mothers) (Thompson 2013b). There may also be within-individual differences across time, as the nature of attachment can also change over time with any partner.

In addition, there may be discernible differences across individuals. Although normative attachment systems are likely to exist in all cultures (even as their characteristics differ from one culture to another), there is also likely to be variation with any cultural system around that norm. Here we explore the research implications for two related issues: In a given culture, does the quality of attachments differ (across individual children, multiple caregivers, contexts, and time)? If so, how does it differ?

One fundamental question that arises at the outset is whether it is more productive to consider such qualitative differences within and across individuals using the infant or the attachment system as the unit of analysis. While attachment security has been conceptualized as relationship-specific in infancy and childhood (i.e., infants can be "secure" with one person and "insecure" with another based on the quality of relationship the child experiences with that person), over time, attachment is characterized as being increasingly person-specific, so that by adulthood, people are sometimes characterized as being secure or insecure (Thompson 2013b). Since the quality of attachment is typically assigned to the individual rather than to the dyadic relationship, and correlates and outcomes of the security of attachment are also focused on the individual in analyses (Sroufe et al. 2010), there is often an implication that attachment is a characteristic of the child from early in life. (Assigning these characteristics to the individual-in-a-cultural-setting would be a further step.) Conventionally, however, attachment researchers believe that these associations occur because aspects of the dyadic relationship have become internalized by the child in the form of "internal working models" (mental representations) based on the child's experiences over time in an attachment relationship (Carlson and Egeland 2004; Weinfield et al. 2008). It is these working models that become more elaborated, complex, and consolidated to characterize individuals eventually as secure or insecure by the time they reach adulthood (Bretherton and Munholland 2008; Dykas and Cassidy 2011).

From the perspective of understanding normative attachment in context, looking more consistently at attachment systems and characterizing their differences may be especially productive for understanding young children's experience, assuming there are consistent differences in the categories of people who function as attachment figures. In particular, studying attachment systems rather than individuals may be a more productive unit of analysis for understanding how variation in multiple caregivers across families influences attachment. We want to understand if the characteristics of attachment systems vary in interesting ways, depending on who participates in an individual system. A child who has two attachment figures (e.g., mother and father) may differ from another child who has five (e.g., mother, father, sister, mother's sister, grandmother). Alternatively, the experiences of children who are attached to child caregivers (e.g., older siblings or cousins) may differ from experiences with solely adult attachment figures (e.g., mothers, fathers, grandmothers). In addition, such co-regulatory systems with multiparty actors need to address not only who takes responsibility for what, but also what happens when one or more parts of the system fail. With the recognition that many children world-wide have multiple attachment figures, and that there likely is variation in how attachment systems are organized, we have introduced not only new sources of individual differences, but also the potential for "subgroup" variation based on the makeup of the attachment system that falls in between individual variation and group variation.

Beyond these issues about unit of analysis and capturing qualitative differences in different configurations of attachment systems, a second fundamental question is how individual differences in quality of attachment can and should be conceptualized and evaluated in a way that respects cultural differences in attachment systems. The current vocabulary for individual differences in attachment uses the basic distinction between secure and insecure, which can be elaborated into a four-way category system (secure, avoidant, ambivalent, and disorganized). This system of categorization has strong inherent traction in current research on attachment, even when looking at cultural differences (e.g., Mesman et al. 2016b) because of the long history of research connected to it (Cassidy and Shaver 2016). However, we feel strongly that these terms are fundamentally inappropriate because they imply a built-in value system imposed on complex patterns of behavior by the judgments of one particular (i.e., Western) culture. When the four-part attachment categories are used in any other culture, their meaning becomes difficult to understand (Gaskins 2013). In addition, these categories are inherently value laden: they are identified by terms that unfortunately present one pattern of behavior as desirable and three patterns of behavior as less desirable. Although we recognize that systematic differences across individual children or across attachment systems may indeed exist (and be describable in culturally meaningful terms), these differences should not be judged *a priori* as being either desirable or undesirable. Thus, contextualizing attachment may require dropping these categories altogether.

Despite our deep concerns with the current system of characterizing individual differences in attachment, we recognize that there is dysfunction and even pathology in the world (indeed, in every culture), and that the full range of attachment relationships observed in a given culture may not be equally adaptive in promoting children's well-being or ensuring their survival. There is value in capturing such individual variation within a culture along with describing culturally normative patterns. There could be a number of sources of this mismatch: children whose temperament does not fit with the expectations laid out by their particular social world; attachment figures who are inconsistent or even deficient in providing support for safety and psychobiological regulation and entry into social learning, either because of their own psychological problems or because of living with personal stressors from their environment; and families or communities that are under great stress (e.g., poverty, illness, social conflict). When caregivers face significant personal, family, or community problems, they must often direct their attention to solving those problems first and, as a consequence, may fail to invest in their relationships with their infants (Scheper-Hughes 1985). Even for a particular child, there could be wide variability in the social ecologies inhabited by different attachment figures (e.g., parent vs. grandparent or home vs. childcare center), which could contribute to important differences in how adaptive each relationship is.

One set of new vocabulary we considered are the terms "culturally adaptive" and "culturally nonadaptive," which we conceptualize as opposite ends of a continuum. Judging whether a given attachment system (or the relationship between a child and a particular attachment figure) is adaptive would include considering some basic concepts consistent with current attachment research: whether there are consistent responses to the needs and concerns of the infant which keep the infant safe and content, whether the attachment figures help a child achieve a culturally desirable affective state, and whether the coordinated actions within the relationship of attachment figures and child are well regulated and predictable. To this list we would add whether an attachment figure and infant are operating together in a manner consistent with the expectations of the cultural community and whether the attachment figure is serving as an effective entry point for the infant's culturally appropriate social engagement.

While this terminology has clear advantages over the four-part categorical system currently in use, some potentially problematic issues arise as well. For example, who can serve as a legitimate judge of the evidence for whether a system is adaptive or not, and on whose terms? Can there be more than one adaptive pattern of interaction in a given culture, and is it important to capture those differences? How can researchers (and practitioners) accurately observe and evaluate a relationship in cultures that are proactive about attending to children's needs so there are many fewer and less extreme displays of distress? Is "adaptivity" the right term or would "well-being," "social trust," or "social competence" be better choices? How can reasons that stand outside the nature of the relationship (e.g., endemic malnutrition causing parental or child unresponsiveness) be best addressed? Also, to return to a fundamental issue of unit of analysis discussed above, does "adaptation" characterize the individual child, the dyadic relationship, a particular social setting or activity, or the entire attachment system provided by the child's social community?

These questions are fundamental and should shape how we think about individual differences in attachment systems across cultures. Although we could not resolve all issues, we did agree that for outsiders to make any valid judgment about adaptivity for an individual infant, they would have to understand, at a minimum, what the cultural norms are for social relationships between infants and caregivers and the indigenous ways by which children are assessed (by their attachment figures and others). They would also need to understand the cultural beliefs about relationships, distress, and security, as well as the terms that are used in that community to define and describe the "attachment system." Outsiders would then need to reflect on how they are judging the adaptivity of the attachment system, relying on both their own conceptualizations about attachment and those of the social group they are studying. To make a judgment about a particular child, an outsider would want to know what that child's attachment figures think and what they value (to be able to understand what they are doing and why) and perhaps consult with local informants about their judgments about the functioning of the infant's social world as well.

Some of the central conceptualizations from current approaches to attachment that are candidates for use in other cultures include proximity seeking, exploration, and stress regulation, all standard measures of attachment-related behavior used in European and American cultures (Ainsworth et al. 1978). We have already discussed some of our concerns about using stress regulation as a measure across cultures. There are also reasons to be concerned about the paired measures of proximity seeking and exploration. According to Ainsworth and her colleagues, in a secure infant, the balance between the attachment and exploratory behavioral systems tips toward the attachment system in the presence of threat or danger, leading the infant to seek out the caregiver for protection and reassurance. Once the infant is comforted and the event that activated the attachment system is no longer perceived as threatening, the balance tips back toward the exploratory system, whereby the infant feels comfortable exploring their surroundings in the presence of her/his attachment figure, who serves as the infant's secure base. Ainsworth argued that this kind of dynamic interplay between attachment and exploration, in which infants can rely on their caregivers for protection and comfort when needed and as a base to engage in competent exploration of their environment, is seen as highly adaptive and conducive, in an evolutionary sense, to individual and species survival (Ainsworth et al. 1978).

In Ainsworth's model, "insecure" infants are similarly defined in terms of a balance between attachment and exploration, but for such infants, this dynamic balance is not seen as adaptive. Infants deemed "insecure-avoidant," for example, are likely to explore new territory in their caregivers' presence in an almost compulsory fashion, tend not to respond to or look toward their caregivers when called, and tend not to seek out their caregivers at times when doing so would afford the infants needed protection and security from threat or potential danger. In such infants, the attachment-exploration balance is tipped predominantly toward the exploratory system and appears to do so at the expense of the infant's safety. "Insecure-resistant/ambivalent" infants, by contrast, often appear to have difficulty separating from their caregivers, with the attachment-exploration balance tipped toward attachment at the expense of exploration (Ainsworth et al. 1978).

We believe that significant cultural issues affect the conception of attachment as a balance between proximity seeking and exploration and how that balance might be organized in everyday activities. For instance, in cultures where children are carried by caregivers, even as they reach their first birthday and beyond (e.g., Konner 1976), proximity seeking, in a physical sense, is limited by the culture's childcare practices. More fundamentally, the concept of "security" and the attachment-exploration balance presume that children are expected to be independent actors that modulate their movements toward and away from their caregivers. Exploration appears to have very different characteristics across cultures: in many cultures, young children do not rely as much on social referencing and are not used to organizing social interactions with others around objects (Gaskins 2006). Moreover, as stated before, stress regulation is much less visible in cultures that proactively adjust circumstances to avoid displays of stress (Keller and Otto 2009). Emotional display rules are socialized differently across cultures, just as are other behaviors, including in infancy (Ainsworth 1967; Otto 2014). Additionally, the coordination of caregiver-infant behavior could be based on other models. The concept of accommodation and responding to the needs of others as an integrated social unit could provide a more useful framework for describing the shared goals of an attachment dyad across cultures than the concepts of stress regulation, security or proximity/exploration. Kochanska (2002:192), for example, describes the parent-child relationship in dyadic terms, as a "mutually responsive orientation" between parent and infant, and Tronick and Beeghly (2011) describe this relationship as a dynamic open system.

In studies with older children in WEIRD cultures, the attachment-exploration balance is assessed in a manner very different than in infancy (Solomon and George 2008). In contrast to a focus on proximity seeking in infancy, researchers of attachment in older children focus instead on the quality of verbal discourse between caregiver and child: essentially, do children seek or regain psychological closeness to caregivers by sharing what the child was doing when the caregiver was out of the room, or in other ways, for instance, talking about personal matters (e.g., how the child was feeling) rather than other topics (e.g., admiring a wall decoration)? The quality of verbal discourse could be characterized as a kind of representational proximity seeking rather than physical proximity seeking. Although this example is based on attachment constructs of WEIRD families, we suggest that this approach be explored to see whether it could be adapted for thinking about proximity seeking in diverse cultural contexts as well.

Despite their limitations, there may be value in keeping the traditional behaviors in mind when studying attachment in other cultures as long as researchers are open minded about the cultural relevance of such behaviors, which need to be considered in the context of understanding beliefs, values, and behaviors that are locally relevant. Research balanced between the "insider" and "outsider" perspectives would gradually produce a list of potential attachment behaviors that might not all be seen in any one culture but which, as a whole, would provide guidance for how to recognize attachment behavior in any culture and how to judge whether or not it is well organized and serves the needs of the individuals in the attachment system.

There are many difficulties in shifting gears to begin a study of attachment grounded within the cultural meanings for the groups being studied, as we advocate here. These difficulties are illustrated by the mixed messages found in the study by Mesman et al. (2016b), in which they review studies of individual differences in attachment in several cultures. In their introduction, they echo many of our concerns about the need for more of a cultural balance in attachment research, a greater recognition of the reality and importance of multiple caregivers (including siblings) in children's lives, and acknowledgment that there is cultural variation in how "sensitive caregiving" is understood and expressed (including rates of contingent responding and modalities of responsiveness) in diverse groups. Thereafter, however, they proceed to conduct a meta-analysis on existing studies of attachment in other cultures based on Western concepts of individual differences and rely on measures of individual differences (the Strange Situation or the Attachment Q-Set) that have many cultural assumptions built into them. Unfortunately, many of the studies included in their review were informed by only minimal information about cultural beliefs and practices; linkages between caregiving behaviors and secure attachment were frequently indirect and inferred. The measures themselves were at best only slightly modified to ensure cultural validity. Moreover, most made little attempt to link attachment classifications to child competencies in the wider world. Despite the inadequacies of the available data, Mesmen et al. conclude with confidence that "…the available cross-cultural studies have not refuted the bold conjectures of attachment theory about the universality of attachment, the normativity of secure attachment, the link between sensitive caregiving and attachment security, and the competent child outcomes of secure attachment. In fact, taken as a whole, the studies are remarkably consistent with the theory. Until further notice, attachment theory may therefore claim cross-cultural validity" (Mesman et al. 2016b:809). We suggest that this conclusion is premature for the reasons outlined above.

Measuring individual differences in attachment figure behaviors across cultures raises the same issues of cultural appropriateness as measuring infant behaviors. Traditionally, maternal sensitivity and responsiveness (and effectiveness) have been used to evaluate the quality of the attachment figure's behavior (Ainsworth et al. 1974). "Responsiveness" means different things in contexts in which caregivers are in continuous physical contact with infants compared to contexts in which caregivers and infants are physically separate. It also means different things in contexts where there are multiple caregivers present to anticipate or respond to infants' needs compared to contexts in which a single caregiver has primary responsibility. Similarly, some in the group felt that it was important to employ conventional measures of attachment figures' behavior (after adapting them as much as possible for culturally variable contexts) (e.g., Mesman et al. 2015) as a reasonable first step, while others stressed that such an approach should be balanced by a careful reflection on what those measures mean in the particular culture in question while at the same time working to develop additional culturally specific measures that may be more meaningful for capturing the characteristics of caregiver behavior (e.g., Harwood et al. 1995; Yovsi et al. 2009).

Procuring certain kinds of additional information about cultural beliefs and practices would support the ability to judge the validity of measures of individual differences in infant and caregiver attachment behaviors and therefore the confidence in the interpretation of research findings. Cultural beliefs about age-salient child stages, competence, interdependence, trust, and security need to be gathered from key informants in a community and selected sample members in a study. The type of family system in which the child and caretakers are embedded also needs to be specified (e.g., joint, extended, single mother, conjugal, commuting, child sharing/lending practices). In general, the insights gained about a specific culture's beliefs, values and practices that we discussed above would all be crucial for identifying appropriate dimensions of behavior to use to evaluate individual differences, both those that fall within the cultural norms and those that fall outside them.

A Methodological Tool Kit

To understand attachment in context, a suite of new methods is needed to investigate how attachment systems function in the daily lives of human infants and young children, as well as in infants and young juveniles from other primate groups, when they interact with those around them. To enable comprehensive study, both qualitative and quantitative methods are needed (Table 8.1). Rather than proposing a fixed research agenda, we favor a methods tool kit to direct enquiry into attachment within particular cultural contexts.

We contend that using mixed methods is essential. The epistemological assumption underlying the use of mixed methods is that in scientific endeavors, the world can be represented through both numbers and words, and that numbers and words (as well as photos and videos) should be given equal status in developmental science (Yoshikawa et al. 2008). Behaviors or contexts relevant to human development are not inherently qualitative or quantitative, but the methods of representation through which behaviors or contexts are recorded in research are. It is important to remember that these can be complementary, and at times even overlapping, methodologies. In their study of six European and U.S. communities that compared four major daily routine activities (meals, family time, play, and school- or developmentally related activities), Harkness et al. (2011:811) state that "qualities can be counted, and quantities can be described."

The world in which young children and caregivers maintain safety, manage psychobiological regulation, and engage in social learning is complex; it

Table 8.1   Overview of a tool kit designed to aid in understanding attachment in context.

	Type of Data	Specific Methods
To assess cultural context and cultural interpretation of behavior	Observe the everyday life of children	Participant observation Ethological observation Videotaping and subsequent coding
	Talk about beliefs and behaviors	Focused interviews Elicitations (pictures, stories, videos)
To assess individual behavior	Systematically code attachment behavior in individuals	Generic coding systems Culturally specific coding systems
	Characterize psychobiological functioning	Genetics Physiological assessments
	Assess attachment outcomes in children	Ensure measures are culturally valid Redefine level of outcome measures

certainly is not linear, decontextualized, nor additive (Weisner and Duncan 2014). Families, the central contexts for the development of attachment systems, "are not frozen dioramas: they are alive, active and changing." An intensive repeated-measures approach reaches beyond static representations of the family toward more dynamic models that depict "life as it is lived" (Repetti et al. 2015:126). Of course, at the same time, the world of attachment systems and their contexts can be represented *as if it were* linear, additive, and decontextualized. Thus, for very good *analytic* reasons, we need both numbers and text, both algorithms and photos or videos, whenever possible.

Qualitative and quantitative measures have different strengths and weaknesses. For instance, although quantitative methods are easier to analyze, they often require a commitment to predetermined categories. For quantitative methods, it is important to have a way of factoring in cultural differences in the interpretation of results. Qualitative methods often produce a very rich understanding, but it can be difficult to make comparisons across contexts or cultural groups. In addition, many people who come from one discipline may find it difficult to understand and value methods from other disciplines.

For these reasons, to study the cultural nature of attachment systems, our tool kit aims to balance qualitative and quantitative methods, including methods that emphasize a culturally derived, emic perspective as well as normed assessments that emphasize an etic perspective. For researchers who are less familiar with qualitative methods, useful criteria are available to assess qualitative and mixed methods work (Creswell and Plano-Clark 2007; Weisner and Fiese 2011). Partnerships and team research can make a multi-methods approach easier; this can and should include community members where appropriate and possible.

The research practice we advocate for developmental scientists interested in studying attachment in cultural context is one that relies on this multi-method approach. Below, we describe a suite of measurement tools that we believe will produce more useful information about cultural beliefs and practices, and thus provide a meaningful context for the interpretation of observed behavior. For any given study, researchers could select from among these tools—the measures of context, the assessments of attachment behavior, and the description of outcomes—that they judge to be appropriate and feasible given the resources, sample, and time available.

Tools for Understanding Cultural Context and Cultural Interpretation of Behavior

Observing the Everyday Lives of Children

Observations of infants in their everyday contexts are a good place to start in understanding the cultural organization of their interactions with attachment figures. Observing a particular infant across a number of events in typical daily life ensures a sampling of behavior from which to draw generalities. The particular events may differ across cultures or species, but their choice should be informed by the contextual issues listed above. A range of events could include potentially caregiving-rich occurrences (e.g., bathing, eating, nursing, and sleeping; opportunities for body contact; exploration of the environment; intentional social engagement) and other occasions when the functions of attachment (e.g., safety, psychobiological regulation, and privileged entry into the social world) are likely to be activated.

The age of the infant should also factor into deciding who, and when, to observe. Typically, research indicates that attachment systems develop in the first year of life in humans and great apes; thus, assessments often occur around an infant's first birthday. Clear-cut attachment could, in principle, occur between nine and 12 months of age (or during an equivalent infancy period in other species). It is also possible that cultural differences in age (in terms of achieving developmental milestones) might affect the trajectory of the attachment systems. For instance, there are theoretical reasons to believe that attachment may be linked to the onset of locomotion (Campos et al. 1992), which is known to vary across cultures (Adolph et al. 2010). There also may be cultural differences in the perception of the age at which attachment systems develop or the range of ages for which attachment behavior is considered appropriate; for instance, parental beliefs about infant recognition and memory of relationships may be related to lay perspectives on attachment and be the basis for organizing children's social worlds (Liu et al., this volume). Different attachment behaviors might also be expressed at different ages; for example, among the

Yucatec Maya, seeking comfort from an attachment figure is usually expressed by the infant's first birthday (if not before), but it is not until around the second birthday that fear is expressed when a stranger arrives at the infant's home. For each culture (or species) under study, the ages of observation of a sample should be set only after preliminary observations have allowed exploration of the normative ages of expression of attachment systems through observable behavior. Thus, one might choose a broader age range (e.g., three months to two years) to include early attachment behaviors and more sophisticated examples of the systems in practice.

Participant observation.   Participant observation occurs while the researcher engages in the activities being observed, as part of the group. This is a central tool in basic ethnographic research (Dewalt and DeWalt 2010). The researcher may record entries about ongoing behavior if the activity allows it, or entries may be made soon after the events when there is time. Informal questions can be asked during the observed activity for clarification purposes. This kind of data tends to be qualitative, open-ended narrative descriptions rather than recording of specific behaviors. However, as with all observation, it is guided by research questions and interests. When partnered with other data, it may be the first data collected (along with open-ended interviews) because it does not impose a preconceived structure on the type and range of information that can be recorded. Once this type of research has been conducted for a given culture, it can be used as the foundation for the development of future research tools (bearing in mind that cultural change may make it necessary to repeat this step).

Ethological techniques for observation in naturalistic environments.   Similar to, but distinct from, participant observation are ethological methods of observation (Lehner 1998). Both techniques observe and record ongoing behavior in everyday contexts. An important difference between the two is whether the observer participates while observing or is an external observer to the event. Many ethological methods are designed to quantify either the frequency or duration of specific behaviors. The data collected with this kind of observation is often organized by predetermined categories. There are a variety of ethological/observational methods that can be used to obtain samples and arrive at an overview of a "typical" day: spot observations, diary methods, time sampling, and all-day observations. Selection of length and frequency of observations may vary but should be broadly similar for all infants who are the focus of observation in a specific study of attachment and should represent the full range of children's experiences in their daily lives. As with participant observation, it is important to collect observations surrounding different kinds of events in the children's lives that are relevant for attachment.

Videotaping child interactions with caregivers and others, and subsequent coding of behaviors.   While participant observation and ethological methods provide valuable insights into children's everyday experiences through the observation of live behavior, video recording of naturalistically occurring behavior allows a wider range of methods to be used to describe the interactions. Videos can be analyzed using quantitative and qualitative procedures that might be used in other types of observation. Analysis of sequences of behavior is also a powerful tool for understanding how behaviors are related (Bakeman and Quera 2011). Capturing behavior on video means that it can be viewed multiple times and analyzed frame-by-frame. This also allows a focused microanalysis of interaction behavior between child and caregiver or others, including talk, which is difficult to observe and capture in real time (Erickson 1995). Because of the time-consuming nature of analyzing behavior at this level, it is used primarily for targeted examples of behaviors of interest. Sometimes, these examples are identified after a longer session, by identifying behavior that signals the beginning of an event of interest to the researcher. Other times, videotaping is done to capture short segments of behavior that is structured or elicited at the time it occurs. For quantitative measures, coding for occurrences of behavior, including inter-observer reliability, can be done at a time and place outside the context of the behavior occurring. In addition, unlike live ethological observations, the system of categories of codes can be developed or revised post hoc if new distinctions come to light.

Talking about Beliefs and Behaviors

Once observations have been conducted to learn about everyday activities, researchers can move on to asking for information directly from participants. In addition to interviews that seek to shed light on general cultural practices, there are a number of tools that can be used to talk about children's everyday worlds and their social interactions. Interviews may be profitably used to learn about the culturally specific meaning of attachment systems, figures, and behaviors.

Focused interviews. Interviews are a basic tool of the ethnographer (Spradley 1979), and there are many types of interviews (e.g., open-ended or structured). Just as with observations, minimal structure is often the best place to start for learning about a new culture, because as an outsider, one might not even know the important questions to ask. A significant advantage to focused interviews which hone in on details of children's everyday lives is that they can produce responses that are more comparable across respondents and across cultural groups. (The same issues discussed for observations should be considered when deciding what age range of infant to include in the research sample or in the materials to be used in the elicitations.) The ecocultural family interview (Weisner 2011b, c, 2016a) is a good example of a more focused interview for learning about young children's everyday lives and the cultural context of attachment. It utilizes a daily routine of activities as a universal frame for beginning a conversational interview with caretakers about their child's activities and socially significant caregivers and others. The interview provides a conversational framework: "Walk us through your day. From the time you get up, what are the first activities for the day? And then the next activities…?" With this framework, the topics of focus depend on the study. For attachment, one might focus on identifying significant caretakers, distress, and experiences with multiple caregivers and strangers, as well as related topics. Common features of activities and settings can be identified that can then be explored across the day and compared across families, communities, and cultures, such as the ones we have proposed earlier.

Elicitations.   Many of the beliefs and practices relevant for attachment are not explicit types of knowledge that can be easily accessed by asking direct questions about children's behavior and its cultural context. In contrast to interviews, elicitations can support conversations that tap into implicit knowledge which may not emerge in a straightforward interview. Since attachment systems often operate below the surface of caregiver awareness, this method is a particularly important tool for learning about attachment. Elicitations can either be done with individual respondents or focus groups. Typically, they use some type of prop to introduce a concrete behavior or event into the interview so as to elicit information and opinions about it. The targeted behaviors and events may be drawn from the variation observed in the specific culture under study or from variation across cultures. When drawn from a specific culture, individuals represented in the stimuli should be unknown as individuals to the respondents but recognizably members of the same culture. When drawn from a range of cultures (chosen for their diversity in geographic location, economic system, etc., to provide some representativeness), elicitations are a promising cross-cultural tool that can be used to assess meaningful across-group comparisons. Below we describe three types of elicitations that can be useful for learning information about beliefs and practices surrounding attachment systems. A partial list of potential events/sequences might include nursing/feeding, response to infant crying, approach of caregiver(s) to infant (or initiation of social interaction), approach of toddler to caregiver(s) (or initiation of social interaction), putting a child to sleep, bathing, playing a social game, demonstration of affection, demonstration of anger/rejection, reaction to stranger, and reaction to attachment figure leaving child:

1. Picture Cards: To draw the attention of the participants to specific topics of behaviors and to elicit answers more easily, a semi-structured procedure using picture cards which depict caregivers (from the sociocultural community or outside of it) interacting with children is a helpful tool (Keller et al. 2004a). These picture cards may represent diverse child states (e.g., distress) and typical responses from others, or they may depict diverse contexts for children's everyday behavior (e.g., where and how children sleep). Respondents are invited to describe what they see, and their interpretation of it: What would you

think about this? What would you do? Why is it important for the child? Responses can be analyzed qualitatively or coded by categories and analyzed quantitatively.

2. Vignettes and story stems: Short vignettes can be developed to illustrate a variety of circumstances that are likely to produce insight into the beliefs, motivations, behaviors of subjects being interviewed (Finch 1987). Like pictures, vignettes should capture context and behavior that produce opinions reflecting cultural values and practices. Likewise, the interviewer can initiate a story, providing context, characters, and circumstances in a story stem (Emde et al. 2003), but stop at some point of tension to ask the respondent to complete the story.

3. Videos of children's social interaction with attachment figures: These can be used to ask respondents to describe, interpret, and comment on what they perceive is happening in the video. Respondents can be shown videos of their own behavior, of others who are unknown to them but from the same cultural group, or of people outside their cultural group. This method, developed by Tobin et al. (1989), has been particularly successful in producing illuminating conversations where participants analyze the meaning and value of complex behavior such as that found in attachment systems.

Developing New Measures of Individual Behavior

Past attachment research has not inspired cultural confidence in its ability to identify which specific behaviors should be observed and used to measure individual differences in adaptiveness. As a result, we cannot advocate for Western-based, structured observations of attachment, such as the Strange Situation, for at least two reasons. First, the empirical base of the Strange Situation derives primarily from Western, industrialized cultures. It is unclear to what extent the psychological experience of such a structured observation, and the classifications derived from it, are valid in non-WEIRD cultures, especially when multiple caregivers are involved. Second, even in WEIRD cultures, the caregiving antecedent of Strange Situation classifications (i.e., parental sensitivity) is, at best, controversial, both in infancy and during the preschool period (Cassidy et al. 2005; Verhage et al. 2015). We believe that using naturalistic observations to examine how children and their caregivers direct their attachment behaviors to each other will reveal more useful information about the nature and quality of attachments in a given culture than a structured observation. We propose the following methods for characterizing the quality of interaction.

Generic Coding System for Individual Attachment Behavior

We are optimistic that the Attachment Q-Set (AQS)—a 90-item q-sort procedure developed by Waters (1995) to assess quality of attachment behavior in the home—could be adapted for use across a broad range of cultures. Although this assessment was developed for and used primarily in Western cultures, we discuss it as a general methodology that could be adapted for a particular culture or even for cross-cultural comparison. We also note that a q-sort has been developed for rhesus monkeys (Kondo-Ikemura and Waters 1995).[2]

AQS measures quality of attachment by rating observed behavior along a continuous security dimension, rather than classifying behaviors into specific categories. In the current measure, individual AQS items are pre-rated by attachment experts along a security dimension ranging from "very much like a secure child" to "very much unlike a secure child." As discussed above, we would propose that the dimension should be modified to reflect a more general concept such as "adaptiveness," "regulation," or "well-being," rather than "security." The individual items would have to be rated on this new dimension according to cultural understandings of what the concept means and what behaviors reflect it.

In the current measure, the evaluation of a particular child consists of sorting these items in order of how much they characterize the behavior observed, and then using the ratings to determine where on the continuum of attachment security the child falls. AQS has been used to assess quality of attachment with nonmaternal attachment figures and in environments external to the home (e.g., in daycare environments) to examine secure base behavior to a daycare provider (Waters et al. 2017). It also provides flexibility in terms of examining specific child behaviors (with specific items), and thus enables a derivation of a specific attachment profile per child, rather than just a simple "score."

To be adapted for use in other cultures, such an approach would need to articulate clearly what a hypothetical child, at both ends of the continuum, would look like in a given culture, based on a strong working knowledge of core attachment constructs and on how attachment behavior is manifested in the specific culture. Based on this cultural understanding of normative behavior, specific items could be developed for rating individual children. Part of this process would involve determining to which degree and in what ways a criterion sort for a specific culture is similar to or deviates from the Western criterion sort that is currently available.[3] We anticipate that there will be significant differences for cultures that limit infant exploration, that have ways of demonstrating affection and closeness that differ from the West, that rely more on close body contact and less on distal face-to-face contact, or that have multiple caregivers with highly differentiated functions. For example, there may be a need to include items related to concepts other than secure base, because that concept may put too much

---

[2] Although the Strange Situation Procedure is not appropriate for use across cultures, it may be appropriate for assessing attachment in nonhuman primates, but only in certain settings, such as human-based laboratory nurseries set in WEIRD cultures (e.g., van IJzendoorn et al. 2009).

[3] Since this measure was developed for use in WEIRD cultures, our discussion is phrased in terms of how it might differ if redeveloped for use in other cultures. This is not to assert that the WEIRD characteristics and behaviors captured in the AQS should serve as a standard for use in other cultures, merely that it is by default the starting point for redesigning this measure.

emphasis on the child as an individual (being primarily characterized as seeking distance and returning when there is need). Here, one could look for other systems in which social trust is displayed (e.g., co-participation and collaboration), and items describing these systems could then be developed. To achieve a level of confidence in the ratings of the items, one could ask members of the culture to rank the items according to an idealized concept of adaptiveness (or regulation or well-being) in attachment relationships.

Classifying the current items in the AQS in terms of the constructs to which they refer is an important step toward illuminating the cultural organization of the measure. This step would also potentially facilitate the development of a new rating system for another culture, by allowing the developers to identify existing constructs that might be similar to constructs in the culture being studied. Where a construct is found that is similar for the two groups, the items that represent that construct would become candidates for being appropriate for the non-Western culture. For example, if "reaction to strangers" is a culturally appropriate construct, then the items related to the infant's response to strangers could be included in those that are considered for use, either as is or modified to reflect cultural practices. If it is not an appropriate concept, they could be dropped altogether. We also envision that such an approach could be used to assess children's attachment quality to an overall caregiving system, not just to specific individuals. If it turns out that there is a subset of items that apply equally well across a range of cultures, then it could become a valid tool for comparison.

To ensure that the full range of attachment behaviors in a given culture is being measured, any significant constructs that are not represented in the Western-based measure would need to be identified. Items based on these constructs would also need to be developed. These items could then become a resource for working in yet another culture, along with the current types of behavior measured in the AQS. They could also, in fact, be used to enlarge the range of constructs studied in WEIRD populations.

For those not familiar with the AQS, we list below a few examples of current items (both high and low on the security dimension) that may be found to have relevance for understanding attachment systems across a range of cultures, even if they need to be modified in their particulars. AQS items that are high on the security dimension it is designed to measure include:

- Child often hugs or cuddles against caregiver, without the caregiver asking or inviting the child to do so.
- When caregiver says to follow, the child does so. (Do not count refusals or delays that are playful or part of a game unless the child clearly becomes disobedient.)
- Child recognizes when caregiver is upset, becomes quiet or upset, and tries to comfort the caregiver.
- If held in caregiver's arms, the child stops crying and quickly recovers after being frightened or upset.

AQS items low on the security dimension include:

- Child often cries or resists when caregiver puts the child to bed for naps or at night.
- When child is upset about caregiver leaving him, the child sits right where he is and cries. Doesn't go after the caregiver.
- Child is easily upset when caregiver makes the child change from one activity to another, even if the new activity is something that the child often enjoys.
- When something upsets the child, the child stays put and cries.

Culturally Specific Coding Systems for Individual Attachment Behavior

Capturing the meaning of ethnic and cultural differences in caregiver-child interactions may be subtle, yet significant. For instance, in a large multiethnic sampling of preschoolers and their primary caregivers, factor scores for caregiver and child behavioral ratings exhibited different patterns of correlations, with independent measures of family environment and child social behavior. This suggests that existing measures may not capture parent-child interaction patterns across different groups (Bernstein et al. 2005). Culturally specific tools provide an important balance to more general ones, such as the AQS, to ensure that cultural patterns of behavior are adequately represented. Constructing a tool from the bottom up increases the chances that the tool is culturally appropriate and that unrecognized cultural biases in terms of values or priorities have not been imported into the study through the measure. These culturally specific rating tools can be used to characterize children's behavior, caregiver behavior, or the attachment system as a whole. By providing a different perspective, they can be used to inform the process of developing tools that could be used across cultures.

A good example of such a measure has been reported by Yovsi et al. (2009). They study caregiver-infant interaction in two cultural groups, Cameroonian Nso and German middle class, using one measure constructed on the Western concept of sensitivity (Ainsworth et al. 1974) and another based on the Nso concept of responsive control (defined by emotional involvement and bodily closeness in interactions with a goal of obedience and responsibility). Perhaps not surprisingly, Yovsi et al. found that each group scores higher than the other on their own culture's measure of interaction style. The Nso measure is not only a useful tool for highlighting the cultural values of that group in contrast to other groups—that is, it measures behaviors that the caregivers of that culture value, not those valued by Western culture—but it also could be used to identify caregiving behavior that would be considered maladaptive for children forming healthy attachments in that culture.

At face value, these two types of methods may appear similar. Both methods can be used to characterize children's behavior, caregiver behavior, or the attachment system as a whole. Both are intended to be culturally sensitive, but they differ in one very significant way: the culturally specific observational tool is informed initially by the understanding of cultural meaning that organizes the infant's social world and the model of caregiving. Only secondarily is it concerned with comparison across cultures.

In contrast, a sorting tool based on the AQS would begin from the current Western model of attachment and work toward a more inclusive characterization of other cultural models by filtering out or modifying inappropriate items. Culturally specific rating systems are inherently less likely to be biased, whereas the AQS is more likely to be able to be used across cultures for comparison. The goal for both approaches ideally would be to end up with measures that are culturally valid in a single culture, but also allow legitimate comparison across cultures. If thoughtfully designed and evaluated, they produce very similar methods. A culturally modified AQS can represent a culture's perspective quite accurately, and culturally specific rating tools can be developed that allow generalizations so that they can be meaningfully used across cultures. Both could focus on behaviors that reflect the two functions of attachment proposed in this chapter.

Characterizing Psychobiological Functioning

In our reconceptualization of the functions of attachment, the first function concerns the regulation of psychobiological functioning. To assess an infant's more biologically based functioning, we include here tools that measure aspects of physiology and/or genetics.

Gene-environment interactions are associated with cognitive, affective, and behavioral outcomes in humans and other animals (e.g., Coll et al. 2004), and there is every reason to think that they also are related to the development of attachment systems (e.g., Bakermans-Kranenburg and van IJzendoorn 2006). DNA samples can easily (albeit perhaps somewhat expensively) be obtained from saliva using cheek swabs (e.g., Cicchetti and Rogosch 2012), from which genotypes, the presence of risk alleles (e.g., for neurotransmitters such as serotonin or dopamine), and neuropeptides (e.g., oxytocin) relevant to attachment can be obtained. In studies where such genetic analyses have provided evidence for differential susceptibility among infants based on their unique physiological or genetic profiles, some infants may be more reactive to environmental variables than others; that is, they may be more likely to show negative outcomes in adverse environments and positive outcomes in good environments (e.g., Bakermans-Kranenburg and van IJzendoorn 2007). It appears, however, that few of these genetic studies of attachment have included non-WEIRD samples.

Physiological assessments of stress reactivity (e.g., salivary cortisol) may be useful in a tool kit. Evaluating changes in cortisol before and after exposure to stressful events could be used to evaluate how the attachment system functions to regulate stress in different cultures. Individual differences in stress reactivity via the functioning of the autonomic nervous system could also be measured, by using cardiac measures such as heart rate or heart rate variability. This requires that individuals be fitted with heart rate and respiration rate monitors. Other related cardiac measures include vagal tone and respiratory sinus arrhythmia. Feldman et al. (2014) demonstrated that early synchrony in mother-infant interactions, including skin-to-skin contact, facilitates infants' biobehavioral regulation with long-term consequences. Other studies have even administered small "doses" of oxytocin to ascertain the effect of this hormone on attachment-relevant behaviors, such as trust (e.g., Kosfeld et al. 2005).

Assessing Attachment Outcomes in Children

Outcome variables are measures of behavior or capacities in behavioral systems that one might expect to be related to the quality of attachment. These outcomes could be conceptualized as cross-cultural differences based on normative patterns of attachment in two or more cultures. Alternatively, they could be conceptualized as within-culture differences based on individual patterns of attachment. We would not expect all behaviors to be influenced by the quality of attachment systems' functioning. We do, however, have some confidence in choosing a set of behaviors that appear to be closely related to attachment in WEIRD groups as potential candidates of where there might be important outcomes based on the qualities of attachment in other cultures (either at the cultural group or individual level). The following outcome domains are identified as being potentially related to differences in infant attachment systems:

- The quality of children's other relationships (e.g., other family members, other children).
- A child's socioemotional and sociocognitive competence (e.g., empathy and prosocial behavior).
- Cognitive competence (and relatedly, language competence).
- A child's level of competence with emotional regulation and adjustment. Problems in this area have been categorized by psychologists as consisting of both internalizing (e.g., depression, anxiety) and externalizing (e.g., disruptive behavior) problems.
- Substance and patterns in children's play.

There are a number of measures of specific variables for each of these outcome domains that exist for use with small children in Western cultures. The problem with almost all of the existing measures is that they have been conceived from and developed to work in one particular (WEIRD) culture. All of the concerns expressed above about using standardized measures of attachment in other cultures apply equally to all of these measures of behaviors in these five outcome domains. For each measure, the cultural values and practices would have to be understood before the measure could be adapted to reflect them. To be useful in a culturally informed study of the outcomes of attachment, measures would need to be examined for their cultural appropriateness and validity, building on the cultural knowledge that was produced in the service of studying attachment itself.

At a practical level, to achieve cultural validity in terms of the meaning of the activities involved, each measure would have to be piloted on children in each particular culture and adjusted or redesigned to ensure that the social assumptions of engagement in the activity and the domains of responses are appropriate. For instance, in some cultures, it would be inappropriate for a child to play "a game" with an adult. In others, expecting a child to provide a fluid verbal answer to an adult might be inappropriate. As we have suggested for attachment itself, it would be methodologically less problematic if these abilities could be observed in more naturalistic settings, by defining everyday behaviors that would be evidence of the same constructs that are usually measured in an assessment activity. If naturalistic observation is not feasible, the next best option would be to develop new assessments (e.g., based on observing and analyzing relevant naturalistic behavior) that would be more appropriate (in setting and activity) than many laboratory tasks.

More centrally, in addition to the methods being culturally appropriate, the constructs being measured need to be culturally meaningful. What it means to negotiate a social relationship, to be competent in the areas of socioemotional or sociocognitive functioning, to engage in specific kinds of problem solving or other cognitive behavior, to regulate emotions, or to participate in social play are all highly culturally specific (Gaskins 2017). The problem is more complex than merely developing tools that rely on culturally appropriate rules of engagement. The concepts themselves and their categories also need to be locally grounded for each culture to ensure that they are meaningful and representative. This issue makes comparison across cultures particularly difficult to obtain.

Another strategy that avoids problems, which occur when one culture's concepts and measures are used to assess and evaluate individuals from other cultures, is to define the outcome of attachment at a more abstract level—one that is less culturally specific in terms of its meaning. For example, one candidate could be the claim that certain qualities of attachment enhance the child's psychological well-being. To evaluate this claim, one could define well-being as the engaged participation of the child in the activities deemed desirable by the child's and family's cultural communities. Such an approach puts competence, initiative, and social trust in context by basing their meanings on the psychological experiences of children as they engage in meaningful cultural activities, leaving the specific measures to be tailored for each culture. In this model, the goal of research on outcomes in any culture would be to discover how the qualities of attachment systems are related to young children's psychological well-being.

An important caveat is that outcomes should only be measured after the age when attachment systems are established, and we do not know if the timing of the development of attachment systems is the same across cultures (or across contexts in other species). If we assume that developmental timing is the same and that attachment systems develop in the months surrounding the first birthday (or equivalent life stage), then measuring outcomes in the second year of life would be an appropriate time frame (for human infants). However, as argued above for observations and interviews, the assessment time points are best informed not only by evidence of attachment systems in the infants' and caregivers' behavior, but also by the culture's beliefs regarding the development of attachment behaviors. Cultural practices about changing children's caregiving arrangements suggest that some cultures recognize flexibility and adaptiveness in the attachment system far into early childhood (Lancy 2014).

One potentially interesting research question is whether attachment systems come online at more or less the same age, regardless of cultural understandings about such systems, or do they become observable in behavior in accordance with cultural expectations? If there are differences in the developmental timeline of attachment across cultures, then we must ask: Are there also differences in how attachment is related to other developmental milestones and abilities? From research in European and American contexts, stronger relationships are found with attachment at a more proximal age; over time, predictive power weakens (United Nations 1989; Thompson 2008b). These relationships may also vary across cultures.

Summary

As originally conceived and still practiced today, traditional attachment theory does not recognize and is unable to describe adequately, or account for, significant variations in attachment systems across cultures and across species. In this chapter, we have proposed ways to theorize and measure attachment systems that will respect and be informed by cross-cultural and cross-species perspectives. By necessity, the complex and often obscure nature of the cultural and ecological contexts that organize attachment systems must be studied in detail if we are to understand and describe accurately the group specific nature of healthy attachment systems around the world. At the same time, we recognize the importance of measuring individual differences within a culture and have suggested specific research strategies to permit them to be measured, evaluated and compared in culturally valid ways.

To enable effective research, different tools are needed to studying attachment systems within and across diverse cultures as well as in other primates. We have proposed tools that focus on understanding the cultural context and the cultural interpretation of behavior, including various kinds of observations and interview methods. Such methods are often missing in cross-cultural studies of attachment, but we feel they are essential in providing an accurate and informative context for understanding attachment systems and their meanings. We have also proposed a number of measures of individual behavior, including the qualities of interactions between infants and their attachment figures, psychobiological functioning, and outcomes in multiple domains of children's development that might be related to the qualities of their attachment systems. For all of these, we have emphasized the importance of using culturally informed, appropriate measures, even while recognizing the value of measures that permit valid comparisons across groups.

As a whole, the list of research tools presented here exceeds the capacity of any one research project, let alone any one researcher. We have thus used the model of a tool box to refer to a wide range of approaches to measure the meaning and behavior involved in the cultural organization of attachment systems. We hope that this will inspire researchers to think more broadly about the limitation of traditional approaches, to consider what new approaches are needed to study attachment systems across cultures and species, and to seek cross-disciplinary resources to conduct their investigations using multiple kind of methods. We firmly believe that by widening the lens, theoretically and methodologically, researchers will come to a richer and more accurate understanding of attachment, both as a universal system structuring human infants' experience (and the experience of infants in related species) and as culturally and contextually organized systems that demonstrate attachment's variation and flexibility.

Finally, our discussions included attention to real-world applications in the areas of policy and practice. This discussion is presented separately in Chapter 13 (this volume).

9

Neural Consequences of Infant Attachment

Margaret A. Sheridan and Kim A. Bard

Abstract

Typical studies of the impact of the quality and presence of attachment relationships on child development have focused on the child's safe-base behavior. In terms of neurobiology, this has primarily led to investigations of the child's control over negative affect. In nonhuman primates, early investigations into the neurobiological consequences of attachment used models where attachment relationships were absent or severely curtailed. Institutionalization of infants, a common practice, mirrors these early primate studies since attachment relationships are limited or absent. These investigations are based on models of disruptions in attachment and used here to illustrate the impact of attachment relationships on two neural systems not typically considered: the neural substrates of reward learning and the neural substrates supporting complex cognitive function such as executive function. While attachment is central to the development of negative affect regulation, it is argued that the context in which the brain develops can also serve as an additional focus of early attachment relationships. This offers insight into the multiple functions served by attachment, and thus the role it plays in the development of other neural systems.

Introduction

In this chapter, we review evidence of the neurodevelopmental consequences of attachment. We discuss the neural mechanisms that support the emotionally positive bond between caregivers and infants, and the neural mechanisms which support the regulation of infant distress. We suggest that these two components act together in support of the homeostatic functioning of the secure base phenomenon. According to traditional attachment theory, this allows the infant to explore while in the presence of the attachment figure, and to seek safety in close contact with the caregiver when distressed (Ainsworth

1985). The emotional bond and the regulation of negative emotion functions develop from birth. Thus, attachment relationships are a constant source of experiences which are likely to shape neurodevelopment during infancy, i.e., periods of peak developmental plasticity (Greenough et al. 1987; Fox et al. 2010; Nelson and Sheridan 2011). Attachment theory focuses primarily on one function of attachment (i.e., regulation of distress). We propose that the neural correlates of attachment also include positive emotional bonds, which undoubtedly are present earlier in development and influence neurodevelopmental outcomes. These attributes of attachment relationships give them the potential to influence neural development profoundly and thus impact developmental outcomes.

Much has been written about the importance of attachment relationships for many developmental outcomes but, to date, relatively little is known about the impact of attachment relationships on neural development. Moreover, this relative paucity of information neglects due consideration of cross-cultural perspectives of attachment, including attachment as it naturally occurs in well-functioning environments. Much of the information we have about the impact of attachment on neural development relies on studies of individuals raised in grossly impoverished settings and resultant groups of people with dysfunctional attachments. This means that much of our knowledge is about neural consequences of the lack of an adequately or well-functioning attachment system. Secondarily, most of our knowledge about attachment comes from cultural settings (western, educated, industrialized, rich and democratic settings, or WEIRD; Henrich et al. 2010) where monotropic attachment is the stated norm. While Western readers may find this characterization of attachment relationships familiar, numerous sources indicate that in other settings, attachment networks comprised of multiple attachments (between the infant and a number of important caregivers) are normative (see Keller and Chaudhary as well as Morelli et al., this volume). After reviewing the extant literature, we will advance some suggestions on how current findings might apply to infants with fully functioning attachment systems, including those with an attachment network, rather than a monotropic attachment.

In addition to the issues regarding the sample, there are other issues to consider in the ascertainment of neural substrates and neural pathways of attachment. We distinguish attachment from general mother-infant bonding, and thus exclude much of the rodent work that focuses on the neural mechanisms involved in the basic mammalian mother-infant bond (e.g., Moriceau and Sullivan 2005). As many chapters in this volume attest, attachment differs from and is much more than the initial emotional bond. Most definitions focus on the function served by attachment figures, such as aids for the regulation of negative emotions or for general psychobiological regulation, although new proposals also include aids for privileged access to the social world (see Chapter 8, this volume).

Importantly, relatively little is known about the neural consequences of variations in attachment (e.g., individual differences in security). In part, this is because such individual differences would need to be profound and permanent, and need to be found in well-studied and understood neural systems in order to be observable using current neuroimaging methods (but see Serra et al. 2015). Interestingly, individual differences in maternal caregiving do appear to be accompanied by differences in neural functioning. For example, maternal styles of high interactive responsivity with three- to five-month-old infants (vs. a maternal style of high intrusiveness) were related to significant differences in neural function in support of reward and stress-related action (Atzil et al. 2011). We do not know how maternal styles map onto individual differences in attachment. Using neuroimaging techniques such as MRI with infants is difficult, primarily because of their inability to stay still, lie by themselves in a scanner while awake, or follow directions. To address these difficulties, infants only participate in MRI or fMRI studies while asleep or sedated. Thus, neuroscientists tend to rely on studying the neural substrates of attachment that result from profound disruptions of attachment relationships, such as the absence of any primary caregivers, the presence of maltreatment (neglect or abuse), or disrupted caregiving.

The use of these indirect measures to assess the impact of attachment on neurodevelopment is justified because we know that these adverse early experiences often result in disorders of attachment, including reactive attachment disorder and disinhibited social engagement disorder, or indiscriminate friendliness (Zeanah and Gleason 2015). Reactive attachment disorder is a disorder of attachment characterized by a lack of developmentally appropriate attachment behaviors, including the failure to seek comfort from a caregiver. This disorder is commonly accompanied by disruptions both in emotion regulation and positive affect. Disinhibited social engagement disorder is a disorder characterized by a lack of specificity in attachment behaviors, which includes children exhibiting overly familiar or intimate behaviors with unfamiliar adults. The tight links between severe disruptions in early caregiving environments and disorders of attachment allow neuroscientists to study neurodevelopment in these populations, to gain insight into neural substrates of attachment.

It is not enough to ask whether variation in attachment impacts neural development. We must also investigate the ways in which variation in attachment shapes neural development. Neuroscientists wish to identify the pathways and developmental processes through which attachment, and variations in attachment, impact neural structure and function. Delineating these processes is useful in part to further our knowledge, but also to facilitate the creation of targets for both remedial interventions and prevention of negative outcomes. Ultimately, as a consequence of identifying these targets for intervention and prevention, all children should be able to experience the benefits of satisfying, early caregiving relationships.

In this chapter, we review the evidence that describes the processes and pathways by which severely disrupted attachment relationships impact neurodevelopment. We explore the implications of these findings for understanding the neural foundations for well-functioning attachment, hint at what might be found in studies across cultural contexts, and present our ideas for future directions in the neuroscience of attachment.

Institutional Care: A Model of Disruptions in Early Attachment

Some aspects of the importance of attachment relationships to neural development may be studied by examining instances where children have no primary caregivers as a result of growing up in settings of institutional care. Institutional care, as we use the term, refers to very poor quality care as a function of rotating and overworked caregivers and a high ratio of infants to caregivers. Institutional caregivers are neither consistently present nor frequently able (or willing) to interact positively with infants in activities unrelated to health concerns. Such disrupted caregiving creates a context in which children are unable to form close relationships with any specific adult figure, or even a set of adult figures, and where caregivers are not able (or willing) to form a special bond with a specific child. Although the physical environment of most orphanages is most decidedly not stimulating, it is not consistently unsafe, lacking in nutrition, or without access to medical care. Thus, researchers have concluded that psychosocial deprivation and thus poor or absent early caregiving is the primary adversity to which these children are exposed (McCall et al. 2016). The poor quality of caregiving is measurable both in frequency and quality; in institutionalized settings, significantly fewer child-caretaker interactions occur and these are significantly lower in quality compared to interactions within families (Smyke et al. 2007). Additionally, quality of attachment is higher in children who live outside institutions (Smyke et al. 2010). Although it would be difficult to support the idea that all neural deficits found in children from institutionalized settings can be attributed to a lack of attachment figures, we can support the claims that institutionalized settings are primarily characterized by psychosocial deprivation for infants and young children, in the absence of other forms of adversity (e.g., inadequate nutrition, poor medical care, physical and sexual abuse).

Beginning as early as 1975, many studies have examined the impact of institutionalization on child development. Tizard and Rees (1975), for example, studied children from London orphanages. An issue that arises in studying eventual outcomes is that the reason why any infant has been placed into the institution is often not known. Some infants are placed in orphanages due to illness, failure to thrive, or other perceived deficiencies, whereas others are placed for political reasons. The Bucharest Early Intervention Project (BEIP) is a longitudinal study of a sample of children raised from early infancy in institutions in Bucharest, Romania. BEIP was initiated at the request of the Secretary of State for Child Protection in Romania. Its major advantage over other studies of children from orphanages is that assignment to foster care, as an alternative to institutional rearing, was randomized. Thus, these outcomes are most precisely related to the interventions since interventions were randomly assigned. All study procedures were approved by the local commissions on child protection in Bucharest, the Romanian Ministry of Health, and the institutional review boards of the home institutions of the three principal investigators (Zeanah et al. 2006; Miller 2009). Studies using BEIP, therefore, provide the best available evidence for a causal relationship between lack of an attachment figure during infancy and early toddlerhood and disrupted neurodevelopment.

In BEIP, a sample of 136 children (aged 6–30 months) was recruited from each of the six institutions for young children in Bucharest. An age-matched sample of 72 community-reared children was recruited from pediatric clinics in Bucharest and comprised the never-institutionalized group. Half of the children initially raised in institutional care in Bucharest, Romania, were randomly assigned to high-quality foster care (Smyke et al. 2009) with a primary caregiver. The other half was assigned to care as usual in the institution, with several infrequently available and rotating caregivers and no primary attachment figure (Smyke et al. 2007).

Given the circumstances in which these children were raised, it is not surprising that the BEIP has provided clear evidence for disrupted attachment relationships resulting from institutional care. Children in the foster care group and care-as-usual institutionalized group exhibited increased rates of reactive attachment disorders and indiscriminate friendliness compared to the never-institutionalized group (Zeanah et al. 2005; Gleason et al. 2014). However, placement into foster care earlier than 24 months of age decreased rates of attachment disorder in the foster care group compared to the care-as-usual institutionalized group (Smyke et al. 2010), indicating that the compromised early care in institutions led to the observed disruptions in attachment. In sum, early postnatal exposure to institutionalization can be used to model the impact of a lack of early attachment on neurodevelopment.

Neurocognitive Impact of Disruptions in Early Caregiving

Several neurocognitive domains have been identified as being susceptible to disruptions in early caregiving. These include the neural bases for emotion expression (both positive and negative), emotion regulation, and neural substrates across multiple cognitive domains. In this section we review findings

and explore implications of extending our knowledge about the neural correlates of disruptions attachment across these various domains.

Lack of Early Caregiving Relationships Disrupts
Emotional Control over Negative Affect

There is robust evidence that early exposure to institutional care causes disruptions in (a) emotional reactivity and (b) control over negative affect. The
lack of an attachment relationship in infancy and early toddlerhood, caused
by institutionalization, is strongly related to pathological disruptions in negative affect (Zeanah et al. 2009) and leads to elevated rates of depression and
anxiety. In addition, this exposure is related to the development of neural systems which support emotion regulation and reactivity. Specifically, lack of
early attachment relationships that result from institutionalization has been
linked with medial prefrontal cortex function, amygdala volume and reactivity, as well as the quality and extent of connectivity between the amygdala
and medial prefrontal control regions (Mehta et al. 2009a; Tottenham et al.
2010, 2011). Relatedly, early exposure to institutionalization causes a blunted
stress response to interpersonal stress and rejection. In BEIP, children from
the care-as-usual institutionalized group, compared to foster care and never-
institutionalized children, showed blunted sympathetic and hypothalamic-pituitary-adrenal (HPA) axis responses to a laboratory Trier social stress test and
a social rejection paradigm (McLaughlin et al. 2015). It has been proposed that
the lack of early attachment figures may speed neurodevelopment of negative
affect regulatory systems, shifting limbic/prefrontal connectivity so that the
neural correlates of emotional reactivity and control over emotional responses
look more "adult like" in children exposed to institutionalization at younger
ages (Ganzel et al. 2013; Gee et al. 2013; Tottenham 2014). Supporting this
theoretical model, evidence from BEIP indicates that disruptions in attachment relationships moderate the impact of institutionalization on pathological
disruptions in regulation of negative affect. Specifically, for children in the
BEIP study who were randomly assigned to foster care, improvements in their
attachment relationships (self-reported and reported by their foster parent)
moderated the association between institutionalization and psychopathology
(McLaughlin et al. 2011; Humphreys et al. 2015).

How Does Lack of Early Attachment Lead to
Disruptions in Control over Negative Affect?

Early in development, young infants lack the ability to regulate their own emotions, and attachment relationships play a role in providing emotion regulation. Attachment relationships also play a role in scaffolding their developing
abilities to regulate emotion, especially negative emotion (Ainsworth 1985;

Morton and Browne 1998; Zeanah and Gleason 2015; Chapters 6 and 8, this volume). Indeed, one of the core criteria of secure attachments, as assessed in the Strange Situation Procedure, is seeking proximity in the face of stress and maintaining contact with the attachment figure(s) until the distress is resolved (i.e., securely attached children use primary caregivers as a source of emotion regulation). Even in cultures in which infants have multiple attachment figures, each caregiver engages with distressed infants and contact is maintained until crying and fussing are successfully reduced (e.g., Meehan and Hawks 2013). In some cultures (e.g., when infants are often in cradles or slings), this regulatory function of attachment appears to be so well developed that infants rarely appear distressed (Gaskins 2013).

Given this central role of attachment figures in modulating negative emotions during infancy, it is perhaps not surprising that infants who have not experienced a primary attachment relationship have deficits in emotion regulation, which later results in psychopathology as children and adolescents. Infancy is a period of peak neural plasticity and is thus highly responsive to environmental inputs (Fox et al. 2010). The neural and physiological systems that regulate negative affect are likely "tuned" by early caregiving experiences. If infants do not have external sources of emotion regulation to calm them during infancy, these systems may develop in aberrant ways. For example, infants may be in a constant state of distress with an inability to downregulate their distress, or infants may develop a systematic unresponsiveness to distress which may develop as a "too mature" response, as described above. Conceptually, infants could be thought to be spending much of their time in a context of danger. In this way, the lack of an attachment figure in infancy and early toddlerhood is an environmental signal that lets the developing neural system "know" what the future is likely to hold. The resultant mismatch between the early developmental context with no attachment relationships and future experiences of relatively safe environments in middle childhood and adolescence may affect emotion regulation and stress physiology in ways that can be understood as pathological.

Lack of Early Caregiving Relationships Disrupts Processing of Positive Affect

In the English and Romanian Adoptees study (Mehta et al. 2009b), exposure to institutionalization during early childhood was found to be associated with blunted striatal activation during reward anticipation. Reward anticipation is measured using neuroimaging during the monetary incentive delay task. This task links a previously neutral stimulus (e.g., a circle) with reward (e.g., winning money) by iteratively pairing responses to this stimulus with a reward over time (Knutson et al. 2001). Neural activation in response to reward anticipation is indexed by measuring responses in the brain to the stimulus which predicts reward, before any reward has been administered. This finding suggests that disruptions in early caregiving may affect neural circuitry involved in reward processing and in learning to anticipate reward. In typically developing children and adolescents, reward anticipation is associated with increased activation of the ventral striatum, ventral medial prefrontal cortex, and dorsal anterior cingulate cortex (Knutson et al. 2003; Haber and Knutson 2010). The ventral striatum is activated during reward receipt, reward learning, and processing of secondary reward stimuli (e.g., happy faces) across numerous studies in humans and animals (for reviews, see Schultz et al. 1997; Haber and Knutson 2010). Similar to findings from the English and Romanian Adoptees study, neural activation was assessed using fMRI while adolescents and children viewed happy and fearful faces in another sample of participants exposed to institutionalization during infancy. In this study, when happy faces were viewed, the ventral striatum was less activated in adolescents exposed to institutionalization early in life compared with age-matched controls (Goff et al. 2013). The degree of ventral striatal activation in this study was also associated with symptoms of depression. Finally, in BEIP, adolescents in the care-as-usual group showed a reduced behavioral response to monetary reward, relative to adolescents in the foster care group and never-institutionalized group in the monetary incentive delay task. This suggests that it is the lack of an attachment relationship in childhood—a species-expected caregiving experience—that leads to a disruption in reward processing (Sheridan et al., under review).

Other forms of disrupted early caregiving (including emotional neglect, cumulative adversity, and maltreatment) have also been linked with blunted neural and behavioral responses to reward as measured by the monetary incentive delay task and similar computerized tasks (Guyer et al. 2006; Dillon et al. 2009; Pechtel and Pizzagalli 2013; Hanson et al. 2015a, b). Taken together, existing evidence indicates that disruptions in early caregiving may confer risk for dysfunction in basic learning mechanisms around reward which ultimately support healthy mood function.

How Does Lack of Early Attachment Lead to Disruptions in Control over Positive Affect?

Currently there is no consensus as to the mechanism underlying the association between early caregiving and reward processing (for reviews, see Pechtel and Pizzagalli 2011; Goff and Tottenham 2015). Considering aspects of attachment relationships may, however, shed some light on potential mechanisms. As we mentioned in the introduction, attachment relationships appear to have at least two core properties: secure base and positive emotional bonds. While the "safe base" behavior is the defining feature of attachment relationships, as measured in the Strange Situation Procedure, the positive emotional bond is a defining feature in more naturalistic contexts, and may be the core emotional aspect of the developing system of trust (Keller 2013a; Gaskins 2014). Attachment relationships play several roles in early child development. In addition to providing a source of regulation over negative emotions, the attachment figure also provides initial instances of reward learning. In WEIRD settings, a primary task for a child is to elicit caretaking behaviors, such as provision of food and comfort, through the use of vocalizations and behaviors. An attuned caregiver will use infant's hunger signals to guide their behavior to provide food. Through this process, the child of an attuned caregiver will learn that some behaviors will elicit reward (e.g., food) and will learn to perform these more readily, particularly when hungry. This is just one example, in particular cultural settings, of the manner in which infant behaviors with a caregiver may be linked with reward. In other cultures where constant physical contact is the norm, there is a more immediate pairing of the provision of comfort and food with attachment figures. In the case of severely disrupted or absent attachment in early childhood, it is likely that the infant has requested and received fewer rewards and less clear learning opportunities, because fewer rewards and interaction opportunities were available from which to learn. If reward (e.g., food, comforting, play) occurs randomly with respect to their behavior, infants will not form a strong association between their attachment figures and positive reinforcement. For the institutional infant, it is possible that the neural circuits which underlie reward learning (connectivity between dopamine-rich sites in the prefrontal cortex and ventral striatum) will have less "practice" during this period of peak developmental plasticity, thus resulting in the observed long-term disruptions in both reward learning and activation of the ventral striatum in the context of reward.

Lack of Early Caregiving Relationships Results in Global Cognitive Deficits

Children exposed to institutional care in early childhood exhibit clear neurocognitive deficits. Exposure to institutional care shows reductions in IQ, which can partially be remediated following randomization to foster care (Nelson et al. 2007). This suggests that the presence of an attached early caregiver is important for intellectual development. Exposure to institutional care is also associated with general cognitive deficits, including deficits in executive function, which is defined as the ability to hold in mind rules and ideas no longer present in the environment and to inhibit immediate responses (Bos et al. 2009; Beckett et al. 2010). Relatedly, disrupted early caregiving following institutionalization is associated with disruptions in attention and impulsivity (Zeanah et al. 2009), which is the behavioral manifestation of poor executive function (Tibu et al. 2016). Finally, early institutional care is associated with global reductions in cortical volume, neural function, and cortical thickness across studies (Chugani et al. 2001; Vanderwert et al. 2010; Sheridan et al. 2012a). Importantly, these general reductions in cortical thickness and neural function statistically mediate the association between institutionalization and behavioral manifestations of attention and impulsivity, indicating that it is because of the reduced thickness that attention and impulsivity are elevated (McLaughlin et al. 2010, 2013).

How Does Lack of Early Attachment Lead to Disruptions in Global Cognitive Function?

Elsewhere, it has been argued that this impact of institutionalization on cognitive development results from deprivation in rich cognitive stimulation during early childhood (McLaughlin et al. 2014; Sheridan and McLaughlin 2014). Studies of rodents show that a lack of cognitive stimulation will increase synaptic pruning processes, and thus rodents exposed to very low levels of cognitive stimulation show overall reductions in cortical volume (Diamond et al. 1972). Here we posit that because attachment relationships are the primary source of stimulation in infancy and early toddlerhood, the lack of these relationships are likely to result in an unstimulating environment, which also increases synaptic pruning globally throughout the brain. Pruning is the mechanism by which many environmental childhood experiences (e.g., phonemic retention in the context of multiple language exposure, visual cortex organization) impact neural development (Wiesel and Hubel 1965; Hensch 2005; Morishita and Hensch 2008). The general reductions in cortical thickness and volume observed following institutionalization are likely to yield deficits in higher-order cognitive functions because these functions require coordinated activation of multiple areas of association cortex (e.g., prefrontal and parietal cortex) and rely on late-developing areas of the brain such as the prefrontal cortex.

The context of attachment relationships is the first in which a child can expect to receive cognitive stimulation. In the Western caregiving systems, this is through contingent vocalization and face-to-face play. In other cultures, social, visual, and vestibular stimulation can occur via exposure to multiple sensory and social environments experienced through carrying. In Mayan cultures, for example, the infant might be held by an older child, and together they may sit next to the mother, surrounded by numerous others, while the mother engages in social exchange and prepares food (Morelli et al., this volume). When attachment relationships are disrupted or absent, this cognitive input is grossly reduced. As a result, the typical developmental process of synaptic pruning, which creates the most efficient neural system possible given particular early inputs, may prune connections relating attachment figures with rewarding events in the world. Given that the institutional setting has a reduced sensory environment and a lack of an early attachment figure, the pruning process may tune brain function with this impoverished setting (Greenough et al. 1987; Fox et al. 2010). Unfortunately, large differences are still observable as late as 8 years of age in children raised in institutions versus those raised in families, indicating that these processes may be difficult to reverse (Sheridan et al. 2012a).

Multiple Roles of the Early Attachment Relationship in Shaping Development

Currently the most common manner of assessing attachment style is through the Strange Situation Procedure. In this experimental context, infants use their primary caregiver as a source of regulation over emotions generated by the novelty of the unfamiliar setting and stranger, as well as by separation. Usually this is interpreted as fear, or wariness elicited by novelty, and distress elicited by separation from the attachment figure. The Strange Situation was meant to mimic everyday situations as well as to represent a potentially dangerous situation—being left alone in an unfamiliar environment (Ainsworth 1985). Although this approach has been valuable in linking quality of attachment to child development outcomes in Western or urban cultures, it has focused attention primarily on regulation of negative emotion as outcomes. Here we review literature which documents that a lack of early attachment results in disruptions in neural development, in part in neural systems that support emotion regulation (e.g., the amygdala and ventral medial prefrontal cortex). These types of disruptions in neural development of emotion regulation systems are consistent with our understanding that early caregiving is important in shaping infant learning about the regulation of negative affect.

There are, however, two additional prominent neurocognitive deficits that result from institutional care: problems associated with reward learning and global deficits in cognitive function. These difficulties, which result from a lack of attachment in early childhood, involve neural structures and functions different from those that account for regulation of emotion. Specifically, reward response and anticipation are supported by the functioning of the striatum and dorsal medial prefrontal cortex, whereas complex cognitive function is supported by coordinated activity across association cortex, including lateral prefrontal and superior parietal cortex. In addition, aspects of attachment relationships with a primary caregiver which likely support development of these neural and cognitive functions are proposed here. Evidence from studies of institutionalization indicate that the lack of an early attachment relationship results in severe deficits in cognitive and emotional function, as well as in the neural structures that underlie these functions. The various impacts of disrupted caregiving likely transpire through multiple pathways, representing the multiple important functions of the early caregiving relationship (for further discussion of the functions of attachment, viewed from cross-cultural and cross-species perspectives, see Chapter 8, this volume).

Future Directions

In this review, we have focused on extreme exposures characterized by an almost complete absence of attachment relationships. In at least one study, however, randomly assigned interventions allowed causal inference (Zeanah et al. 2003; Nelson et al. 2007). Unfortunately, access to natural or actual experiments in rearing is limited in human studies. Most investigations of early exposures and subsequent outcomes in humans are correlational in nature and rely on naturally occurring variation in rearing environments (e.g., neglect, abuse). Because natural variation of this kind is unfortunately common in humans, these types of studies are easier to perform and should be viewed as important complements to studies with actual exogenous variation in attachment environments. For example, correlational studies could more carefully delineate and describe observations made in studies with random assignment. Equally important, the number of studies in humans where random assignment is used should increase. While it is unethical to assign children randomly to negative early environments, treatment studies where children are randomly assigned to treatments in which these environments are ameliorated (e.g., parenting interventions) are possible and increasingly common (e.g., Bernard et al. 2012; Caron et al. 2016). Unfortunately, evaluations of these interventions are unlikely to focus on the neural functioning of the child. Future work that evaluates the neurobiological consequences of early parenting interventions could contribute meaningfully to our understanding of the impact of early experience on neural structure and function.

In addition, there is a large body of work in nonhuman primates which can address gaps in current understanding of the effects of early rearing deficits on neurobiology due to a lack of experimental evidence. Indeed, since the very early days of attachment research, primate models have been informative (Suomi et al. 2008). This research often mirrors and further develops the work reviewed above, as primate models of attachment disruptions generally involve random assignment to peer or nursery rearing: a total lack of maternal care. There are many similarities in the downstream consequences of peer rearing and institutionalization. For example, both forms of early maternal deprivation disrupt stress physiology and regulation (Dettmer et al. 2012; McLaughlin et al. 2015). However, the impact of these two experiences on neural structure may differ; indeed, they may go in opposite directions (Spinelli et al. 2009; Sheridan et al. 2012a). The diversity of findings between humans and primates exposed to superficially similar experiences points to the importance of carefully considering species-specific effects and exact nature of the exposures. Importantly, studies in nonhuman primates, with greater access to random assignment, can be used to test specific theories from the human literature about the effect of variation in early caregiving on long-term neurobiological outcomes. Increased collaboration between researchers investigating early rearing exposures in animal and human models has the potential to accelerate our understanding of the impact of attachment disruption on the developing brain.

Finally, we wish to stress the importance of expanding consideration of attachment beyond the mother-infant dyad to include attachment networks (Keller and Chaudhary, this volume). In addition, the proposed functions of attachment need to expand beyond the regulation of negative emotion to include introduction to the world (Chapter 8, this volume). Our consideration of the neural consequences of attachment, or lack thereof, provides strong support for the involvement of two neural mechanisms: one that underpins regulation of negative emotion and another that underpins reward anticipation and reward response. Although most of the evidence is conceptualized from a framework of monotropic attachment, we have purposively indicated that the evidence equally supports the interpretation that an infant's neural substrate "expects" experiences which involve both multiple functions of attachment and an attachment network, in the sense of experience-expectant and experience-dependent processes (Greenough et al. 1987) and probabilistic epigenesis (Gottlieb 2007). Further research is needed (e.g., new treatments for institutionalized children) to test the neural foundations and consequences of attachment, as reconceptualized through this volume.

10

Neural Foundations of Variability in Attachment

Allyson J. Bennett, William D. Hopkins,
Ruth Feldman, Valeria Gazzola, Jay Giedd,
Michael E. Lamb, Dirk Scheele, Margaret A. Sheridan,
Stephen J. Suomi, Akemi Tomoda, and Nim Tottenham

Abstract

Neuroscience offers insight into processes that support the development of the social brain within the cultural contexts that permit attachment relationships to form. Both human and nonhuman animal studies are critical to inform theory development and hypothesis testing via descriptive and experimental studies. A scientifically valid evolutionary theory is necessary to account for the remarkable diversity of parenting systems across human and many nonhuman animals. This chapter examines the neural foundations of attachment and poses critical questions that relate to the initiation of this relationship: How does attachment interface with brain development? What is the interplay between attachment and brain development (including elements of bidirectionality)? Are there negative consequences associated with variation in attachment, and are they reversible? Rather than conceptualizing attachment in terms of a single type of relationship, or a rigid developmental channel, this chapter proposes that an expanded consideration of variation is necessary to understand the neural foundations of infant-caregiver relationships, and the role of those relationships in developing competence across the life span. This approach will permit identification of common neurobiological elements of attachment as well as the remarkable plasticity and diversity within and across individuals, cultures, and species.

Group photos (top left to bottom right) Allyson Bennett, William Hopkins, Margaret Sheridan, Ruth Feldman, Michael Lamb, Jay Giedd, Valeria Gazzola, Stephen Suomi, Nim Tottenham, Akemi Tomoda, Margaret Sheridan, Dirk Scheele, Michael Lamb, Allyson Bennett, Jay Giedd, Nim Tottenham, Valeria Gazzola and Dirk Scheele, Akemi Tomoda, Ruth Feldman, Stephen Suomi, William Hopkins,

Introduction

Neuroscience offers insight into processes that support the development of the social brain within the cultural contexts that permit attachment relationships to form. Both human and nonhuman animal studies are critical to inform theory development and hypothesis testing via descriptive and experimental studies. Understanding the neural foundations of attachment relationships can extend insights beyond the human case. In turn, identifying core cross-species similarities in the neural processes involved in the initiation and maintenance of attachment relationships provides the foundation to better understand individual, cultural, and species variability, both in the development and the outcome of attachment relationships. A scientifically valid evolutionary theory is necessary to account for the remarkable diversity of parenting systems across human and many nonhuman animals. Neuroscience can enrich our understanding of the biological mechanisms that support plasticity in attachment formation and outcomes.

A number of guiding questions shaped our consideration of the neural foundations of attachment. How attachment interfaces with brain development was a primary focus, in part because the infant-caregiver(s) relationship occurs at a period in which neural systems undergo major maturational changes. At the same time, the interplay between attachment and brain development (including elements of bidirectionality) is necessary to frame other research questions and to form testable hypotheses and theory. For instance, uncovering shared and unique neural systems across species and cultures is productively informed by identifying how neural development supports, and is affected by, attachment relationships. This information is requisite to address another major question of relevance to healthy development: If there are negative consequences associated with variation in attachment, are they reversible? Finally, understanding the degree to which the neural circuitry underlying infant-caregiver attachment is the same or different from the neural circuitry underlying other relationships (e.g., pair bonds) is important for a number of reasons. Better understanding of the similarities and differences in neural circuitry involved in different types of relationships is needed to identify the neural processes involved in relationships across an individual's life span, across cultures, and across species. This is particularly true given the differences in infant-caregiver arrangements across the nonhuman animal species that provide important avenues for experimental research aimed at identifying neural foundations of attachment.

We begin with a discussion of attachment in the context of studying neural development. The themes selected highlight both advances in understanding of attachment as well as ongoing challenges related to the range of neural plasticity that supports diverse systems for infant caregiving. We briefly highlight what is known and what remains unknown with respect to the neural processes involved in infant-caregiver relationships, and propose a conceptual framework to guide further consideration of how current evidence provides a foundation for new avenues of research.

Attachment: Neural Development across Cultures and Species

Social relationships occur in diverse configurations across human cultures and different species: no single definitive type of social relationship exists in human and nonhuman animals (see Hawkes et al. and Morelli et al., this volume). While these relationships may emerge from, and indeed depend on, some of the same neurodevelopmental processes and neural circuitry, early relationships between infants and their caregivers vary in a number of key aspects. Our primary focus is on attachment; however information specific to attachment is not always available. Thus, studies of social relationships, considered more broadly, play a key role in developing our knowledge about the neural contributors to attachment.

With respect to differentiating attachment relationships from others, the features that are most readily apparent are primacy and dependency. From the perspective of considering the neural foundations of attachment, the primacy of the relationship differentiates it from other relationships in a number of critically important ways. As a result of developmental timing (i.e., it occurs during a period of rapid neural maturation and integration of neural, behavioral, physiological, and other systems), attachment is fundamentally tied to survival and early competencies during infancy as well as to competence in subsequent life periods. In other words, given the time course of neural development, relationships during infancy are uniquely positioned to affect developmental trajectories for perceptual, motor, emotional, social, and cognitive health.

Asymmetrically dependent social relationships (i.e., those between the infant and the caregiver) occur for infants of many species. For some, the infant-caregiver(s) relationships are characterized in a specific way as attachments. We interpret this to mean that there is a degree of specificity to these relationships: each individual has a relationship with a *particular* other—or *particular* individuals—and the relationship is reciprocated by the other (or others). Whereas this distinction and definition of reciprocity may not be equivalent across human cultures, it is one that can productively organize comparative analyses and considerations of the usefulness of nonhuman animal models. Figure 10.1 illustrates our views of attachment as found across animals, highlighting differences between species both in terms of reciprocity and specificity. As illustrated, in some rodents, reciprocated specificity is not evident, such that the participants are, for example, specifically rewarded by the other's presence (as opposed to that of any other) and are not specifically concerned about the other's absence. By contrast, relationships in chimpanzees and in

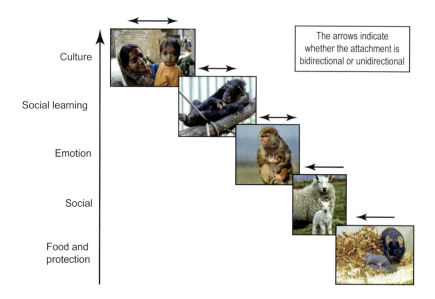

Figure 10.1 Illustration of the bidirectionality of attachment relationships across different model species that are often compared in neurobiological studies relevant to attachment. Complexity of functions and domains involved in the infant-caregiver(s) relationship is shown as increasing across phyla. Photo credits: culture, Nandita Chaudhary; social learning, Magnus Johansson; emotion, ©Kathy West Studios; social, Donald Macleod; food and protection, Indiana University.

humans can—but may not always—be characterized by both bidirectionality and specificity.

Figure 10.1 also highlights the increasing complexity of functions that play a role in, or may be crucial to, such relationships. At the most basic level, this includes the provision of food and protection and extends, with increasing complexity, to incorporate socioemotional, cognitive (social learning), and cultural domains. Infant-caregiver relationships—particularly for the infant—may exhibit common features (including common neural foundations) across different species. Social relationships, including attachments, have multiple components that include perceptual, behavioral, emotional, cognitive, and higher-order representational features that are subject to variation at the level of the individual, culture, and species. Attachment in nonhuman primates, for example, may have a high degree of overlap with humans in terms of many behavioral and emotional processes, but differ in aspects of cognitive representation. In addition, infant-caregiver relationships among mammals may share behavioral features and underlying neural processes, but differ in emotional and cognitive components compared with primates. The core similarities provide the critical foundation for studies aimed at better understanding the neural foundations of attachment, as well as the variation in attachment that occurs across individuals, cultures, and species. As we aim for this, however, we must remember that these differences exist and are of central importance.

Conceptual Framework: Variation and Neural Development

To facilitate better understanding of the neural processes involved in attachment and identify how variation in attachment affects neural development, we put forth the following conceptual framework. As illustrated in Figure 10.2, attachment relationships are initiated over the course of early development as infant-caregiver(s) relationship(s) develop. The initiation is conceptualized as a process separable from the maintenance of the relationship(s) that unfolds over time, across the entire life span. The time course of development is represented across the horizontal area of the figure, on to which the range in variation of outcomes is overlaid. Outcomes are represented as a spectrum related to behavioral, social, emotional, and cognitive "competence" and the associated degree of functionality of the neural systems that support those processes. As illustrated, the model conceptualizes a wide range of variability in which competent functioning may occur, with an optimal level that is defined within the context of the individual's environment and culture (e.g., Keller and Chaudhary, this volume).

The model also illustrates a lower zone in which competence and outcomes are compromised, again within the context of the individual's environment and culture. Finally, the model explicitly includes recognition of plasticity (as

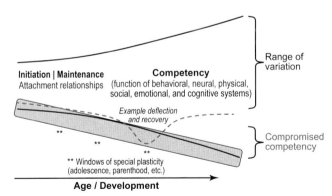

Figure 10.2   Conceptual framework showing the relationship between early attachment and development of competence in a range of systems critical to individual functioning. Variation in outcomes and development is shown as a vertical range; the lower limit represents compromised competency, which significantly impedes the individual. Plasticity across development and following the early attachment relationships is depicted as a dotted line to show deflection and recovery. This underscores an anti-determinist perspective and acknowledges that development of competence can follow suboptimal attachment relationships. Finally, the model incorporates windows of special plasticity (**) throughout development (e.g., adolescence and parenthood).

opposed to determinism). Thus, there is potential for variability in development, across the life span, even following relative similarity in the early attachment network. Moreover, there are "windows of special plasticity," corresponding to developmental points in time, in which neural plasticity is high (e.g., during infancy, adolescence, parenthood).

The development of competence is not specified in terms of specific outcomes or systems; these could include neural, behavioral, emotional, physical, and cognitive outcomes as well as the interplay between them. Our framework does, however, emphasize the area under the threshold: whereas there can be a great deal of variation in competence that remains in the functional range, there is both a relatively smaller range and significant concern for outcomes that have substantial and persistent adverse effects on individual's ability to function competently (including their survival). An individual's trajectory may vary, however, as a function of unstable, disrupted, or absent infant-caregiver relationships, with the predicted trajectory represented in terms of higher likelihood of challenge to attain competency.

For an individual organism, the developmental trajectory is influenced by (a) previous developmental outcomes, (b) environmental and experiential events and conditions, (c) genetic factors, and (d) resulting interactions (Gottlieb 2007). An interaction occurs because an individual's own state, development, and previous experiences all play a role in shaping an individual's experience with a subsequent environment or experience: two individuals can experience the same event and/or environment differently. Plasticity is represented in this model as potential for deflection from a developmental trajectory. In the case of risk, deflection is a movement away from competence. In the case of recovery or resilience, movement is toward greater competence and includes recovery from pathology or compromised function.

A core principle of our framework is the idea that the central functions of different developmental phases—reflected in behavioral, reproductive, affective, and cognitive changes—have a neural basis and can bidirectionally affect their neural substrates. From the neural perspective, therefore, it is particularly important that the developmental trajectory contains "windows of potential" for especially high plasticity (e.g., Dahl 2004). These windows may, for example, be transition periods that correspond to developmental stages known to show high levels of change: infancy, adolescence, and perhaps parenthood.

The framework is inclusive of variation across cultures, populations, and species with respect to typical maturational timelines and parameters for well-being. What is constant is the initiation of the attachment relationship(s), or the infant-caregiver relationship(s), during infancy and the maintenance of that relationship network during the critical infant and juvenile developmental period. Moreover, the framework allows for variation in the configuration of infant-caregiver relationships. This variation is critical as it provides a way to encompass the range of infant caregiving practices for humans and is also inclusive of the range of infant care practices across the many species that may be studied to inform attachment research.

"Flexible Glue" Model

The "flexible glue" model offers a way to describe the initial process of attachment (initiation) and the subsequent maintenance of the relationship. Here, attachment is conceptualized as a flexible glue or bond between an infant and caregiver(s). The glue is strengthened, or becomes more stable over time, but the glue need not be permanent and is subject to change across the life span. This glue, or the bond that initiates attachment, is essential to infant survival. It forms during a period of neural development which involves multiple systems central to social relationships, including regulation of affect and behavior. Then, over the course of child and adult development, attachment grows and it can change, for example, in response to disruptions.

Attachment describes the special relationship between caregiver(s) and a child. Although the unit of analysis most often used to identify an attachment relationship is the child, it is important to stress that the attachment process— that is, the linking of a child and caregiver(s)—can be observed and evaluated in both the child's and caregiver's brain. In addition, the attachment relationship impacts on the child's brain. Thus, for the purposes of considering the "*neurobiology of attachment*," we must consider: (a) the neurobiology (or core processes) within the parental brain that results in an attachment to a child; (b) the neurobiology within the child brain that results in an attachment to a caregiver or group of caregivers; and (c) the impact of that attachment relationship on neural structure and function. The latter is related to the maintenance of the attachment relationship, not the initiation of this relationship. In this way, the neurobiology of attachment is likely to be quite different as a function of which member of the attachment relationship and which stage of attachment (initiation vs. maintenance) is being discussed. Some systems will be shared across these individuals and stages whereas some systems will not.

Three sets of questions organized around the "flexible glue" model can guide enquiry into the neural foundations of attachment. The first centers on the effect of variation in caregiver(s)-infant arrangements. Here, the goal is to understand how the initiation and maintenance of relationships are affected by variation in caregiver(s)-infant arrangements and how these are represented at a neural level. This line of enquiry seeks to identify the neural foundations of the process (including what systems and substrates are required and involved in initiating and maintaining attachment relationships) and the neural outcomes which specify, for instance, the neural consequences and pathways affected by the relationships.

The second set of questions looks at neural foundations of attachment across species. For both the dependent infant and the caregiver(s), the initiation of an attachment relationship relies on reward and salience systems. Reward systems include the dopamine-modulated ventral striatal and prefrontal areas of the brain; salience systems involve the amygdala and striatum. Evidence for this has been obtained in the adult caretaker (Feldman 2015, 2017), but evidence in infants is limited to the salience network and is largely incomplete. Most research has been conducted during the maintenance phase, where impact was on function and structure of the salience and reward networks, as well as higher-order cognitive/emotional processing regions (e.g., prefrontal cortex and potentially association cortex, more generally; see Sheridan and Bard, this volume).

The third set of questions addresses the range of variation in competence and developmental change. Decades of evidence from both human and nonhuman animal studies demonstrate deleterious, wide-ranging, and persistent health outcomes associated with severe impairment (or even absence) of infant and childhood caregiver(s) attachment relationships (Felitti et al. 1998; De Bellis et al. 1999; Machado and Bachevalier 2003; Teicher et al. 2003; Gilbert et al. 2009; Tottenham and Sheridan 2010; Nelson et al. 2011; Callaghan and Tottenham 2016a). These extreme cases include maltreated and neglected children, children raised in severely impoverished orphanages, and nonhuman animal models in which offspring are reared without parental attachment which include seminal studies in rhesus monkeys (e.g., Harlow 1958; Harlow and Harlow 1965; Sackett 1965). Focus on extreme cases offers an important opportunity to address a significant human health challenge; it also can deepen our understanding of how neural and other systems are impacted when development is affected by early attachment relationships. Just as lesion studies have played a role in isolating the specific neural substrates of behavioral and cognitive functions, studies of the outcomes following absence of early attachment relationships have yielded critical insight into pathways and neural substrates involved in healthy development. To identify core systems, one approach is to analyze how widely variable relationships (within and across species) are supported by common neural systems, and whether there are some that depend upon unique neural foundations.

Less well studied is the association between early attachment relationships and the range of variation in development of competence that occupies the center of the range above pathological or deleterious and into optimal states. Pathological states have been of central interest from the perspective of high relevance to minimizing adverse outcomes and challenges to human health. By contrast, the broad swath of variability within the central range has remained relatively underexplored. This is the case both in the context of neurobiologically informed studies of outcomes from particular infant-caregiver(s) attachment arrangements, as well those within a broader cross-cultural or cross-species perspective.

State of Current Knowledge

Neural Foundations of the Attachment System

One of the benefits of neurobiology is that it can be used to identify sensitive periods to assess the impact of experience. In general, data which pinpoint the impact of deficits in the presence or quality of the attachment relationship reveal that early deficits in attachment are always more detrimental to neurobiological outcomes than later deficits. This is consistent with the fact that earlier in development a child's brain is more plastic, and therefore more vulnerable than later in life. Beyond this general principal, we lack ample evidence about when certain kinds of attachment disruptions are most detrimental. This area requires increased investigation. Here we highlight current knowledge about the neurobiology of attachment itself (i.e., the glue).

To facilitate integration of current neurobiological knowledge across both humans and nonhuman animals that serve as model systems, attachment can be represented as a multidimensional system with neural, perceptual, behavioral, and other components which can—and do—overlap with those used for other social bonds. In other words, components of the systems and circuits involved in infant-caregiver networks also subserve other social relationships, including peer, romantic, or sexual relationships. At the same time, components or processes evident in less cognitively complex species may serve as basic common modules that are elaborated upon with increases in neural, behavioral, or cognitive complexity.

We divided our consideration to the following examples of knowledge about neurobiological systems. First we considered sensory modality and oxytocin as they play roles in early bonding, or pre-attachment. Second, we considered lines of neural and physiological evidence about the initiation and maintenance of the bond that can develop into attachment relationships. Understanding early bonding is relevant to illustrate some of the likely foundations upon which initiation of attachment—the initial glue in our model—depends. It also provides a translational bridge to model systems in which infant-caregiver bonds occur but do not develop into attachment, the sustained relationships similar to those found in humans.

Sensory Modality

Research on human attachment has primarily focused on visual and auditory modalities, but the glue that binds individuals is multimodal. For instance, evidence for the importance of tactile stimulation comes from studies documenting the beneficial effects of early skin-to-skin contact between a mother and infant, particularly on increasing cardio-respiratory stability and decreasing infant crying (Moore et al. 2012). In fact, the practice of skin-to-skin contact, early suckling, or both during the first two hours after birth can positively influence mother-infant interaction one year later (Bystrova et al. 2009). Skin-to-skin contact between a mother or father and their preterm infant induces the release of oxytocin, which in turn may mediate the positive anti-stress effects and stimulate the reinforcement system (Uvnas-Moberg et al. 2014; Cong et al. 2015). Further, this contact appears to have long-term effects, as demonstrated in a randomized study which showed that skin-to-skin contact improved the mother-infant relationship and the infant's neurobiological system (respiratory sinus arrhythmia, stress response, sleep, executive functions) for up to ten years (Feldman et al. 2014). Interestingly, people spontaneously stroke their partners and their baby with velocities that are effective in stimulating C tactile fibers in the stroked recipient (Croy et al. 2016); this supports the notion of overlapping affection systems. It is important to note, however, the caveat that much of the evidence is from Western middle-class families in which the mother is the main caretaker. Thus, we do not know, for example, about oxytocin release in babies who are passed from arm to arm after birth, as in multiple caregiving among the Beng (Gottlieb 2014).

While olfaction may not be as important for parent-infant bonding in humans as in other macrosomatic mammals (Levy et al. 2004), chemosensory signals have been found to modulate very early interactions between mother and infant. For instance, natural breast odors are sufficient to attract and guide neonates to the odor source (Varendi and Porter 2001), and the presence of pleasant familiar odors increases infants' attention and reduces crying and mouthing, but increases smiling (Coffield et al. 2014). Furthermore, mothers can use odor cues to identify their infant as early as 2–6 days postpartum (Porter et al. 1983), and it has been observed that the mother's odor can elicit automatic imitation effects in children with autism (Parma et al. 2013).

Neurohormones

The neuropeptide oxytocin (OXT) has been implicated in mediating numerous prosocial effects, ranging from approach/avoidance behavior (Scheele et al. 2013) to interpersonal trust (Kosfeld et al. 2005). Importantly, accumulating evidence from animal and human studies suggest that OXT is a key factor for both mother-infant bonding and pair bonding (Feldman 2012b, 2016; Feldman et al. 2010; Gordon et al. 2010; Weisman et al. 2012). Intracerebroventricular infusions of OXT induce maternal behavior in hormone-primed rats (Pedersen et al. 1982) and sheep (Kendrick et al. 1987). Furthermore, central OXT receptor distribution has been linked to naturally occurring variations in maternal behavior in the rat (Champagne et al. 2001), and it has been shown that OXT in female mice enables pup retrieval behavior by enhancing auditory cortical pup call responses (Marlin et al. 2015). However, OXT not only shapes maternal behavior, it also influences the social behavior of the infant. For instance, OXT increased affiliative affective facial expressions in newborn macaques toward the caregiver and decreased salivary cortisol (Simpson et al. 2014).

Interestingly, the neurobiological mechanisms underlying the long-term persistence of mother-infant bonds in rats and pair bonds in monogamous prairie voles are remarkably similar (Numan and Young 2015). Both bonds are based on OXT and dopamine (DA) action within the nucleus accumbens, which promotes the synaptic plasticity required to make the infant or the mating partner rewarding (Atzil et al. 2011; Feldman 2017).

In humans, viewing the face of the romantic partner produces activity in striatal regions (Bartels and Zeki 2000; Acevedo et al. 2012), and an overlapping set of areas is activated when mothers are confronted with photographs of their children (Bartels and Zeki 2004). New lovers have higher OXT plasma concentrations than people not in a relationship (Schneiderman et al. 2012), and OXT administration significantly increased positive communication behavior during an instructed couple conflict discussion (Ditzen et al. 2009). By using functional magnetic resonance imaging (fMRI), it was shown that OXT may also enhance the bond between romantic partners. Specifically, the intranasal administration of OXT augmented neural responses to the romantic partner compared to a familiar person in the ventral tegmental area and nucleus accumbens in men (Scheele et al. 2013) and women (Scheele et al. 2016). Notably, this OXT effect was evident only in women who did not use hormonal contraception, indicating that bonding-related OXT effects are also influenced by gonadal steroids. To date, however, there is no direct evidence from human studies that these social effects of OXT are mediated by DA. The existence of OXT DA D2 receptor heteromers in the ventral and dorsal striatum have been documented (Romero-Fernandez et al. 2012), but so far the only positron emission tomography (PET) study to use the D2 receptor radioligand [11C] raclopride did not find altered endogenous DA release in the striatum or pallidum following OXT administration (Striepens et al. 2014). In this PET study, highly attractive but unfamiliar faces were presented to the participants, and the absence of bonding-specific stimuli (e.g. the participant's romantic partner or own child) could account for this null finding. Still, it is conceivable that non-dopaminergic actions are more important. In mice, the rewarding properties of social interaction require the coordinated activity of OXT and serotonin in the nucleus accumbens (Dolen et al. 2013), and another human PET study observed a modulatory impact of OXT on serotonin signaling (Mottolese et al. 2014).

Further fMRI studies have revealed that OXT also influences how mothers and fathers respond to photographs of children. Mascaro et al. (2014) reported that fathers have higher plasma OXT concentrations than nonfathers, whereas Wittfoth-Schardt et al. (2012) found that OXT increased activity in the caudate body in fathers in response to photographs of their own child compared to an unfamiliar child. Interestingly, Wittfoth-Schardt et al. also observed a diminished response in the globus pallidus to their own child or an unfamiliar child compared to a familiar child. In postpartum and nulliparous women, OXT enhanced activity in the ventral tegmental area to photographs of unfamiliar crying infants (Gregory et al. 2015). The presentation of an infant crying (Riem et al. 2011) and laughing sounds (Riem et al. 2012) produced activations in the amygdala which were reduced under OXT, thus suggesting that OXT also improves responsiveness to infant cues within parents in WEIRD (Western, educated, industrialized, rich and democratic) populations.

Notably, early life experiences appear to have a long-lasting effect on the OXT system. Current information in WEIRD contexts suggests that OXT interacts with attachment representations in adults' responses to infants. The OXT effect on amygdala reactivity in response to infant crying is particularly pronounced in individuals with insecure attachment representations (Riem et al. 2016). Likewise, the peripheral OXT response to infant contact at seven months is significantly higher in secure mothers compared to insecure/dismissive mothers, and positively correlated with greater activation in the ventral striatum when mothers with a secure attachment viewed their own infant's smiling and crying faces (Strathearn et al. 2009). Moreover, harsh parenting experiences moderate the OXT effect on the use of excessive force while listening to infant cry sounds (Bakermans-Kranenburg et al. 2012). Intriguingly, the exogenous administration of OXT also augments attachment representations later in life. Less anxiously attached individuals remembered their mother as being more caring and close after OXT, but more anxiously attached individuals remembered their mother as being less caring and close after OXT administration (Bartz et al. 2010; see also Buchheim et al. 2009). Thus, current concepts of social OXT effects emphasize the moderating role of interindividual and contextual factors (Olff et al. 2013; Hurlemann and Scheele 2015). For instance, OXT increases the perceived pleasantness of interpersonal touch and its associated neural response in insula, precuneus, orbitofrontal, and pregenual anterior cingulate cortex only when the touch is framed in a positive manner (Scheele et al. 2014). Along these lines, it seems likely that OXT also moderates the pleasant experience of interpersonal touch between parents and infants, as well as romantic partners. Collectively, OXT may contribute to human social bonding by modulating activity in a broad neurocircuitry involving reward-associated brain areas, such as the nucleus accumbens and ventral tegmental area and the amygdala (see also Sheridan and Bard, this volume). Clearly, future studies are warranted to elucidate how the multifaceted social effects of OXT result from the interplay between OXT and other neurotransmitter/hormonal systems.

Parental-Offspring Behavioral and Physiological Synchrony

Synchrony is one process that has been studied in some forms of attachment, albeit primarily in WEIRD populations. We note that synchrony is a general term that can refer to overlapping or joint action, as well as to contingent action—the latter defined as turn-taking and stimulus-response types of activity. Contingent actions are much more common in WEIRD mother-infant interactions, whereas overlapping or joint behaviors are more prevalent in relationally oriented cultures (Gratier 2003; Otto 2014; Morelli et al., this volume). Synchrony unfolds in different ways in more Western and more collectivistic societies (see Feldman et al. 2006). Although it remains for ongoing study to fully identify how synchrony unfolds and the specific aspects of similarity and difference across populations, current knowledge suggests that it is one core process in particular types of attachment relationships.

Early attachment relationships can provide a unique opportunity for "biobehavioral synchrony" or attunement between parent and child's physiological processes (Feldman 2012a). Biobehavioral synchrony is defined as the coordination between a parent and child's physiological (e.g., autonomic, hypothalamic-pituitary axis, alpha band activation in temporoparietal areas, OXT release) and behavioral states. Behavioral synchrony occurs via nonverbal patterns in the gaze, affective expression, co-vocalization, and touch-and-contact modalities. Of note, there is cultural variation with respect to the modalities that are emphasized. Thus, differences may appear in terms of norms for continuous contact or gaze synchrony, predominant in WEIRD cultures, whereas physical contact and movement synchrony may predominate in rural, relationally oriented cultures (Keller 2007; Morelli et al., this volume). Overall, synchrony can occur via any nonverbal modality and it likely shapes culture-specific neural pathways toward species-typical outcomes. For instance, we have behavioral evidence that Western parents express more facial synchrony whereas in other cultures, caregivers express more contact synchrony. In each culture, however, synchrony can predict a child's adaptation to the social group in later childhood and can also relate to a reduction of aggression toward peers.

In Western populations, synchronous interactions (in which behaviors are matched between parents and child) occur in somewhat different ways when the partner is the mother or the father. The maternal and paternal forms of synchrony—with more mutual gazing, affectionate contact, co-vocalization in the mother, and more matching of high positive arousal, exploratory behavior, and stimulatory touch in the father—can predict biological processes, such as OXT response or cortisol modulation. Parent-infant synchrony may play an important role in the three main functions of early relationships. These include the management of stress-regulation-homeostasis of physiological processes. In some cultures, the early relationship also serves to amplify positive arousal. For instance, young infants may experience positive affect primarily in social contexts, and less often when they are alone. Parent-infant synchrony may also function as a mechanism by which infants learn culture-specific rules of social behavior. Such learning may generally include whether, when, and how much expression of affect or emotion is appropriate in social contexts. More specifically, infants may learn how much arousal is appropriate to express, how much gaze to elders is permitted, and to what extent expression of negative affect is appropriate in social contexts.

Links between infant-parent synchrony and a range of outcomes in later life have been found in longitudinal studies from the newborn to late adolescence (18 years of age) (Feldman 2007a, 2012a). These studies show that the experience of synchrony (mother or father) predicts a host of social-emotional outcomes across childhood and adolescence, such as greater empathy, emotion regulation, and socialization in multiple contexts (e.g., with close friends, social group, parents, or siblings). By contrast, disruption of synchrony can occur in ways that are unique to specific conditions that interfere with parent-infant bonding, such as prematurity, postpartum depression, or high contextual stress. Each of these may disrupt different aspects of synchrony. For example, in maternal depression, there is a major reduction in typical maternal behavior. By contrast, in conditions of high anxiety or stress, mothers often overstimulate and thus overload the infant, whereas mothers with substance abuse disorders often oscillate unpredictably between "too much" and "too little" mothering. Of key importance from the perspective of plasticity, it appears that each of these deflections may be repaired by synchrony-focused interventions (Bernard et al. 2013).

The evidence provided here is largely specific to one type of attachment relationship and derives from studies of a limited number of populations. Synchrony is a core process in some forms of attachment, especially between infants and caregivers, but we do not know in the same amount of detail, the forms and roles of synchrony in contexts with other caregiving arrangements. It seems possible that because synchrony requires familiarity with the behavioral pattern of the partners, the number of caregivers with whom an infant might establish synchrony is limited, not infinite. Thus, broadening the study of synchrony to include populations with other forms of infant-caregiver relationship networks remains important in order to identify the extent of similarity in terms of contributing to the development of competence and underlying neural processes.

The Parent/Caregiver Brain

Processes such as synchrony provide one example of how joint consideration of both the infant and the caregiver can illuminate the neural foundations of attachment. The role of the parental, or caregiver, brain is also of inherent interest in understanding how attachment relationships unfold. What we know about the parental brain is largely derived from work in rodents, including that of Fleming et al. (2009). They describe a subcortical system that includes the medial preoptic area, primed by hormones of pregnancy (OXT, prolactin), that projects to both ventral tegmental area (increasing maternal reward from pup stimuli) and amygdala (increasing maternal vigilance). This charts a subcortical system underpinning "motherhood."

Human studies have shown that these structures, particularly striatal and amygdala, are also activated in humans in response to their infants' cues. In humans, these subcortical structures are connected via multiple ascending and descending projections to several cortical systems, particularly those involved in "mentalizing" (temporoparietal junction, superior temporal sulcus), "mirroring" (inferior frontal gyrus, supplementary motor area), "affect sharing or empathy" (anterior cingulate cortex, anterior insula), and "emotion regulation" (medial prefrontal cortex, orbitofrontal cortex) to enable parents to share infant affect, infer intention from action, and plan long-term goals. The maternal and paternal brains chart somewhat different pathways. However, when fathers are primary caregivers, there are mechanisms by which the "paternal" pathway recruits the "maternal" pathways and no differences are found between them.

In Western human contexts, there is evidence that both the mother and father "synchronize" their brain response to their infant cues in real time in mentalizing and mirroring networks (supplementary motor area, temporoparietal junction, superior temporal sulcus, anterior insula). Whether similar rapid online brain-to-brain synchronization may occur or be required to support caregiving in arrangements in which multiple individuals care for the same infant is not yet known. What is known is that disruptions to the parent-infant bonding process (e.g., in situations of abuse, depression, or premature birth) also disrupt the parental brain in multiple ways. As with behavioral synchrony (discussed above), preliminary evidence shows that the parental brain is plastic and that some changes can occur after intervention.

Neural Outcomes of Disrupted Attachment and the Development of Competence

Much of our current knowledge about the neural processes and outcomes associated with attachment is derived from studies of individuals with significant disruption early in life, including children placed in orphanages, those abused as children, and those with diagnosed attachment disorders. Experimental studies of nonhuman animals have also focused on disruption models to identify neural consequences of infant-caregiver relationships. Although these studies do not directly identify the neural processes involved in the initiation and maintenance of infant-caregiver(s) attachments, they have been extremely valuable in identifying neural circuits and processes, brain regions, and functions that are likely involved in attachment.

The consequences of absent and adverse offspring-caregiver relationships are robust, apparent in many neurobehavioral systems, and show some consistency across species. For instance, a relatively old and large body of literature demonstrates that the hypothalamic-pituitary-adrenal (HPA) axis system is profoundly affected by disruption of infant care in multiple species, including humans, monkeys, and rodents (Gunnar et al. 2015a, b). Nonhuman animal studies are critically important to address questions that cannot be answered with human studies. Therefore the convergences between the results of human studies of disrupted early care and those of animal models provide key foundational data for hypothesis-driven experimental studies. Below we high-light, as examples, several areas in which both human and nonhuman animal studies provide such convergent evidence of neurobiological outcomes asso-ciated with absent or adverse offspring-caregiver relationships. The biologi-cal evidence gained from experimental studies in animal models has reliably underscored the persistent effects of early childhood experiences and their sig-nificance for health across the life span. It is important, however, to note the probabilistic nature of the relevant processes. Even the harshest conditions, which can be devastating for most people, can be associated with successful "adaptation" in some cases. This speaks to the multiple pathways to success and the incredible resilience of human development.

Oxytocin

Long-term consequences of negative child-caregiver relationships are found in neuropeptides that are centrally involved in social relationships. For example, lower than normal levels of OXT in cerebrospinal fluid occur in adult women with a history of childhood trauma and abuse (Heim et al. 2009), whereas lower than normal levels of OXT are found in the urine of socially deprived children when they interact with their mothers (Wismer Fries et al. 2005). Oxytocin diurnal secretions are presumably hyperregulated for coping with the environment in order for maltreated children to survive and thrive. Hormonal dysregulation was found by measuring salivary cortisol/OXT diurnal patterns; differences were found in maltreated children living in "settled" environments and "unsettled" environments (Mizushima et al. 2015).

Brain Morphology and Neurofunction

A growing body of evidence demonstrates associations between childhood maltreatment and aspects of brain morphology and function. The volume of some brain structures are reduced in individuals with childhood maltreatment: in the superior temporal gyrus (De Bellis et al. 2002), cerebellum (De Bellis and Kuchibhatla 2006), corpus callosum, and hippocampus (for a review, see Teicher et al. 2003). In children with reactive attachment disorder (RAD), a range of alterations in brain morphology and function are observed. In com-parison with children with secure attachment relationships, children with RAD have a decreased volume of gray matter in the visual cortex and ventral striatum (Shimada et al. 2015). Neural circuits involved in fearfulness are af-fected by maternal care in rodents (Caldji et al. 1998) #and by maltreatment in human children (Gee et al. 2013; Fareri and Tottenham 2016; for a review, see VanTieghem and Tottenham 2017). In humans, maltreatment is associ-ated with deficits in emotion regulation and with difficulties in activating the ventral medial prefrontal cortex in the context of negative affect. In addition, maltreatment is associated with deficits in behavioral and neural responses to rewarding stimuli.

In U.S. study populations, other forms of childhood adversity, such as poverty, are associated with difficulties in higher-order functions, particularly language and executive function. These behavioral differences are likely related to differences in prefrontal cortex function and structure, which are observed in children from lower socioeconomic status (SES) parents compared to higher SES parents (Hackman and Farah 2009; Noble et al. 2012; Sheridan et al. 2012b). In addition, poverty is associated with differences in hippocampal function and structure (Hanson et al. 2011). Whereas the deficits in higher-order cognition may be related to the quantity and quality of parent-child interactions, the differences in hippocampal function and structure are likely the result of chronic stress exposure associated with poverty (Hair et al. 2015). Although consistent with a long tradition of work on the behavioral and educational outcomes of poverty, these claims have yet to be substantiated with experimental studies.

Neurocognitive Development

In humans, there is also robust evidence of the influence of institutional care on child attachment and neurocognitive development. For institutionalized children, the lack of an attachment relationship is associated with risk for dysfunction in neurocognitive deficits: reduction in IQ, cognitive deficits, impulsivity, and attention as well as decreases in cortical volume and function (see Sheridan and Bard, this volume).

Childhood maltreatment also increases the risk for psychiatric disorders throughout childhood and into adulthood (Edwards et al. 2003; Gilbert et al. 2009). Maltreatment encompasses a spectrum of abusive actions (sexual, physical, emotional abuse) or lack of actions (physical, emotional neglect) by the parent or other caregivers. Associated with early life abuse and neglect, RAD is a psychiatric disorder that is characterized by a child's wary, watchful, and emotionally withdrawn behavior (APA 2013). Given the emotional dampening that occurs in RAD, the disorder closely resembles internalizing disorders with depressive and anxiety symptoms. In populations of maltreated children in foster care, 19–40% had signs of RAD based on DSM-IV criteria (Zeanah et al. 2004; Lehmann et al. 2013), in which RAD (inhibited type) and disinhibited social engagement disorder (disinhibited type) were not completely independent. Within a general population, RAD (as defined by DSM-5 criteria) has been reported in 1.4% of children (Minnis et al. 2013; Pritchett et al. 2014).

Despite its high prevalence and clinical importance, there have been very few investigations on the possible neurobiological consequences of RAD except for recent publications by Tomodo and her colleagues (Mizuno et al. 2015; Shimada et al. 2015; Takiguchi et al. 2015). Children with RAD have reduced activity in caudate and nucleus accumbens relative to typical-developing children. Overall, dopaminergic system alterations appear to be associated with RAD, in a manner that provides convergent evidence that attachment disruption is associated with persistent effects on the neural circuitry involved in both salience and reward (Tomoda 2016).

Nonhuman Primates

Decades of research with nonhuman primates have evaluated the neural consequences of variation in maternal care and infant rearing. A long-standing literature demonstrates that experimental manipulations, which include alternation of maternal behavior as a result of variable foraging demands and nursery-rearing in absence of the mother, produce wide-ranging and persistent effects. In these comparisons, individuals who are reared by their mothers are contrasted with individuals reared in a nursery, which parallels institutionalized children in terms of lack of primary attachment relationship (van IJzendoorn et al. 2009). In rhesus monkeys, a range of adverse early social experiences (e.g., maternal deprivation, maternal neglect, chronic low maternal status during the infant's first 6–7 months of life) is associated with both short- and long-term disruption of normal peer relationships (low play, excessive aggression), altered HPA reactivity, chronically low central serotonin metabolism, higher C-reactive protein levels, as well as differences in brain structure and function (Suomi 1987; Kraemer and Bachevalier 1998; Machado and Bachevalier 2003; Lyons et al. 2009; Nelson and Winslow 2009; Bennett and Pierre 2010). Consistent with findings from human studies (see above), monkeys with disrupted attachment exhibit alternations in responding to both rewarding and aversive stimuli (Nelson et al. 2009).

Studies that have addressed neurobiological differences between mother- and nursery-reared macaque monkeys have consistently (with the exception of Ginsberg et al. 1993) demonstrated significant effects of early differential rearing on various measures of brain morphology and composition, including differences in the caudate-putamen (Martin et al. 1991; Ichise et al. 2006), hippocampus (Siegel et al. 1993), cerebellar vermis, dorsomedial prefrontal cortex, dorsal anterior cingulate cortex (Spinelli et al. 2009), and corpus callosum (Sanchez et al. 1998; cf. Spinelli et al. 2009). The pattern of early-rearing group differences in these monkeys parallels the findings from neuroimaging studies of human populations with histories of early stress and trauma. However, some differences in brain morphology, such as hippocampal volume, that are associated with low SES in humans (see above) do not seem to be apparent in nursery-reared monkeys (Spinelli et al. 2009). The divergence in findings suggests that the experimental control possible in nonhuman primate studies (including group equivalence in adequate nutrition, environment, and clinical care) allows for a disentangling of the effects of the infant-caregiver relationship from other factors that can be confounded in human studies.

In chimpanzees, we know that nursery-reared individuals have lower gray matter volumes than mother-reared individuals; however, no difference in white matter volume, total gyrification, or overall gray matter thickness has been found (Bogart et al. 2014). We also know that early-rearing experiences have a significant impact on the heritability of personality and dimensions of psychopathy, including meanness, boldness, and disinhibition. These traits are significantly heritable in mother-reared but not nursery-reared chimpanzees (Latzman et al. 2015). In rhesus monkeys, nursery rearing results in major changes in genome-wide patterns of mRNA expression (Cole et al. 2012) and DNA methylation (in lymphocytes and in prefrontal cortex; Provencal et al. 2012; Massart et al. 2014). As is true in other neurobiological systems associated with attachment, these patterns appear to be at least partially reversible with subsequent targeted interventions (Dettmer and Suomi 2014).

Genes, Epigenetics, and Plasticity

Evidence of plasticity, as well as knowledge about the biological mechanistic pathways that underlie plasticity, continues to emerge, as noted in many of the findings discussed above. There is a remarkable range of outcomes associated with variation in early experiences, including the nature of infant-caregiver relationships and, potentially, the diversity of relationship networks. The findings underscore a critical cautionary note: there is not a *single* structure, system, or gene responsible for overall outcomes with respect to the development of competence. Likewise, we should not expect that a *single* variable will account for an overwhelming proportion of variance in development. Such a cautionary note may appear obvious and simple. It is worth remembering, however, that as scientific findings are conveyed to the public and policy-makers, interpretative errors may convey a determinism that is not warranted (see Chapters 13 and 14, this volume) Both the exaggeration of the magnitude of effects as well as implications that a *particular* effect or mechanism is the *only* possibility can give the impression of determinism that is neither consistent with the multiple, redundant, developmental pathways, nor with our current knowledge of plasticity and epigenetic change.

Caveats and Limitations

In reviewing what is currently known, it is important to note that relatively narrow definitions of attachment have been employed to date, and that the majority of findings on the neural bases of attachment have been derived from research that has been conducted on a narrow range of model species. These limitations have most likely hindered the identification of universal core processes, which together permit variation and adaptation in offspring-caregiver relationships. That progress has relied on studies from relatively few species is not surprising, since laboratory studies in biological psychology, neuroscience, and other fields which contribute to this research focus primarily on a few model species. Arguably, this very focus may have enabled the detailed knowledge that we currently have. Nonetheless, because the diversity of offspring-caregiver relationships among other species is great (Hawkes et al., this volume), research should expand to include more diverse species. This is needed to uncover both similarity and variation in neural systems that contribute to attachment relationships.

It is critical to note that constraints involving the range of species used in neurobiological studies do not result from a narrowness in scientific inquiry, but rather from practical, political, economic, and sociocultural factors imposed on science. These constraints may occur across different types of study and pose unique challenges to research that is geared toward understanding the neural foundations of attachment: some assessments are noninvasive (e.g., OXT administration, cortisol from saliva), but detailed mechanistic experimental studies require different protocols. Much of what we know at the molecular level is from invasive research, terminal, and large N studies conducted with rodents (e.g., mice, rats, voles). This type of mechanistic work has been largely conducted in rodents not only because they are common laboratory model species, but also because their designation as such reflects particular societal views and practical considerations.

There is a wide range of nonhuman animals that are more similar to humans than rodents, in terms of neurobiology, behavior, or offspring-caregiver relationships—features which would enable comparative studies and increase our understanding of the neural foundations of attachment. Such animals include primates (particularly apes), dogs, cats, dolphins, and elephants. However, due to societal and practical considerations, it is highly unlikely that new (or continuing) experimental or invasive studies will occur using these animals. Thus, while it is crucial to include a greater range of species that can represent a greater variation in infant-caregiver relationships, there is also a great need for sensitivity to the broader sociopolitical reality in which the work occurs. In parallel, there is a corresponding need for thoughtful consideration of how sociopolitical reality affects the scope of scientific questions, the probability of advances in understanding causal mechanisms and the neural foundations of behavior, and in turn, the implications and consequences of those effects on both human and nonhuman animals (Bennett and Panicker 2016).

Open Questions

Many questions about the neural foundations of attachment remain as challenges. Some of these require attention from researchers working at points of intersection with basic neurobiological research and often involve nonhuman animal studies. Increased understanding of how culture impacts attachment would be beneficial, since a more specific understanding of the diversity of infant-caregiver(s) relationship networks is needed to inform the basic assumptions and definitions which underlie the selection of questions, animal models, specific hypotheses, and design of studies in attachment neuroscience research. Here we provide a summary of several sets of open questions that we believe are essential to advancing knowledge about the neural foundations of attachment.

Generalization, Cross-Cultural, and Cross-Species Considerations

Basic descriptive information about the processes and time course of different infant-caregiver arrangements is needed to generate hypotheses and design studies that can illuminate variation in neurobiological foundations related to attachment. For instance, all things being equal, does attachment manifest in the same way in monomatric and other forms of child-rearing? If one were to measure HPA axis in one-year-old children in different child-rearing cultures, would both the stress response and time course to return to baseline be the same across cultures?

A parallel set of open questions surrounds cross-species comparisons. Similar to the cross-cultural consideration, questions about the neurobiology of attachment could be informed by comparative analyses of species that vary in infant-caregiver arrangements. For instance, do the infants of mothers from primate species with and without alloparenting differ in HPA axis in response to separation? Do they differ in response to presentation of a stranger? To take another example, some prosimian species "park" their infants while the parent forages. What is the experience for the offspring during these times? Do they show any stress response over repeated separations?

When thinking about variation across species as well as across cultures, it is helpful to pay close attention to the ecology—broadly conceived (e.g., Keller 2007)—and other species characteristics, because this will help us define the challenges faced by both infants and parents. For example, the challenge for sheep (group-living mammals who deliver precocial young) is for mother and lamb to identify one another and learn to recognize each other very quickly; otherwise they will become separated, thus threatening the survival of the lamb. For primates, there is wide variation across species with respect to where they live, what they eat, whether or not they have multiple births, and group composition (size, stability, and structure). For humans, we tend to pay too little attention to these issues, because all human babies are born immature and are highly dependent for a long period of time. Nonetheless, ecology and subsistence patterns obviously play a role in determining childcare patterns. For example, within hunter-gatherer groups, there are differences between forest-dwelling, desert-living, savannah, and fishing-oriented groups in terms of what adults need to do to care for children (Hewlett and Lamb 2005). Do these environmental differences have implications or correlates at the neural level? If so, are these correlates differentially associated with the development of competency, with risk, or with plasticity?

Individual Differences, Stability over Time/ Development, Range of Variation in Competency

As discussed above, much of what we know about the neural bases involved in infant-caregiver(s) relationships derives from studies focused on disruption and aimed at understanding the effects of adversity and compromised outcomes. As a result, many open questions remain about individual differences in behavioral, emotional, and cognitive development; biological functioning; brain development; and patterns of gene expression or methylation. Additional questions exist concerning how these individual differences and child-rearing circumstances interact within different cultures, given the diversity of infant-caregiver(s) relationships. From the perspective of developmental systems theory, the need for basic knowledge about individual variation encompasses an appreciation of complex multidirectional processes, with interplay between culture, environment, behavior, brain, and genes, that unfolds across the life span. Better understanding of the range of variation within the spectrum of competency and underlying neural functioning is needed to guide the identification of core processes that are basic and common across the diversity of infant-caregiver(s) arrangements. How variation in developmental trajectories is associated with infant experiences and how experiences influence individuals during periods of special plasticity (i.e., adolescence, parenthood) also remain open. At the same time, they are questions that can be guided by integration of rapidly increasing knowledge about neural development during those life stages.

Evaluation of Domain-Specific Sensory Systems

It is clear that both the initiation and the maintenance of the infant-caregiver(s) relationship involve multiple and interacting systems. What is currently unclear, however, is whether evidence from one system can inform knowledge about others. For instance, to what extent can the neurobiological substrates of visual and auditory attachment cues be extrapolated to tactile and olfactory stimuli? Further research is needed to evaluate both the generalizability and the specificity of the role of different systems and domains in the development and maintenance of attachment.

Animal Models

Studies of nonhuman animals as model species are crucial to progress in understanding the neural foundations of attachment. Identifying the molecular mechanisms for plasticity, for instance, would most likely depend on experimental study, including invasive research, that is prohibited in humans. As a result, open questions surrounding what aspects of attachment can and cannot be modeled in other animals remain as large and ongoing challenges. A key differentiation here is in identifying what we know (and do not know) about behavioral, social, emotional, cognitive, and higher-order representational processes of diverse species. In turn, open questions about the implication of these similarities and differences for processes relevant to attachment and infant-caregiver(s) relationships must also be addressed to advance hypotheses about underlying neural systems.

Possible Research Avenues

We have identified a broad range of gaps in knowledge about the neural foundations of attachment, particularly as concerns variance across cultures, species, and infant-caregiver arrangements. The number of avenues for possible research is extensive and diverse. Still, we believe that core areas for research have the potential to advance our understanding of neural contributions to attachment and may thus provide a more robust platform for identifying similarities and differences across culture, species, and types of infant-caregiver relationships. Here we outline several of those areas.

Neural Bases of Initiation and Maintenance of Attachment

With respect to the flexible glue model proposed above, several foundational questions need to be addressed. Identifying the core processes involved in both the initiation and the maintenance of attachment—and doing so at the level of neural, behavioral, and hormonal processes—is important. Mirror neurons, OXT, and amygdala-prefrontal cortex connectivity are potential areas to target: they have been identified in studies of disrupted relationships and in social neuroscience, and provide good examples for hypothesis generation and testing. It is of critical importance, however, for hypothesis-driven study to evaluate explicitly whether—and how—these core processes are unique to infant-caregiver attachment relationships. Fully identifying the neural and molecular elements of these processes will likely depend on experimental research with nonhuman animals.

In terms of the initiation phase specified in the flexible glue model, we believe that it will be difficult to pull apart the specific attachment pathway from other systems, particularly early in life. It may, however, be feasible in the future to demonstrate individual variations in attachment through longitudinal studies. At the same time, continued studies that focus on disruption of the developing attachment system are anticipated to contribute important data, particularly when the form or type of "adverse" experience is increasingly specified or constrained (e.g., Sheridan and McLaughlin 2014).

Regarding the maintenance phase, there is a need to focus on the neural basis of children's attachment during this phase. There is a good amount of knowledge about how parents (mostly the mother or father in WEIRD populations) respond when they see pictures of their own versus other children. Less is known, however, about other modalities, the neural activity in children, and the range of variation. Studies directed toward understanding caregivers' response to hearing their own versus another child crying or to a response to an infant's body pheromones would, for example, be of interest.

Variability across Individuals, Culture, and Species

Future research needs to address variation in attachment across individuals, cultures, and species. Identifying species similarity and differences is crucial if we are to understand the neural systems involved in attachment and connect neural systems with different aspects of the infant-caregiver relationship. Identifying variations is also central to theory refinement and hypothesis testing. In addition, investigating the neural correlates of cultural variation in infant-caregiver(s) attachment networks is needed. Neuroscientific enquiry, however, must depend on, and be informed by, basic research on cultural variation in attachment. For example, research to identify the range of variation in the development of competency across cultures and how attachments are related to cultural variation is the necessary foundation to identify the underlying neural correlates of core processes, as well as variation. Thus, neurobiological research will continue to be dependent on research from other disciplines which assess variations in the quality of care, attachment, and outcomes in other countries. Absent that knowledge, neuroscientific studies will remain in jeopardy of applying inappropriate lenses.

Consequences of Adversity across Cultures and Generations

Despite substantial knowledge about the neural consequences of childhood adversity and disrupted attachment relationships, large gaps remain that need to be informed by cross-cultural study. Avenues for future research include identifying the neural consequences of adverse experiences that occur across cultures (e.g., infant and child abuse). Does cultural context modulate neural outcome? If so, how does this occur? It is critical to note that research must begin with a determination of the basic phenomena under investigation, as this is necessary to ensure comparability of data.

Alongside cross-cultural evaluation of the neural consequences of adversity, attachment across generations should also be examined. Research is needed, for instance, to illuminate how parenting and the underlying neural processes or "parent brain" may differ in individuals whose own early attachment was disrupted. There is good evidence that there is often cross-generational continuity. There is also good evidence that this is not even remotely *inevitable*. Thus, while making the case that an individual's own early experiences affects how they engage in infant-caregiver relationships, it is important to emphasize that, in many cases, adverse early-life experiences do not result in compromised competency. It is critically important to communicate this nuance and to avoid conveying an element of early-experience determinism. Evaluation of the neural consequences of variation in infant-caregiver(s) experiences, integrated with consideration of multiple systems which play a role in competency and the full range of outcomes, provides a path away from determinism and toward a more eclectic understanding.

Integration: Identifying Core Processes to Further Understanding

An important yet still unanswered area of enquiry concerns the need to identify and delineate better those systems and processes that facilitate the initiation of the attachment system, as well as those which contribute to its maintenance. Over the past decades, areas of the brain used to act upon, feel, and sense the environment have been observed to reactivate when these actions are observed in other individuals. This reactivation of the motor, emotional, and sensory systems has been named "shared circuits." The association between attachment and shared circuits is likely to be bi- or multidirectional. Shared circuits may help initiate and maintain attachment, and attachment may help wire up shared circuits.

Attachment may contribute to the development of shared circuits. Although we still know little about how shared circuits develop in infants, an influential theory suggests that congruence between (a) infants' motor programs, sensations, and emotions and (b) the sight and sound of these, as perceived by sensory systems, is key for the brain to connect neurons in sensory systems selectively with matching representations in the motor, somatosensory, and emotional systems, respectively (Keysers and Gazzola 2014). In some cases, such Hebbian learning does not rely on social interactions. For instance, when a child learns to grasp objects, we know that the child will look intensely at his/her own hand. This means that the child will simultaneously activate neurons encoding motor programs to perform these actions and neurons encoding the sight of the action that the child now sees him/herself perform.

These simultaneous activations mean that synaptic plasticity will reinforce the connections between neurons responding to the sight of grasping with those responsible for the motor program for grasping. After this training, the sight of grasping will trigger activity in neurons responsible for performing the action. When a child sees someone else perform this action, the same connections will trigger the child's own motor program: a shared circuit has been wired. However, for other actions where we know shared circuits exist, the child cannot see him/herself perform the action. When a child is happy, for instance, the child will smile, but the child cannot see the smile. In Western societies, child-parent dyads show patterns of behavior in which the parent will avidly imitate the facial expressions of the baby. A happy baby will thus see the parent return a smile. The better this interpersonal synchrony, the more the brain will be able, according to Hebbian learning theories, to connect neurons involved in smiling and being happy with neurons representing the sight and sound of a smile and giggle. After repeated synchronized imitative interactions, the baby's brain will activate his/her own smile and happiness when witnessing happiness in others. A shared circuit now emerges from the synchronized social interactions which characterize parents generating healthy attachment styles in Western contexts.

To test whether attachment really favors the development of shared circuits via interpersonal synchrony, a promising research line might involve comparing societies in which parents imitate the facial expressions of their young babies with societies in which parents do not. With this type of study, we could explore whether an experience of imitation of facial expressions would lead to systematic differences in levels of motor and limbic activations in children when witnessing the facial expressions of others.

The visual-motor association described above could also be complemented and/or replaced by other sensory modalities-motor association. For instance, a happy state could be associated with an increase in vocalization, which would then allow the recognition of other's emotional states primarily through the auditory domain. Cultural and species differences in the preferred modality, therefore, would not necessarily impair the development of shared circuits or their relationship with attachment.

In addition to attachment being the basis for the development of shared circuits, shared circuits might be the basis for the development of attachment. This could take two forms. First, individual differences in shared circuits have been associated with differences in reported empathy (Singer et al. 2004; Jabbi et al. 2007), as well as with levels of prosocial motivation (Hein et al. 2010). Individuals who demonstrate stronger activations in shared circuits for emotions have been found to show higher levels of empathy for others and greater willingness to help. Accordingly, shared circuits in the parent may be a motivator for engaging in the type of interactions with the child that are the basis for the generation of healthy attachment. Second, shared circuits have been associated with the ability to engage in interpersonal synchrony—the ability to generate congruent actions in real time, in response to the actions of others (Kokal et al. 2009). As has been shown, Western parents who synchronize their actions to those of the child have children with more secure attachments; shared circuits in the parents may thus be critical in establishing and maintaining attachment. Finally, because they develop in part via attachment (see above), shared circuits enable the child to share the emotions of the caregiver and attune his/her own actions to those of the parent. Accordingly, the synchrony that is so critical for strong attachments is then no longer entirely the burden of the parent, but increasingly becomes the result of fine-tuned bidirectional synchronization via shared circuits in the child.

Although this approach would allow a certain degree of variability in the amount of synchronization required and the agent offering such synchronization, the presence of at least another agent and a certain degree of synchronization seem to be necessary conditions. Investigating shared circuits across different cultural settings or other species will expand understanding of the impact that the amount or type of synchronization and the number of agents might have on the relationship between shared circuits and attachment. Studying those cases in which shared circuits are impaired would then further facilitate understanding of the causality between shared circuits and attachment. Cases of poor or absent attachment could be used to investigate whether alterations in attachment cause a reduction in shared activity.

Conclusion

In this chapter, we have summarized important aspects of what is known about the neurobiology of attachment and posed critical questions that surround understanding of how attachment relationships are initiated. Much of what is known about the neural foundations of attachment derive from research that has been conducted on a limited range of species, within limited cultural contexts, and has focused on limited types of infant-caregiver relationships. Increased biological evidence about specific neural components of infant-caregiver relationships has, however, produced a fertile platform for hypothesis-driven and descriptive research, and can advance understanding of how brain development interacts with experiences from a diversity of caregiving relationships. Furthermore, it has become increasingly apparent that plasticity (and thus, resilience) constitutes a common theme in neurobiological, epigenetic, and behavioral findings. Observed plasticity illuminates what we believe to be a centrally important point in contextualizing attachment for neuroscientific study. Rather than conceptualize attachment in terms of a single type of relationship, or a rigid developmental channel, an expanded consideration of variation is needed to understand the neural foundations of infant-caregiver relationships and the role these relationships play in developing competence across the life span. This approach should enable the identification of common neurobiological elements of attachment, as well as the remarkable plasticity and diversity within and across individuals, cultures, and species.

11

How Attachment Gave Rise to Culture

James S. Chisholm

Past loves shadow present attachments, and take up residence within them.
—Martha Nussbaum (2001)

Abstract

This chapter reviews advances in evolutionary theory since Bowlby and proposes that our capacity for culture emerged with the evolution of human attachment by means of selection for increased mother-infant cooperation in the resolution of parent-offspring conflict. It outlines the evolutionary-developmental logic of attachment, parent-offspring conflict, and the view of culture as "extended embodied minds." It describes how the embodied mind and its attachments might have been extended beyond the mammalian mother-infant dyad to include expanding circles of cooperative individuals and groups. It argues that because attachment came before and gave rise to culture, no culture could long exist that did not accommodate the attachment needs of its infants. On this view, all the myriad cultural contexts of attachment foster secure-enough attachment—except when they cannot. Theory and evidence show that when mothers and others are unable to buffer their children against environmental risk and uncertainty, insecure attachment can be (or once was) evolutionarily rational. The major source of risk and uncertainty today are the causes and consequences of intergenerational poverty or inequality. It concludes that an attachment theory fully informed by twenty-first century evolutionary theory is fully consilient with normative emic perspectives on the nature of the child and appropriate child care, in both favorable and unfavorable environments.

Introduction

Evolutionary theory is our only scientific theory of life, and attachment theory, as formulated by John Bowlby, is the predominant evolutionary theory of social-emotional-cognitive development. But for attachment theory to mature, it must fully incorporate the cultural contexts of attachment, especially the role of cooperative breeding and alloparenting. I believe that the many advances in evolutionary theory since Bowlby show very clearly that it can.

The most productive of these advances is the emerging "extended evolutionary synthesis" (EES), which is a convergence of ideas from information science, evolutionary-developmental ("evo-devo") biology, and behavioral ecology. Bowlby had limited knowledge of each, which by today's standards were primitive anyway, but he was the first to conceive of emotional development in terms of "control systems" that regulate "feedback" between mother and infant (Bowlby 1969:65). I will argue that the EES provides *a priori* logico-mathematical grounds for rejecting nature-culture, mind-body, and individual-group dualisms and shows how our capacity to cooperate (*cooperārī*, "work together") and develop culture might have evolved from our infant ancestors' ancient mammalian motivation to form attachments.

Because my focus is on the evolution of the development of human attachment, I view "mother" and "attachment" from the perspective of a generic newborn mammal's body. Like all forms of life, infants are complex adaptive systems: they use the information encoded in their genes to act, expressing it in their attempts to adapt to life outside the womb. I begin at the beginning because the human attachment process evolved from the stem mammalian attachment process (e.g., Broad et al. 2006; Royle et al. 2012). All mammalian newborns are motivated to approach species-specific patterns of sign stimuli; in the environment of evolutionary adaptedness of mammals, these stimuli would essentially always emanate from their biological mothers. The primary adaptive function of being close to one's mother is to survive infancy. Parent-offspring conflict theory (Trivers 1974), however, holds that mothers and infants are naturally conflicted. Although mothers share 50% of their genes with each offspring, offspring also share 50% of their genes with their father; thus, mother-offspring conflict is inevitable. Infants are expected to seek more resources (material and socioemotional) than mothers are willing to provide because infants seek to benefit themselves (copies of *both* parents' genes). At the same time, mothers are expected to provide fewer resources than their infants would like because mothers, too, seek to benefit themselves (copies of their genes in current or future offspring). For Trivers, "socialization is a process by which parents attempt to mold each offspring…while each offspring [is expected] to resist…and to attempt to mold the behavior of its parents" (Trivers 1974:260).

At the very least, mammalian mothers must accept or tolerate their infants' proximity-seeking behavior and attempts to nurse. In doing so, mammalian mothers keep their infants alive by regulating their new physiological functions for them. By adjusting their behavior to that of their infants, mammalian mothers help the infant's body learn how to function as it has to as an adult, in terms of temperature maintenance, blood sugar level, arousal level (hypothalamic-pituitary-adrenal [HPA] reactivity), and much more. For a generic mammalian infant, a "mother" would thus be anybody who first does no harm and at least tolerates it long enough for its homeostatic control systems to mature. Corresponding minimalist views of attachment are the "mutual regulation" model (Tronick 2007), the "biobehavioral synchrony" model (Feldman 2007a, 2014), and the "psychobiological attunement" model (Field 1985), in which "each partner provides meaningful stimulation for the other and has a modulating influence on the other's arousal level" (Field 1985:415; see also Schore 1994, 2013; Polan and Hofer 2008; Beebe and Lachmann 2014; Leclère et al. 2014). A good minimalist definition of secure attachment was provided by Gunnar et al. (1996:200): "secure attachment relationships protect or buffer infants from elevations in cortisol."

The EES views natural selection and development as mechanisms for acquiring information and organisms as matter and energy that have been organized by information. Accordingly, it views "culture" as minds that have been organized by shared information, analogous to the view of culture as cultural models or shared cognitive schemas (Quinn and Holland 1987; D'Andrade 1992; Strauss and Quinn 1997). These minimal definitions of "mother," "attachment," and "culture" help set the stage for the proposition that the bodily connection between infant and mother leads to their cognitive connection.

The Extended Evolutionary Synthesis

The dominant paradigm in evolutionary theory in Bowlby's day was the Modern Synthesis (MS), which unified Darwin's theory of natural selection with population genetics in the 1930s and 1940s. Since Bowlby, the MS has incorporated numerous advances in evolutionary theory, and it remains the dominant paradigm because of its powerfully predictive mathematical models of what gene selection should favor and how gene frequencies change over time. While the EES constitutes a major shift in evolutionary theory, it is a shift in emphasis, not a Kuhnian paradigm shift. Mutation and recombination still generate genetic variability, gene variants are still tested by natural selection, and variants that pass the test are still copied into the next generation more often than those that do not. What the EES is shifting are the old notions that mutation and recombination are the only or even major source of variability, that genes are the only mode of inheritance, and that selection operates in a single mode or at a single level.

Dissatisfaction with the MS goes back to Waddington (1942) and has been building ever since. Tinbergen (1963) expressed his dissatisfaction in the form of his famous "four questions" and Stearns (1982) with his distinction between biology's "adaptationist" and "mechanist" perspectives (Table 11.1). Stearns (1982:238) was dissatisfied with the way the thoroughly adaptationist MS "made a series of simplifying assumptions that had the effect of reducing the objects of study to changes in gene frequencies: the organism disappeared from view, and with it went the phenotype, the ecological interactions of the phenotype with the environment that determine fitness, and the developmental interactions with the environment that produce the phenotype."

Table 11.1   The structure of an evolutionary explanation of human behavior.

Stearns's Perspectives	Tinbergen's Levels of Explanation
Mechanist:	Proximate: What causes the expression of behavior X?
How organisms work Fitness as work	Ontogenetic: What is the developmental/cultural history of behavior X?
Adaptationist:	Phylogenetic: What is the evolutionary history of behavior X?
How evolution works Fitness as measure	Ultimate (natural selection): What is the adaptive value of behavior X?

Flush with the success of its powerfully predictive mathematical models of how gene frequencies change over time, the MS was not very interested in biology's mechanist perspective. The adaptationist perspective focuses on how evolution works and views fitness as a measure: relative reproductive success. The mechanist perspective focuses on how organisms work and views fitness as work: the work they have to do to stay alive, grow, and develop in order to reproduce. If not always explicitly, the EES uses Tinbergen's four questions to bridge the adaptationist-mechanist gap by tracing the developmental pathways from genes to behavior, and back, when behavior changes the focus or strength of selection (Pigliucci and Müller 2010; Stotz 2010, 2014; Sterelny 2013; Pigliucci and Finkelman 2014; Laland et al. 2015).

A major step toward bridging these gaps was the development of life history theory (see also Hawkes et al., this volume). Its key insight was Waddington's: evolution and development are two sides of the same coin. Genotypes "push" phenotypes into one generation; phenotypes "pull" genotypes into the next. Selection operates on flesh, blood, and behavioral phenotypes, not the raw DNA in genotypes. Selection cannot "see" the information represented in genotypes until it has been embodied into a phenotype during development. The life cycles of all sexually reproducing organisms begin with a single-celled zygote, but zygotes must develop into adults before they can reproduce. Development is thus an adaptation for reproduction; life cycles *are* reproductive strategies (Bonner 1965; Stearns 1992; West-Eberhard 2003; Konner 2010) that consist of life history traits (Table 11.2) organized by information about their particular "developmental niche" (Super and Harkness 1986). (More on life history theory later.)

In the EES, information has the quality of "aboutness" or "intentionality" (*intendere*, "to stretch toward, aim at"): it "points at" that which it is about or represents. Evolution and development are both information acquisition mechanisms. In evolution, the information acquired is about an organism's environment of evolutionary adaptedness, represented in its DNA. The laws of aerodynamics, for example, are represented or embodied in the shape of birds' wings, whereas the laws of optics are embodied in eyes. In development, the

Table 11.2   The major dimension of differences in life history strategies (adapted from Pianka 1970; Reznick et al. 2002; see also Hawkes et al., this volume).

Strategy	Current ("Short-fast")	Future ("Long-slow")
	Minimize chance of maximal possible fitness loss (extinction)	Maximize chance of minimum necessary fitness gain (continuation)
Ecology	More variable and/or unpredictable	More constant and/or predictable
Mortality rates	Often catastrophic, non-directed, density dependent	More constant, directed, density independent
Survivorship	Low in early life	High in early life
Population size	More variable	More constant
Intra- and inter-specific competition	More variable, lax	More constant, intense
Selection favors:	Rapid development	Slow development
	Early reproduction	Delayed reproduction
	High reproductive rate	Low reproductive rate
	Low parental investment	High parental investment
	Small body size	Large body size
	Semelparity (large litters)	Iteroparity (small litters or single birth)
	Short life span	Long life span

information acquired is about an organism's developmental niche. The attachment process, for example, enables mammalian infants to acquire information about mothers.

The principal difference between the MS and the EES is that the former is gene-centric whereas the latter is information-centric. The MS holds that with rare exceptions genes are the source of the variability on which selection operates and the only medium of inheritance (Dawkins 1976). In contrast, the EES holds that because selection acts on phenotypes, not genotypes, what matters is the information that organizes the phenotype *regardless* of its source or medium of inheritance—as reflected in the title of the best account of this perspective by Jablonka and Lamb (2005), *Evolution in Four Dimensions: Genetic, Epigenetic, Behavioral, and Symbolic Variation in the History of Life*.

The four dimensions of evolution are four levels in the evolution of the complexity of life; each emerges from the preceding lower level (Table 11.3). In the beginning, at the bottom level, organisms were (and still are) organized by genetic information. At the next level, some organisms evolved the capacity to be organized by the epigenetic effects of information about their own or their parents' environment, acquired by their own exposure or inherited from

Table 11.3 Information in the four dimensions of evolution (Jablonka and Lamb 2005).

Information	Context	Acquisition Mode	Transfer Mode
4. Symbolic	Cultural environment	"Mind reading"	Cultural environment
3. Behavioral	Social environment	Learning, imitation	Social environment
2. Epigenetic	Physical or social environment	Parents or exposure	Parents or exposure
1. Genetic	Environment of evolutionary adaptedness	Natural selection	DNA

their parents without exposure.[1] At the third level, animals evolved the capacity (nervous systems) to have their behavior organized by information about the behavior of other animals, acquired and inherited from their social environment by learning and imitation, giving rise to animal behavioral traditions. At the top, most recent level, humans evolved the capacity to have their minds organized by the symbolic, cultural information in other minds, acquired and inherited by "mind reading," giving rise to history.

In the EES, information science includes biosemiotics, complex adaptive systems theory, and game theory. Each is reviewed briefly below, after which I will introduce the concepts of group selection and niche construction. This will set the stage for showing how infant attachment motivations give rise to the key EES concepts of the "embodied" and "extended" mind.

Biosemiotics

Biosemiotics ("signs of life") follows from the premise that organisms consist of matter and energy that have been organized by information (Harms 2004; Skyrms 2010; Emmeche and Kull 2011; Deacon 2012; Witzany 2014). This information is acquired from, and points at, two periods and one point in time, which I will call "old," "newer," and "now." "Old" information is about an organism's phylogenetic history, "newer" information is about its ontogenetic history, and "now" information is about its moment-to-moment sensory experience. "Old" information is about what the state of the organism's body *should* be (in order ultimately to reproduce) at a given age in a given circumstance (e.g., its species' optimal[2] body temperature, blood sugar level, or state of arousal). "Newer" information is about what the state of the organism's body *has been* during development. "Now" information is about what the state of the organism's body *is*; that is, what happens to its body

---

[1] Epigenetic inheritance has been identified in 27 human studies (Turecki and Meaney 2016) and over 100 other species, with effects observed (so far) for three generations in humans and 46 in long-term breeding experiments with the silver fox (Jablonka and Raz 2009).

[2] Best possible, not imaginable.

(consciously or not) as it interacts with its environment. In principle, with information about two periods and one point in time, organisms can "triangulate" on the future.

In biosemiotics, meaning, value, and intention are facts of nature. Meaning exists in nature in the form of intentional information that points at what it means; it is about something in an organism's internal or external environment. Following Peirce (1958), biosemiotics emphasizes the triadic relationship among signs, objects, and interpreters. Signs have meaning only to interpreters. Signs acquire meaning only in terms of what they mean to, or entail for, an interpreter. For instance, even brainless, single-cell bacteria are able to interpret the meaning of the molecular trail left by a decaying organism just by acting on it (without intent: "*if* molecule X, *then* follow them"), implicitly "predicting" that food lies where the sign points. Likewise, mammalian newborns interpret the meaning of signs of mother by approaching (at first with no intent) the source of the signs. Molecule X and signs of mother mean, entail, or "predict" approach behavior because natural selection discovered the natural, "bio-logical" contingency between signs (of molecule and mother) and objects (food source and security).

Value exists in nature in the form of reproductive success; life exists because organisms reproduce. Value exists in nature when a sign points at something of intrinsic value to an organism—any resource that the organism needs to survive, grow, develop, and reproduce (e.g., Chisholm 2012). Bacteria have no feelings; they just act. For infants, detecting contingencies feels good intrinsically; it makes them smile and want to explore, play, and thereby learn how objects, events, and people are connected in ways that naturally feel good (approach) or bad (avoid) (Watson 1972, 1994, 2001; Gergely and Watson 1996; Gergely et al. 2010).[3] Natural selection endowed contingency detection with good feelings as an adaptation for development, for infants to acquire information about their developmental niche.

Complex Adaptive Systems

Evolution is a complex adaptive system, a natural process that keeps itself going by reproduction. According to complex adaptive system theory, "intention" refers to "specific information acting on the dynamics [of the complex adaptive system], attracting the system toward the intended pattern" (Kelso 1995:141). (In terms of attachment, mothers are "attractors" in the "design space," or developmental niche, of infants.) Evolution's "intention" is that life should continue to evolve. Its "intended pattern" is the pattern by which life continues: "The persistence of the whole over time—the global behavior that outlasts any of its component parts—is one of the defining characteristics of complex systems" (Johnson 2001:82).

---

[3] See also Gopnik (2000) on "explanation as orgasm."

Complex adaptive systems are hierarchically nested networks of interacting agents that receive signals from their environment and send signals to agents at their own and other levels (Holland 1992b, 1995, 2012; Capra and Luisi 2014). The study of complex adaptive systems "…involves understanding how cooperation, coalitions, and networks of interaction emerge from [agents'] individual behaviors and feed back to influence those behaviors" (Levin 2003:3). Natural selection is the ultimate *bricoleur*: it constructed life, bottom-up, piecing together simple agents into ever more complex wholes. It organized molecules into DNA, then DNA into cells, which opened the door for cell-cell communication, cooperation, coalitions, and networks of interaction, and the emergence of tissues, organs, organ systems, behavior, behavioral control systems (e.g., attachment), societies, and most recently, cultural models or shared cognitive schemas.

Agents at each level set up the environment for the agents at the next level. Agents at the bottom level are the "sensory organs" of the complex adaptive system. They search the environment, both internal and external, for certain kinds of information. If/when they detect a certain kind, they interpret its meaning by sending signals up to agents at the next level. Agents at that level, in turn, search their environment (signs from the preceding level) for certain patterns of information. If/when they detect that pattern, they send signals up to the next level, where the process is repeated until signals reach the top level, which interprets the output of the penultimate level and sends signals back down the hierarchy, effecting top-down control of agents at lower levels and the system (e.g., the body) as a whole.

Each level of complexity is "self-similar" to the level from which it emerged. At all levels agents use the same conditional "if-then" logic to interpret the meaning of the signals they receive. Agents "learn or adapt in response to their interactions with other agents….the actions of a typical agent are conditionally dependent on what other agents are doing" (Holland 2012:24–25). Complex adaptive systems "…change and reorganize their component parts to adapt themselves to the problems posed by their surroundings" (Holland 1995:18). Sometimes permutations and combinations of these conditional "if-then" interactions spontaneously generate a higher, more adaptive level of complexity (e.g., Konner 2010). Game theory shows, in principle, how the "top-down" behavioral control system of shared cultural models or cognitive schemas might have emerged with the evolution of human attachment by means of selection for increased cooperation in the resolution of parent-offspring conflict.

Game Theory

The emergence of game theory after World War II was a major impetus for the EES, as it raised questions about how the logico-mathematical operations that produce cooperation, in theory, could be embodied, in fact, in the behavioral phenotypes that produce it. Game theory is the study of mathematical models of conflict and cooperation between rational "agents" in which each agent has to choose a behavior in an attempt to obtain some "utility," and the success of one agent's choice depends on those of the others. In the classic iterated prisoner's dilemma game, for example, two players must either "cooperate" with or "defect" from the other to obtain their utility. Game theory has shown that cooperation can emerge (in the iterated prisoner's dilemma game) when the agents are connected by a shared utility or common cause (e.g., Axelrod and Hamilton 1981; Maynard Smith 1982; Axelrod 1984, 1997). Hume provided an early example in his *Treatise on Human Nature* (quoted in Skyrms 2010:21):

> Two men, who pull the oars of a boat, do it by an agreement or convention, tho' they have never given promises to each other. Nor is the rule concerning the stability of possession the less deriv'd from human convention, that it arises gradually, and acquires force by a slow progression, and by our repeated experiences of the inconveniences of transgressing it…

In game theory, the oarsmen are the agents and the utility they seek is what they intend to gain by rowing. Let us assume first that these oarsmen have different intentions. Whatever they are, the "repeated inconvenience" of bumping into each other could serve to draw them together. All they need to do is to detect the contingency between a bump (the signal of asynchrony) and what it feels like (what the signal means) to have their selfish intentions frustrated by the other oarsman. Pure self-interest would then motivate each oarsman to synchronize the timing of their own strokes with that of the other, which would help both to work together to achieve their individually selfish intentions. So it is with mothers and infants, when their intentions conflict. Mother-infant interaction is fundamentally the exchange of information (immaterial signals), the meaning of which can have material effects on their bodies, resolving their conflict and gradually drawing them (and their minds) together.

Alternatively, let us assume that the oarsmen have the same intention. Now, in addition to their self-interested motivation to avoid a collision of oars, they are connected by their shared intention: their motivation to work together to achieve a common good. All they need to do is detect the contingency between their mutually regulated strokes (no collision: synchrony) and the pleasurable feeling (the meaning of not colliding) of realizing their shared intention more quickly or efficiently than either could manage alone or without synchrony. Again, so it is with mothers and infants as they gradually learn that they can trust the other to help them (their group) maintain synchrony (connectedness), which feels better than asynchrony.

Nowak (2006) showed mathematically that there are five processes that can "connect" organisms and favor the evolution of cooperation: kin (shared genes) selection, direct reciprocity, indirect reciprocity, network reciprocity, and group selection. All five can "connect" infants, mothers, alloparents, and others in expanding circles of group identities. Nowak even concluded that "we might add 'natural cooperation' as a third fundamental principle of evolution beside mutation and natural selection" (Nowak 2006:1563). I suggest that increasing the "connection" between our ancestral mothers and infants would have transformed their natural conflict into "natural cooperation" and opened the door for them to co-construct a new niche, a new unit of selection at a higher level of complexity, the mother-infant group. Put differently, I suggest that our infant ancestors' motivation to "connect" or "attach" to mother-like people gave rise to the *feeling* of belonging to a group (and cultural group selection, as we'll see).

Group Selection

Group or multilevel selection theory maintains that when the cost of a prosocial act to an agent at one level is less than the benefit of that act to its group, group selection trumps the selfish agents at the preceding, lower level (Sober and Wilson 1998; Nowak 2006; Wilson and Wilson 2007; Nowak et al. 2010; Wade et al. 2010; Richerson et al. 2016). Thus, for example, because the cost of synchrony to an individual oarsman is minimal, two connected oarsmen would win a race against two selfish oarsmen. The idea that selection operates at the level of groups is, however, contentious. The MS, being gene-centric, rejects it; the EES, being information-centric, endorses it. The details of the kin selection-group selection debate need not, however, concern us.[4] What is of interest here is the key EES concept of niche construction: how the emergence of more connected mother-infant groups constructed a new niche with new selection pressures.

Niche Construction

According to the MS, environments exert selection on organisms. Under the EES, organisms can also select their environments; they can select (occupy) or construct a new niche by modifying an old one, either way potentially exposing themselves to new selection pressures (Odling-Smee et al. 2003; Stotz 2010; Laland et al. 2016). I speculate that the evolution of increasingly connected maternal and infant bodies would have constructed a new developmental niche with a new selection pressure: group selection for deeper, more intimate cognitive connections between mother and infant through mutual "mind reading."

---

[4] In any event, kin selection and group selection may just be two ways of thinking about the same thing. Their math is equivalent and the choice between them may depend more on particular interests, available data, or philosophical inclination (Nowak 2006; Nowak et al. 2010; Birch and Okasha 2015).

The Embodied Mind

Cognition is embodied when the body affects the brain (Varela et al. 1991; Gallese and Lakoff 2005; Fonagy and Target 2007; Gallese 2007; Johnson 2007; Adams 2010; Niedenthal et al. 2014; Clark 2016). It is embodied in infant brains by the effect of each bout of mother-infant interaction on the infant's body. As evolution embodied the immaterial laws of aerodynamics into wings and those of optics into eyes, the attachment process embodies the immaterial principles and logico-mathematical operations of game theory, biosemiotics, and complex adaptive system theory into infant brains. These "immaterialities" ("pure reason") are embodied in newborns through exteroception and interoception. Exteroception provides the infant with information about its mother. Interoception provides it with information about its *milieu inté-rieur*—the "feeling of what happens" (Damasio 1999) to its body—when it detects a sign that points at mother. The infant's interpretation of the contingency between mother's behavior and the way it feels at that moment[5] *is* the way it feels; the way it feels points at and means "mother," and vice versa: the contingency between the way the infant feels at a given moment and mother's response points at and means "me." The way it feels "now" is the phenotypic expression of "old," genetic information about the way mammalian infants should feel in that context. The feeling associated with each "now" experience is stored in memory as "newer" information.

Tomasello et al. (2005:680) proposed that the "origin of cultural cognition" was in the evolution of our capacity to share intentions: "Shared intentions, sometimes called 'we' intentionality, refers to collaborative interactions in which participants have a shared goal (shared commitment) and coordinate action roles for pursuing that shared goal." They believe that this capacity evolved from a common great ape's "understanding [of] others as animate, goal-directed and intentional agents" and a human "species-specific motivation to share emotions, experience, and activities with other persons" (Tomasello et al. 2005:675). However, they do not specify what this motivation was. What did selection "see" such that our infant ancestors evolved the "species-specific motive" to share the work of pursuing a shared intention?

I believe that what it "saw" was our infant ancestors' ancient mammalian motivation to form attachments.

Complex Adaptive Nervous Systems

Nervous systems are themselves complex adaptive systems—networks of hierarchically organized neural agents working together to control the body and its movements. They are the embodiment of the conditional, "if-then" logic of biosemiotics, complex adaptive system theory, and game theory. They become

---

[5] Infants can detect contingencies in less than a third of a second (Beebe and Lachmann 2014).

organized through cell-cell communication: if a neuron receives a certain kind of signal, it then interprets its meaning by sending a signal to other neurons. Nervous systems are the embodiment of Hebb's "law" (Hebb 1949): "neurons that fire together, wire together." That is, neurons become connected by communicating within and between levels.

At the bottom of the hierarchy are the most inferior brain regions (spinal cord and brain stem), which generate what Panksepp and Bevin (2012:96) call "intentions-*in*-action," as opposed to "intentions-*to*-act" at the top, the prefrontal cortex (PFC). Intentions-in-action include unmotivated reflexes and the impulsive "acting out" of feelings. Like bacteria, the deeply subcortical brain's interpretation of a stimulus *is* its response to the stimulus. If there's a discrepancy between "now," interoceptive information about the state of the body, and "old" information about what it should be, the deeply subcortical brain interprets the mismatch by activating innate autonomic or motor reflexes for maintaining homeostasis.

The separation between intention and action begins to emerge at the second level, where a higher subcortical region (cerebellum) interprets a stimulus to move, not by doing so, but as an instruction to consult motor memories ("newer" information) before acting. Inserting information about past motor behavior between "if" and "then" makes for more complex, coordinated, and clearer behavioral signals that are more effective because they are easier for mother to interpret.

The third level is the limbic system (amygdala, hippocampus, thalamus, hypothalamus), which is the interface between the subcortical brain and the PFC. The limbic system generates mammalian social emotions and motivations and inserts them as "value judgments" into the information it passes up to the PFC; that is, what this information means (the feeling it entails), first for the infant's immediate survival, then, as it continues to survive, for its growth and development and ultimate reproductive success. The limbic system is the origin of intention: emotions make the body want to do what its "old" mammalian emotions signal it to do. Panksepp identified seven "primordial" motivational systems in the brains of all mammals (Panksepp 1998; Panksepp and Bevin 2012). I will follow his lead in capitalizing their names to emphasize that, unlike everyday emotion terms, they refer to well-defined neuroanatomical structures and functions: SEEKING (wanting, expectancy), FEAR (anxiety, insecurity), RAGE (anger), LUST (sexual excitement), CARE (nurturance), PANIC/GRIEF (loss, sadness), and PLAY (the joy of play and exploration). I focus on SEEKING because it is the body's prime mover, the "generic appetitive force" that drives all the others. According to Panksepp and Bevin (2012:103), SEEKING is

> ...a general-purpose system for obtaining all kinds of resources that exist in the world, from nuts to knowledge, so to speak. In short, it participates in all appetitive behaviors that precede consummation: it generates the urge to search for any

and all of the "fruits" of the environment; it energizes the dynamic eagerness for positive experiences from tasty food to sexual possibilities to political power; it galvanizes people and animals to overcome dangers either by opposing them or by escaping to safety; it invigorates humans and prompts us to engage in the grand task of creating civilizations. But in the beginning, at birth, it is just "a goad without a goal"…that opens up the gateways to engagement with the world, and hence knowledge.

In fact, the goad does have a goal; its adaptive function is just that—to engage the world. Newborns acquire information about people by engaging with their bodies, much as Hume's oarsmen acquire information when their oars collide. Newborns engage mothers and others when their SEEKING systems energize an intention-in-action simply to move, thereby acquiring contingent feedback with which to construct internal motor models of how it feels to engage with them.

Damasio's "somatic marker hypothesis" holds that the limbic system is the source of mental images of past feelings which the PFC can access and "hold in mind" while deciding what to do next (Damasio 1994). He sees the connection between the limbic system and PFC as an "as-if loop" whereby a mental image of a past feeling is inserted into an action plan "as if" it had already been completed. This "as-if loop" makes it possible to evaluate images of the past and future. It enables us "to use a part of our mind's operation to monitor the operation of other parts" (Damasio 2010:28). The feeling tone connected to the imagined future points at or "marks" it as good, bad, or uncertain for the body's fitness, "now" and in increasingly distant futures.

The evolution of our capacity to "monitor" our subcortical emotions (to feel them) gave us a sixth sense, so to speak: our sense of value, the subcortical, inherently subjective feeling of SEEKING (even if we are not always sure what it points at). Natural selection "built the apparatus of rationality (PFC) not just on top of the apparatus of biological regulation (subcortex), but also from it and with it" (Damasio 1994:18), and because of it, Damasio might have added. The entire adaptive point of having a brain is to use it for the good (survival, growth and development, reproduction) of the body. The amygdala plays a critical role in keeping the body alive by looking for signs of danger in the information it receives from the body and inserting "fear" into the signals it sends up to the PFC (Gee et al. 2013; Callaghan and Tottenham 2016b). When it detects danger (risk or uncertainty), it attempts to avoid it by activating the infant's HPA system, initiating the release of cortisol, with its short-term benefit to survival (but potential long-term costs in the form of stress-related disease). Nervous systems are complex adaptive systems, of which the "fundamental attribute" is that "an internal model allows [the system] to look ahead to the future consequences of current actions, without actually committing itself to those actions. In particular, the system can avoid acts that would set it irretrievably down some road to future disaster ("stepping off a fitness cliff")" (Holland 1992b:25).

Because newborns are all intentions-in-action, they are unable to look ahead to anything. Their optimal developmental strategy is thus to pay particular attention to downside protection against risk and uncertainty. Continuing to survive gives them time to construct internal models for looking ahead to future consequences. Newborns construct internal models of how it feels to engage with mother by applying the scientific method (e.g., Reddy 2008; Gopnik 2009). Each engagement provides the infant with inductive "now" information about the contingencies between the "feeling of what happens" when mother responds to its intention-in-action. This information is passed into memory as "newer" information, adding to what is, in effect, a kind of "correlation matrix of contingencies," a "database" of experience for the PFC to "analyze" for patterns and construct (deduce) models of experience and expectations.

The PFC is a neurobiological adaptation that enables us, among other things, to use information about the logical contingencies among beliefs, desires, and intentions to manage our social relations. (It does the deductive work of internal working models, theory of mind, mentalizing, intersubjectivity, etc.) Tomasello et al. (2005:675) argue that sharing intentions ("we" intentionality) is the "origin of cultural cognition" and the "foundational skill" for theory of mind "because it provides the interpretive matrix for deciding precisely what it is that someone is doing," or did, intends to do, or means. A newborn nervous system has the capacity to make primitive meaning because its interpretive matrix came with some information "wired-in," in the form of "old" information about the environment of evolutionary adaptedness of mammalian infants. With maturation and repeated interactions with mothers and others, infant nervous systems construct larger, multimodal, cross-temporal interpretive matrices. As Bowlby observed, each bout of interaction is an iteration of the feedback cycle that controls infant social-emotional-cognitive development. Each iteration of this "attachment cycle" (Figure 11.1) embodies information that the infant nervous system evaluates in terms of its own self-ish SEEKING for security, and uses to construct beliefs about mother's beliefs, desires, and intentions regarding the infant (Sroufe and Waters 1977). It goes without saying, of course, that while the attachment cycle itself is universal, the beliefs that infants construct are inherently specific to their culture.

Attachment and Cooperation

The attachment cycle integrates information from each of the four levels of complexity in the evolution of life (Table 11.3). First, genetic information establishes the infant's "set-goal" in the form of "old" information about the way mammalian infants should feel, i.e., secure (see Figure 11.1, "Felt security"). Second, epigenetic information may be embodied through exposure to stress or the inheritance of the effects of stress on the infant's parents (e.g., Weaver et al. 2004; Turecki and Meaney 2016), especially the stress of intergenerational poverty (McEwen and McEwen 2017). Information about stress (risk

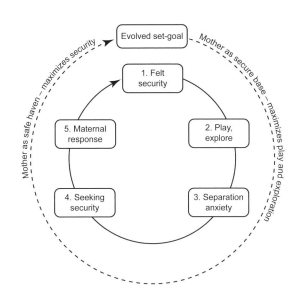

Figure 11.1 Schematic diagram of "attachment as an organizational construct."

and uncertainty) is particularly important to newborn bodies because they are unable to feel the way they should by themselves; they need tolerance and help from mothers or others. Third, behavioral information is embodied through each iteration of the attachment cycle. Information about stress (the feeling of insecurity) is particularly valuable to newborns because the adaptive function of insecurity is to stay alive by SEEKING security. The only way newborns can SEEK security is to signal their needs (intentions-in-action), in the evolution-arily rational expectation that someone will provide a safe haven. (I'll discuss the fourth level of complexity in the following section on "Shared Intentions.")

But there's more to life than safe havens. To acquire information about their developmental niche, infants must explore it. The adaptive function of security is to SEEK information, which, as Panksepp and Bevin (2012:8) pointed out "opens up the gateways to engagement with the world, and hence knowledge." The only way infants can SEEK information is to signal their need for it by the intention-in-action of engaging their world, probing and playing with the people and things around them, in the evolutionarily rational expectation of a secure base to which they can return. Sooner or later, the infant's probes will produce painful contingencies that activate its HPA stress-response system, release cortisol, separation anxiety, and the intention-in-action of SEEKING security. Then comes the nub of the attachment cycle: how its mother (*qua* minimalist mammalian mother) responds to its signal of need.

Each iteration of the attachment cycle provides the infant with an opportunity to detect the contingency between a mother's response to its signal and the feeling of what happens to its body when she does (or does not) respond. Internal models of experience are constructed from "backward" contingencies: they attempt to explain the past ("after X happened, I felt Y"). Internal models of expectations are constructed from "forward" contingencies: they attempt to predict the future on the basis of the past ("because X happened in the past, I will feel Y if it happens again"). In the stark binary terms of propositional logic, biosemiotics, and complex adaptive systems theory, a mother's response either downregulates the infant's HPA system or it does not; it either makes the infant feel secure or insecure. Each iteration of the attachment cycle provides new "now" information that is stored as "newer" information about its attachment history in its "correlation matrix of contingencies" or "interpretive matrix for deciding precisely what it is that someone is doing," or did, intends to do, or means.

Animal behavioral traditions are maintained by learning. Infants learn their mammalian behavioral traditions, first by learning from the mother's body, then through her mind. Infants begin learning from and about their mothers by detecting the contingencies that "connect" their bodies—the feeling of what happens before, during, and after they come in contact, move, and bump into each other. In the stark binary terms of game theory, each contact (Figure 11.1, "Maternal response") gives mothers an opportunity to "cooperate" with their child or "defect" from it. As with Hume's oarsmen, pure self-interest can motivate each to adjust its movements to the other. In addition, generic mammalian mothers are predisposed to cooperate because they share 50% of their genes with their children and are motivated to care for them. It is in their genetic self-interest to tolerate, at least, the child and its attempts to nurse (see, however, further discussion of parent-offspring conflict below.) By "working together" the mother-infant group achieves a degree of mutual regulation, biobehavioral synchrony, and psychobiological attunement. To paraphrase Hebb, "bodies that move together, wire together." Infants also begin to learn from and about mother in the first few weeks of life through imitation (Meltzoff and Moore 1977; Meltzoff 2005).

Compared to young chimpanzees, young children are "hyper-imitators," prone to "overimitation"[6] (Whiten et al. 2009). This may be due to the human mirror neuron system (Gallese 2007). The capacity for imitation was obviously important in the evolution of our capacity for culture (Tomasello 1999; Richerson and Boyd 2005; Lyons et al. 2007; Burkart et al. 2014). As Whiten et al. (2009:280) stated, "we are such a thorough-going cultural species that it pays children, as a kind of default [adaptive learning] strategy, to copy willy-nilly much of the behavioral repertoire they see enacted before them"—but not by just anyone. Henrich and McElreath (2003) argue that the evolution of culture depended on "biased imitation"—our well-documented propensity to imitate others because of their prestige (higher status, fame), success (power, wealth), similarity ("like me": identity), and/or conformity (proximity: do what people around me are doing).

---

[6] Unlike chimpanzees, children also have a strong tendency to imitate the actions of a model that are extraneous to the goal of the action being modeled.

It is not clear, however, where these imitation biases came from. Because selection operates only on phenotypes, not genotypes, the EES would like to know (a) what phenotypic mechanisms motivate anyone to imitate anyone else, (b) how they develop, and (c) what their phylogenetic precursors might have been. In other words, what did selection "see" such that it evolved into our children's capacity for overimitation? I believe the most parsimonious explanation is that newborn infants imitate mother-like people because they are innately motivated to SEEK the species-specific patterns of sign stimuli and contingencies which point at "mother." Mothers are "attractors" in the design space of generic mammalian newborns because, from the infant's perspective, they have higher status, all the power and resources, are "like me" (belong to my group), and are usually in close proximity. This is a parsimonious explanation because it posits a single adaptive learning strategy—imitate whoever acts like a mother—rather than a number of separate adaptive functions or motivations for imitating different people with different qualities; mothers embody them all. Mother-like people are important for the infant's later cognitive development because their "power" includes epistemic authority; infants cannot help but have "epistemic trust" that a mother is modeling appropriate behavior (Csibra and Gergely 2011).

Shared Intentions

At the top of the four levels of complexity (Table 11.3) is the processing of symbolic (i.e., cultural) information, accomplished in the PFC, the brain's "higher association area." Accounting for twice the total brain volume in humans as in the other apes, the PFC embodies the brain's executive functions and working memory. Within it is the medial prefrontal cortex (mPFC), which, with input from working memory, executes top-down cultural control over our generic mammalian social emotions. With the onset of mPFC myelination at about nine months (Paus et al. 2001), infants begin the transition from intentions-in-action to intentions-to-act. They become able to form forward contingencies—expectations about mother's behavior—based on their "interpretive matrix" of backward contingencies in their attachment history. They develop the ability to "hold in mind" the limbic feeling of what happened to their bodies in the past and form internal models of how it would feel (via Damasio's "as-if loop") if they had executed some intention. If the expected feeling does not match its "old" mammalian information about how it should feel (secure), their mPFC can inhibit that intention.

Chronic early stress (HPA activation) can affect the developing connections between the limbic system and the mPFC, with potentially severe consequences for a child's self-control, exploratory behavior, and HPA reactivity. In theory, if mothers consistently demonstrate their ability and willingness to invest, infants will form positive expectations about future interactions. If not, they will have negative expectations, in one of two ways. First, in theory, if mothers have been too inconsistent in responding to their signals of need, infants will be insecure (uncertain, anxious, ambivalent, preoccupied) about their mother's ability to meet their needs. Second, if mothers have too often ignored, rejected, or punished their infants' signals of need, they will be insecure about their mothers' motivation and mirror her unwillingness back to her by not signaling, thereby avoiding, not so much her, as the risk of pain from an expected rejection (Main 1981). Indeed, Behne et al. (2005) showed that by the age of nine months, infants can detect the difference between an adult's ability and willingness (motivation, intention) to perform an act.

The nine-month-old's ability to interpret its mother's intentions and generate expectations is the essence of theory of mind (internal working model, mentalizing, etc.). Belief-desire-intention reasoning is practical (Bratman 1987): it is used to program artificial intelligent agents as well as to model successfully reasoning in three- to ten-year-old children (Wahl and Spada 2000). A secure one-year-old, for example, would (a) *believe* that his mother cared for him, (b) so she must *desire* to do so, (c) therefore she *intends* to do so in the future (assuming she is able), and (d) therefore he *expects* her to do so.

Positive expectations maximize the infant's desire to play and explore. An infant will engage its mother and others by SEEKING information through their minds to make the most of its secure base. Positive expectations emerge from the infant's mammalian sense of being connected to mother through mutual regulation, biobehavioral synchrony, and psychobiological attunement. Each good connection leads to the next. Each is associated with the pleasant feeling of "working with" mother, adding to her reputation for cooperation. Positive connections transform the natural parent-offspring conflict into Nowak's "natural cooperation" (Atkinson et al., in preparation), giving rise to the mother-infant group, shared or "we intentionality," and the feeling of consensus, group identity, or "we-ness" with mother. The infant's capacity for "we-ness" emerges, bottom-up, with the maturation of the connections between her limbic system and mPFC, giving her increasing top-down control over her mammalian infant social emotions. In a "self-similar" way, the mother-infant group emerges, bottom-up, from nine months of postnatal mutual regulation, synchrony, and attunement, gradually increasing its top-down control over their individual identities and desires. If we project this model of the ontogeny of "we-ness" backward as an "evo-devo" model of its phylogeny, it is not hard to imagine that early hominin mother-infant groups whose minds were better-connected would have had an evolutionary edge over less-connected mother-infant groups. Better-connected infants would inherit not only their animal behavioral traditions from mother's body, but also her emerging symbolic, cultural traditions through her mind. On this view, natural mother-infant cooperation created a new niche, culture, at a higher level of complexity, the cultural group, and a new selection pressure, cultural group selection,[7] for mind reading and increasingly well-connected, hypercooperative mother-infant groups.

Negative expectations, on the other hand, maximize the infant's desire for a safe haven, SEEKING security in hope of staying alive just long enough to have a life and avoid a fitness cliff. Building on the pioneering work of Draper and Harpending (1982), Belsky et al. (1991) proposed that the attachment process was an evolved mechanism for entraining alternative reproductive (life history) strategies. It enabled infants to predict the quality of the environment in which they will reproduce from their experience of that into which they were born. Writing in support of their proposal, I argued that it was fully consistent with life history theory (Chisholm 1999b). When mothers are unable to buffer their offspring against environmental risk and uncertainty—above all that of death—it is evolutionarily rational to grow, develop, and reproduce as early as possible to avoid stepping off Holland's "fitness cliff" (Table 11.2). But before looking at the effects of infants' expectations on the development of their reproductive strategies, I need to say a word about how they go from sharing intentions with mother to sharing meanings. In other words, how the infant's expectations draw it into the mother's mind, where it can "read" her intended meanings and learn how to think.

Shared Meanings

Because there is no reason to think unless one is SEEKING a "utility" to think about, emotion is inseparable from cognition. Mammalian emotions were phylogenetically prior to human cognition, gave rise to it, and give rise to it ontogenetically. As Hobson put it, "symbolizing, language and thought are possible only because of the nature of the emotional connection between one person and another, and because of each person's involvement with a shared world" (Hobson 2002:94). Before humans evolved the capacity to think, there "was *social engagement with each other*. The links that can join one person's mind with the mind of someone else—especially, to begin with, emotional links—are the very links that draw us into thought. To put it crudely: the foundations of thinking were laid at the point when ancestral primates began to connect with each other emotionally in the same ways that human babies connect with their caregivers" (Hobson 2002:2, original emphasis).[8]

The link that first connects mothers and infants is bodily: mutual regulation, biobehavioral synchrony, and psychobiological attunement. As Fonagy and Target (2007:428) state:

> Since the mind never, properly speaking, separates from the body, the very nature of thought will be influenced by characteristics of the primary object relation.…

---

[7] For cultural group selection theory, see Richerson et al. (2016).

[8] Baron-Cohen (2002) captured Hobson's argument perfectly in the title of his *Nature* review of Hobson's book: "I am loved, therefore I think."

> The origin of symbolic representation is thought to be biologically significant actions tied to survival and adaptation. Such actions are steeped in somatosensory [interoceptive] experiences and salience and are perceptually guided. Thus, implicit in the use of a symbolic representation is the history of bodily and social experience of actions related to the symbol.

Or, in the words of Lakoff and Johnson (1999:555):

> We can only form concepts through the body. Therefore, every understanding that we can have of the world, ourselves, and others can only be framed in terms of concepts shaped by our bodies.

Infants begin life with the intention-in-action of SEEKING information from and about their caregiver's body through physical interactions. To the embodied mind, symbols are material objects, out there in the world, perceived through exteroception *and* they are the material, interoceptive representation of the object in the neural networks that stand for or point at the object (Trevarthen 2009, 2011; Vogeley and Roepstorff 2009; Clark 2016). Roepstorff (2008:2051) reviewed neuroimaging evidence which showed that "once words are understood by a person, they become material instantiations in some form in the brain." For example, action words with specific targets ("the man goes into the house") activate motor areas of the brain; however, when the target of the action is abstract ("the man goes into politics"), motor areas are silent and language areas active. Thus, he suggests that "the overall neural resonance created by words interacts with non-linguistic brain areas involved in representing processes that the word represents" (Roepstorff 2008:2051). In a self-similar way, the "neural resonance" between mother and infant replicates the mother's symbols in the infant's brain. To paraphrase Hebb again, "minds that fire together, wire together," producing the pleasant feeling of being understood. Everyone's first experience of culture is with people who act like mothers. To paraphrase Geertz (1973:5), mothers and infants are "animals suspended in webs of significance they themselves have spun."

The Extended Embodied Mind

Cultures are the extension of embodied minds in which one person's mental state is extended to or constructed in others (Clark and Chalmers 1998; Moll and de Oliveira-Souza 2009; Menary 2010; Kendal 2011; Shea 2012). Preceding Tomasello's concept of "we intentionality"—he even used the term "we-ness"—by 59 years, Bowlby (1946:62) was concerned with "the psychological problem of ensuring persistent co-operative behavior" in groups of any kind. He maintained that "the principal conditions for willing co-operation are thus that there should be a common aim, apprehended to some degree at least as being of value both to the self and to others who are loved, and that the individual or individuals who present this common aim and the plan for achieving it should do so in such a way that they are respected and trusted" (Bowlby 1946:63). Bowlby believed that the adult motivation to cooperate originated in the infant's desire "to be held in good esteem by the people he values" (Bowlby 1946:65), and that the capacity to value people was the capacity to libidinize,"[9] which "originates in infancy in the child's feeling for his mother" (Bowlby 1946:64). He goes on to describe how the infant's capacity to libidinize mother is the origin of adults' motivation to belong to valued groups and to place emotional value onto group leaders, the group itself, and the group's "policy" (ethos or belief system). I would argue that it's also the origin of the *infant's* capacity to emotionally value groups—groups of alloparents.

The leader of the infant's first group is mother. The infant SEEKS to be held in good esteem by her because he depends on her for his life. His weakness makes him subordinate to her; he needs someone "stronger and/or wiser" (Bowlby 1988a:3). After nine months of good-enough mothering, he begins to develop a sense of group identity, the feeling of "we-ness" about the self-mother group itself. "We-ness" opens the door for "we" intentionality, the feeling of sharing a common aim. Understanding the leader's plans for approaching the group's aim, however, is not easy. It is easier to form feelings of love or pleasure toward a person than toward a plan or policy. It is also easier for infants (and many adults) to identify with a group's leader—and trust that God has a good plan—than to trust the reasons why it is a good plan. People SEEK security in groups more than they seek wisdom because "the group is thought of *as though it were an individual*, and feelings of personal affection are evoked" (Bowlby 1946:61, emphasis added; see also Ein-Dor and Hirschberger 2016).

Like Bowlby, the game theorist Michael Bacharach was concerned with the psychological problem of ensuring persistent cooperative behavior in groups; that is, how "early man managed to function well in groups, by doing things that we are inclined to call 'cooperation'..." (Bacharach 2006:98). He began by emphasizing the huge diversity in the types of game theoretic games that people can play with each other. While cooperation can evolve in the iterated prisoner's dilemma game, Bacharach argues that this cannot explain altruistic or prosocial behavior in other types of games. After a technical explanation of why this is so, he concludes that prisoner's dilemma games "do nothing to explain, *psychologically*, cooperative behavior in common-interest interactions, or in organized interactions" (Bacharach 2006:111, emphasis added). In other words, the iterated prisoner's dilemma cannot explain the emergence of "we-ness" or "group identification":

> ...group identification is the key proximate mechanism in sustaining cooperative behaviour in man. More fully, I conjecture this: dispositions to cooperate in a range of types of game have evolved in man, group identification has evolved

---

9   Libidinize, or the capacity to form a libido, was restricted in early Freudian theory to the formation of sexual drive, but later expanded to include all expressions of love and pleasure.

in man, and group identification is the key proximate mechanism for the former. The main virtue of this hypothesis over that of altruism and other contenders [for explaining human cooperativeness] is that group identification is a more powerful *explanans* of the diversity of cooperative behaviors we see. Group identity implies affective attitudes which are behaviorally equivalent to altruism in Dilemmas, and it can explain what altruism cannot, notably human success in common-interest encounters (Bacharach 2006:111).

Attempting to cast new light on the evolution of human cooperation, Moffett (2013) conducted a phylogenetic analysis of hunter-gatherer and vertebrate social groups. Reasoning that human groups are characterized by an especially high degree of cooperation and the other vertebrate groups are not, Moffet defined "society" minimally as "cooperation beyond mere sexual activity," so as to include as many vertebrate phyla as possible. He found that "individual-recognition" societies (e.g., nonhuman primates, social carnivores) rarely exceeded 200 individuals, whereas human societies were exponentially larger (millions, if not billions). What made the difference, he suggests, is that we evolved the capacity to have a *concept* of identity. We do not have to recognize individuals because we evolved the capacity to identify people by the signs of the group with which they are identified. This concept of identity enabled our ancestors to construct "nested hierarchies" of group identities (e.g., kinship systems) based on "degrees of intimacy." At the bottom of the hierarchy is the most intimate: the mother-infant group.[10] Then, in order of decreasing intimacy (and rough prehistorical order of emergence) came the nested hierarchy of allomothers, nuclear and extended families, bands, clans, tribes, and so forth, up to more than a billion people in modern states and religions, each exerting a degree of top-down control over the preceding levels. Our cortical capacity for cultural concepts of group identity emerged phylogenetically, and does so ontogenetically, from our mammalian limbic resonance or sense of "we-ness" with mothers and others—Hobson's emotional link that draws us into thought.

Alternative Life Histories

Life History Theory

As argued above, the Belsky-Steinberg-Draper attachment model of the development of alternative reproductive strategies is thoroughly consistent with life history theory. Life cycles (Table 11.2) are reproductive strategies that have been organized by information about an organism's developmental niche. The most telling information is about the probability of dying at a given age (Promislow and Harvey 1990, 1991; Stearns 1992; Charnov 1993). When mortality rates are low and predictable, it is relatively easy for mothers to buffer their children against environmental risk and uncertainty. Under such

---

[10] Moffett, a zoologist, mistakenly identifies "married couples" as the most intimate.

conditions, the future or "long-slow" strategy is optimal because it gives organisms time to grow bigger bodies and brains, gain more experience, become more competitive (by cooperating, if they're human), find a good mate, produce a few offspring, and invest heavily in each. This maximizes the probability of the minimum gain necessary for lineage continuation by maximizing offspring "quality" (reproductive value: potential for providing grandchildren). It fosters what Daniel Kahneman refers to as "thinking slow" (Kahneman 2011).

Alternatively, when mortality rates are high or unpredictable, it is harder for mothers or alloparents to buffer children from the causes and consequences of high mortality rates (e.g., intergenerational poverty, inequality). Under such conditions, the current or "short-fast" strategy is evolutionarily rational because it enables organisms to reproduce as early and often as possible. Reproducing early maximizes the chance of reproducing before mortality strikes; reproducing often maximizes the chance that at least one offspring will survive. This minimizes the probability of the maximum possible loss, lineage extinction, by maximizing offspring quantity, even at the cost of future morbidity and shortened lives. It fosters Kahneman's notion of "thinking fast."

Survival, growth, and development are necessary for reproduction but not sufficient. Evolution does not "care" about organisms' quality or length of life; all it cares about is reproduction. To maximize the chance of reproduction under adverse conditions, natural selection favors mechanisms that enable organisms to "make the best out of a bad bargain": available resources from growth and development are reallocated to permit early and/or frequent reproduction. In humans, chronic early adversity (e.g., psychosocial stress, HPA activity) is associated with early puberty in boys (Mendle and Ferrero 2012) and girls (Coall and Chisholm 2003; Chisholm et al. 2005; Chisholm and Coall 2008; Ellis et al. 2009; Belsky et al. 2015). In turn, early puberty is linked to increased risk for obesity, elevated blood pressure, heart disease, type 2 diabetes, and, in women, breast cancer (Collaborative Group on Hormonal Factors in Breast Cancer 2012; Hanson and Gluckman 2016). Early adversity is also linked to a suite of behavior problems. As Bowlby said of adolescents and young adults who had suffered disturbed early family relations, "persistent stealing, violence, egotism, and sexual misdemeanours were among their less pleasant characteristics" (Bowlby 1951:380). The same traits also comprise the "absent father syndrome" (Draper and Harpending 1982), the "young male syndrome" (Wilson and Daly 1985), the "male supremacist complex" (Divale and Harris 1976), and "cultures of risk" (Quinlan and Quinlan 2007) and coping (Burbank 2011). Early adversity is also associated with impaired ability to delay gratification (Chisholm 1999a; Coccaro et al. 2015; Sturge-Apple et al. 2016) and increased psychopathology (Del Giudice 2014; Hurst and Kavanagh 2017), in particular borderline personality disorder. Brüne (2016:52) characterizes borderline personality disorder as "unstable interpersonal relationships, fear of abandonment, difficulties in emotional regulation, feelings of emptiness, chronic dysphoria or depression, as well as impulsivity and heightened risk-taking behaviors." Interpreting borderline personality disorder in terms of life history theory, he describes the lives of sufferers as "fast and furious." Insecure attachment is specifically implicated in its development (Fonagy et al. 2000; Debbané et al. 2016).

The mechanisms by which early adversity affects later health and behavior are not fully understood, but chronic HPA activity can modify the expression of genes involved in neurodevelopment (McGowan et al. 2009; Turecki and Meaney 2016). Early adversity has been linked to enlarged amygdalae (Tottenham et al. 2010) and accelerated amygdala-mPFC connectivity (Gee et al. 2013; Callaghan and Tottenham 2016a, b). In keeping with life history theory's "short-fast" ("fast and furious") reproductive strategy, Callaghan and Tottenham (2016a:79) propose that:

> ...accelerated [fearful] phenotypes emerge because stress experienced early in life may prematurely activate the core circuitry of emotional learning and re-activity. That is, the acceleration of limbic development following early stress may rely on an activity-dependent process. Importantly, we hypothesize that this accelerated development, while meeting immediate [short-term, downside risk protection] emotional demands (i.e., emotional regulation in parental absence), may have long-term consequences...on emotion regulation in adulthood.

Acting out the bodily sensation of fear may be (or once have been) an evolutionarily rational response to chronic risk and uncertainty, but it is not conducive to mind reading, trust, or cooperation.

Cooperative Breeding

Throughout human evolution, the probability of death, and thus the force of selection, has been highest in infancy and early childhood (Jones 2009). Keeping their increasingly needy children alive through the intense selection of their first few years could not have been easy for early mothers. The greater their children's need, the more intensely they would have been selected to meet it. This created a demographic dilemma: the more that mothers worked to keep one child alive, the harder it became to have another and keep it alive as well. The trade-off for rearing quality children was reducing their quantity, making it harder to maintain reproductive rates at replacement level and increasing the threat of lineage extinction (Lovejoy 1981). But our ancestors did increase their reproductive rate: "Humans, who of all apes produce the largest, slowest-maturing, and most costly babies, also breed the fastest" (Hrdy 2009:101). Theory, cross-taxa, and cross-cultural evidence indicate that reducing inter-birth intervals without compromising child survival was possible only because mothers got help with child care.

There is no question that cooperative breeding was critical for the survival of our species (Burkart et al. 2009; Hrdy 2009; Sear 2016), or that grand-mothers in particular radically changed the developmental niche of our infant ancestors (Hawkes 2004, 2014). Long, slow development gave our ancestors more time to grow big brains, with well-developed association areas, and to develop complex social, emotional, and cognitive skills—but it also increased the cost of rearing them. Cooperative breeding solved the demographic dilemma by spreading this cost among alloparents such that our ancestors were more likely to live long enough to benefit from their enlarging cerebral cortices and "learn more, know more, become more efficient at food procurement, outcompete others for mates, and so forth" (Hrdy 2009:277). For Hrdy (2009:277, original emphasis) "cooperative breeding had to come first" because it was "the *preexisting condition* that permitted the evolution of these traits in the hominin line." Likewise, Hawkes (2014:29) argues that "grandmothering sets up the novel selection pressures on mothers and infants identified by Hrdy."

But why would grandmothers or anyone feel like helping a needy mother? Since selection operates only on phenotypes, not genotypes, the EES would like to know what phenotypic mechanism motivates anyone to cooperate with anyone else. Hrdy's model of the role of cooperative breeding in human evolution as well as Hawkes's of grandmothers in particular, are compelling but say little about the role of the infant in the evolution of cooperative breeding or culture. Recently, Hrdy has argued that our capacity for prosocial, "other-regarding" feelings were "predictable corollaries of [cooperative breeding] and as a byproduct of it, preadapted apes in the hominin line for greater social co-ordination" (Hrdy 2016a:43). But what was the preexisting, phylogenetic precursor of prosocial emotion? What did selection "see" such that it evolved into Tomasello's "species-specific motivation" to make common cause with one another? I believe that our prosociality was more than a byproduct of cooperative breeding; it was intimately involved in its emergence and that of culture itself.

Parent-Offspring Conflict

Each iteration of the attachment cycle provides mothers an opportunity for parental investment—a chance to respond cooperatively to their infants' signals of need. Whether and how they respond depends on their ability and motivation. When mothers are materially, socially, and emotionally secure, parent-offspring conflict tends to be minimal and constructive. The constructive resolution of conflict builds trust. It repairs the "messiness" of breakdowns in Tronick's mutual regulation model, restores the synchrony in Feldman's biobehavioral model, the attunement in Fields' model, and buffers the infant against HPA hyperreactivity and elevated cortisol in Gunnar's model. However, when mothers are under stress from the causes and consequences of intergenerational poverty or inequality, they may well be less able or willing to invest: more of their interactions are likely to be messy and end badly, leaving the infant and/or mother feeling insecure (hungry, tired, frustrated, confused, sad). Instead of buffering infants against the stress their environment imposes on them, mothers transduce it to their infants.

Conflict is unpleasant and inevitable, but without it there is no reason to co-operate. Likewise, without parent-offspring conflict there would have been no reason for mother-infant cooperation or cooperative breeding. The evolution of our prolonged helplessness escalated our existing ape level of parent-offspring conflict into an early human mother-infant "arms race." Infants would have exerted continuous selection on mothers for their ability and motivation to respond effectively to signals for care and attention. Mothers would have con-tinuously resisted by allocating their limited time and energy to (selecting for) infants with the ability and motivation to send more persuasive signals. The result would be positive feedback between the effects of mothers' selection on infants and vice versa (Chisholm 2003; Kilner and Hinde 2012). When this feedback cycled to the point that mothers could no longer provide enough by themselves, it would have been evolutionarily wise for them to get help with child care. Those with sufficient social skills and/or relationships to recruit or attract alloparental care would be more likely to produce another child before the previous one was independent. Shorter birth intervals, however, opened a new arena for parent-offspring conflict—the "dark side of cooperative breed-ing" (Hrdy 2009:100). Except for the other cooperatively breeding primates, the callitrichids, only human mothers can have a child before the preceding one is independent. And, like the callitrichids, only human mothers have ever had to decide that one child is a better "investment" than another, and to neglect, reject, and even kill those judged less likely to provide grandchildren (Hrdy 1999, 2009). Understanding a mother's intentions would have been an evolu-tionarily wise basis for attempting to "mold" her into providing more invest-ment or avoiding its termination with prejudice.

Conclusion

I believe that an attachment theory informed by the EES can readily incor-porate the concept of culture and the role of alloparents—those to whom the infant is psychobiologically attuned as if to its biological mother. The key is to focus on the role of the "feeling of what happens" to the body during the acquisition and reproduction of information.

From biosemiotics comes the notion that organisms are matter and energy that have been organized—embodied—by information, and that meaning, value (feelings), and intention are facts of nature represented by signs. This understanding will help us resist mind-body and nature-culture dualism. From complex adaptive systems theory comes the concepts of self-replication and emergence—the idea that complex systems not only reproduce themselves but can acquire information such that higher levels of complexity can emerge, bottom-up, to exert top-down control over lower levels. Evolution is a complex adaptive system; it embodied the information that organized organisms into increasingly complex wholes. Organisms are also complex adaptive systems:

they die, but the information by which they were organized is reproduced, outlasting the matter and energy they embodied. Infants, too, are complex adaptive systems: they acquire the information that organizes their nervous systems into the increasingly complex wholes by which they exert culturally appropriate top-down control over their behavior.

Game theory offers the notion that cooperation can emerge from conflict. I have argued that biology, in the form of (a) mammalian infants' motivation to form attachments and (b) parent-offspring conflict, came before and gave rise to "natural cooperation" and culture: minds that "work together" or "cooperate." This, in turn, opened the door for mothers and infants to co-construct a new niche—a "web of meaning"—a new unit of selection at a higher level of complexity, and for selection to take human evolution in its hypercooperative direction. From life history theory comes the notion that life cycles are reproductive strategies. Evolution is a complex adaptive system that keeps itself going through the acquisition and reproduction of the information that organizes organisms, even when the going gets tough and organisms suffer. Cultures are complex adaptive systems that keep themselves going through the acquisition and reproduction—the extension—of cognitive schemas. As Scheper-Hughes (1992) observed so well, when people suffer from chronic poverty and inequality, cultures of condolence (*condolere*, "to suffer together") are likely to emerge.

Acknowledgments

I'm grateful to the Ernst Strüngmann Forum for enabling Kim Bard, Heidi Keller, and Julia Lupp to organize so many cooperative minds. I'm also grateful to them and Leslie Atkinson, David Butler, David Coall, Barbara Finlay, Kristen Hawkes, Sarah Hrdy, Joyce Klein, Naomi Quinn, Volker Sommer, Nim Tottenham and two anonymous reviewers for their diverse ways of helping me realize what I wanted to say. My thanks as well to the National Evolutionary Synthesis Center at Duke University for the Senior Sabbatical Fellowship from which this chapter emerged.

12

Twenty-First Century Attachment Theory

Challenges and Opportunities

Ross A. Thompson

Abstract

Attachment theory is the focus of considerable contemporary developmental research. Formulated by Bowlby more than fifty years ago, it has been the subject of ongoing critique, particularly in terms of its relevance in non-Western settings. Attachment theorists have modified the theory in response to empirical findings, advances in allied fields, and further ideas. Yet, as evidenced by this Forum, work still remains. This chapter summarizes changes to some of the central areas of attachment theory as well as remaining points of contention: To whom do infants become attached? How should differences in attachment relationships be characterized? What influences lead to differences in attachment relationships? What are the outcomes of differences in attachment? Its intent is to sharpen the ways that culturally informed research can contribute to a better understanding of the attachment process and its consequences. Discussion concludes with broad reflections on attachment and culture.

Introduction

Theory has an uneven place in contemporary psychology. Theoretical concepts from connectionism, constructivism, social-cognitive learning theory, and other formulations provide researchers with broad interpretive frameworks and testable hypotheses. More commonly, however, psychologists use less comprehensive domain-specific theories that offer guidance to particular research fields, such as Bem's self-perception theory or Marcia's identity status theory. As further evidence that contemporary psychology has moved beyond grand theories, many contemporary psychologists pride themselves on being "data driven": research conclusions are derived inductively from statistics and theoretical assumptions are minimized or, at least, are implicit rather than explicit

(Thompson and Goodman 2011). In this contemporary context, "data mining" is an approbation rather than a criticism. Contemporary research in many fields (e.g., behavioral and molecular genetics, cognitive neuroscience, developmental biology) seems to be influenced minimally by theoretical hypotheses and maximally by emergent patterns in large data sets. Even contemporary introductory psychology textbooks commonly refer to the "perspectives" rather than the theories that guide psychological inquiry.

In this context, attachment theory is a notable exception. Inaugurated with Bowlby's seminal writings on the nature of the child's emotional tie to caregivers (Bowlby 1958, 1969), attachment theory has been a preeminent catalyst to developmental research for more than half a century. Its scope has expanded over this time from a primary concern with infant-parent bonding to encompass issues concerning adult romantic relationships, social and personality development, developmental psychopathology, evolutionary adaptation, and public policy problems in divorce and custody, foster care, child care, and grandparents' rights. Further, ideas associated with attachment theory have influenced pediatric practice, marriage and family therapy, parent counseling, family law, and clinical intervention, both nationally and internationally. During this period, theory has been important as attachment researchers have debated these expansive applications of Bowlby's theoretical claims, their consistency with amassing empirical evidence, and the generalizability of attachment ideas (see, e.g., Sroufe and Waters 1977; Lamb et al. 1985; Roisman and Fraley 2013; Thompson 2016).

Attachment theory has changed over the past 50 years. This should be expected. After all, Bowlby formulated the theory at a time when scientific understanding of infancy and early childhood underestimated the cognitive and behavioral sophistication of the young child. There have been significant advances in behavioral genetics, evolutionary biology, and developmental neuroscience since his time, as well as growing sophistication in research methodology. Family relationships in the Western contexts on which he based his ideas are different now than they were in Bowlby's era: the rise of dual-career families, greater recognition of the importance of the role of fathers, increased single parenting, and many other influences. For all these reasons, it has been necessary to update, elucidate, and expand Bowlby's formulations in ways that he could not have anticipated.

As attachment theory has also expanded over the years, it has served as a conceptual umbrella for broad and narrow constructions of the developmental impact of early parent-child relationships. Under this umbrella there has arisen a variety of attachment mini-theories that concern, for example, the organization of personality development (Sroufe 2005), bioevolutionary adaptation (Chisholm 1996), adult relationships (Mikulincer and Shaver 2007), and other topics. Attachment theory has also changed as attachment researchers have had their own ideas about the influence of early attachment security which they have sought to harmonize with Bowlby's formulations.

These changes in attachment theory during the past half-century have been largely beneficial, but they present a problem for contemporary scholars. With its adaptation to new research findings, developments in allied fields, theoretical extensions, and expanding applications, what are the important ways that attachment theory has changed, and why has it done so? How does twenty-first century attachment theory compare with Bowlby's formulations of some fifty years ago? These questions are the focus of this chapter, which discusses challenges, changes, and continuing debate over issues such as the range of partners to whom infants become attached, how to characterize variability in the quality of those attachments, the origins of these differences, and the developmental outcomes with which they are associated.

A portrayal of twenty-first century attachment theory is challenging not only because of the many changes that have occurred in the theory over the past fifty years, but also for two other reasons. First, while some theoretical issues have been resolved as the result of new data and new thinking, others remain open, and opinions currently differ among attachment researchers concerning some of the theory's central claims, as I describe below. On these issues, it is difficult to indicate definitively what attachment theory currently claims. Second, Bowlby's theory was not always clear. His ideas evolved between the first and second editions of his seminal *Attachment* volume, and his other writings modified some claims and expanded others. Moreover, some central concepts in his theory are not well defined and are therefore subject to multiple interpretations as the theory has developed.

Consider, for example, the nature and influence of developing "internal working models," the mental representations deriving from attachment relationships that influence how children approach new relationships, view themselves, and interpret social information. There are currently a variety of interpretations deriving from Bowlby's theory of what internal working models are and how they develop and function (Grossmann 1999; Thompson 2008a):

- formulations that are similar to psychoanalytic concepts of the dynamic unconscious and the introjection of good and bad objects,
- conceptualizations that draw on the prelinguistic cognitive-perceptual schemas of infancy,
- ideas describing emotion biases that become incorporated into preattentive processing, and
- proposals that connect working models to the social-cognitive achievements of early childhood, and other formulations.

There is currently no consensus among attachment researchers on how internal working models develop and function beyond Bowlby's general ideas about their influence on relationships, self, and social information processing. This poses a problem not only for those who are trying to understand the central claims of contemporary attachment theory, but also for the coherence of the theory itself. After all, it is easy to interpret almost any research results in terms of the functioning of internal working models if the construct is so vaguely defined that it can accommodate nearly any empirical findings (Thompson and Raikes 2003). In short, although attachment theory has changed considerably during the past half-century, theoretical clarity is still lacking in some important areas, and this is challenging for those who wish to understand the central claims of the theory.

The effort to clarify how attachment theory has changed, what are continuing points of theoretical uncertainty, and how this compares with Bowlby's original formulations is important to a discussion of the cultural context of attachment. In the years leading up to Bowlby's theory, and more vigorously since then, substantial research has provided evidence that (a) the caregiving conditions of young children are diverse, (b) children develop significant relationships with people other than the biological mother, and (c) complex developmental outcomes are associated with these relationships. This research has engendered considerable debate about its relevance to the central claims of attachment theory, which was developed to understand species-typical characteristics of infant attachments to caregivers. Reconciling the central claims of attachment theory with the meaning and significance of culturally diverse forms of care is thus important. However, this discussion must focus on the claims of contemporary attachment theory rather than only on claims made by Bowlby over a half-century ago. Doing so helps to clarify relevant cultural critiques of the theory and continuing challenges to be addressed as attachment theorists respond to the evidence of cultural research.

In this chapter, therefore, my goal is to discuss some of the characteristics of twenty-first century attachment theory, especially those elements of the theory that are most relevant to cultural critiques of the theory, and in which contemporary ideas have evolved from those originally formulated by Bowlby. The discussion that follows is organized around four central questions:

1. To whom do infants become attached? This relates to Bowlby's concept of monotropy and the influence of multiple attachments.
2. How should differences in attachment relationships be characterized? Here we consider new thinking about the meaning of these differences, especially in the context of Bowlby's evolutionary formulations about the importance of security.
3. What influences lead to differences in attachment relationships? This issue relates to the nature and significance of caregiver sensitivity.
4. What are the outcomes of differences in attachment? Here the range of competencies and liabilities that might derive from early secure and insecure relationships is considered.

The chapter closes with some concluding reflections about future discourse of attachment and culture, as well as the challenges and opportunities of contemporary attachment theory.

To Whom Do Infants Become Attached?

During the middle of the twentieth century, conventional thinking about families in the Western world was that infants developed emotional attachments to their mothers who were biologically and motivationally prepared to provide love, nurturance, and protection. Such a view accorded with normative patterns of family life in industrialized Western societies (especially in middle- and upper-income homes) as well as with prevalent portrayals of how the young were cared for in other mammalian species and throughout human evolution. In this context, Bowlby (1969) contributed a view of infant-parent attachment that was both familiar and novel. He argued that "almost from the first, many children have more than one figure to whom they direct attachment behavior" (Bowlby 1969:304) and that attachment figures may be biological kin (such as a grandparent) but need not be so (e.g., a regular care provider could be, depending on the context, a nanny, babysitter, or alloparent). His explicit use of the term "mother-figure" simultaneously emphasized that person's behavior rather than relatedness while also (perhaps unfortunately) tying that behavior to portrayals of traditional mothering. Bowlby observed that young children treat these attachment figures differently from one another. His view—that there is likely to be a principal attachment figure who is preferred for comfort and security—is the basis for his use of the term "monotropy." Attachment figures are thus differentiated from a broader cohort of social figures in the child's world, and there exists a preferential hierarchy among those people.

Bowlby's original theory thus incorporated a tension between a view of multiple attachment figures and a view of mothers as uniquely important and influential. In the research that followed, attachment researchers in Western industrialized countries focused their attention predominantly on relationships between infants and those who were typically their principal attachment figures, the child's mother. A handful of researchers, however, also examined the nature and developmental influences of other attachment relationships, including those with fathers, siblings, childcare providers, and others (for a review, see Howes and Spieker 2016). This research partially supported Bowlby's hierarchical model of the influence of multiple attachments on children's development in Western contexts, especially the strong influence of the child's relationships with primary caregivers. However, it also supported a model of domain-specific developmental influences, in which different attachment relationships have preeminent influence for children's behavior in certain domains of behavior relevant to that relationship. For example, although mother-child attachment is a robust predictor of children's peer competence in studies conducted in Western industrialized countries (Groh et al. 2014), researchers have found that measures of teacher-child attachment in childcare programs predict peer competence in those programs and elsewhere better than mother-child attachment (Howes et al. 1994; Ahnert et al. 2006). Similar findings have been reported for young children living on Israeli *kibbutzim,* in which relationships with *metaplot* (communal caregivers) predicted children's peer competence five years later whereas the security of mother-child and father-child attachments did not (Oppenheim et al. 1988).

Such research on the independent and combined influences of multiple attachment relationships is far less common in the attachment literature than studies of mother-child attachment alone, which may help to explain why Bowlby's concept of monotropy is identified with a matricentric orientation in attachment theory. Most of what is known about the immediate and longer-term consequences of the security of attachment, for example, derives from studies of mother-child attachment in Western countries (for a review, see Thompson 2016). The uneasy tension between the view of the mother as primary and the influence of different attachments extends to the theory's international influence on public policy. In the United States, for instance, attachment theory was an important influence in moving child custody adjudication from a maternal presumption to a broader focus on the child's primary caretaker (e.g., Neely 1984), but Australian researcher Jennifer McIntosh (2011) and others have enlisted attachment theory to argue for a more exclusive maternal preference.

Despite these tensions in the theory and its applications, twenty-first century attachment theory recognizes, far more than did Bowlby's original theory, the fact that young children normatively form multiple attachments within the family and outside of it, and that these attachments are developmentally important. Attachment theory would benefit from greater attention to the nature and influence of other nonmaternal attachments in young children's lives, such as those in the extended family setting as well as in formal and informal caregiving arrangements (including family, friend, and neighbor care in small communities). Building on research that documents the direct and indirect effects of family relationships on child development (e.g., Parke and Buriel 2006), such studies could expand understanding of the system of attachment relationships that influence a child's development by examining how attachment figures facilitate (or impair) the child's interactions with other figures; the complementary roles that attachment figures assume in offering nurturance, support, and protection; and how these relationships become internalized by the child in terms of mental working models that incorporate hierarchical, domain-specific, or integrated influences. Some efforts have been made by attachment researchers to situate multiple attachment relationships within broader social networks in which partners who are not necessarily attachment figures also provide care, initiate play, offer support, and influence children's development (see, e.g., Lewis 2005; van IJzendoorn 2005). However, much more is needed if attachment theory is to remain relevant to the conditions of children's care around the world.

Here the work of culturally oriented researchers can make an important contribution by studying the child's experience of the diverse environments of relationships characterizing different developmental contexts. To advance the theory further, however, requires more than just documenting the range of social partners with whom an infant interacts, because not all social partners are necessarily attachment figures. It is also essential to understand the significance of different adults to the child.

One illustration of how this might be accomplished stems from the work by Meehan and Hawks (2013) with the Aka in the Congo Basin Rain Forest. Using focal sampling, they sought to describe not only which adults provided care for infants and young children but also children's differential display of attachment behaviors (which they defined in terms of proximity- and contact-seeking, distance interaction, affectional actions, and related behaviors) and adults' responsiveness to these behaviors. They found that infants had an average of six attachment figures, but that the number of adults to whom children displayed attachment behaviors was relatively small compared to the size of children's caregiving networks. For example, not all allomothers were attachment figures, and even though mothers and allomothers responded comparably to children's distress, they responded very differently to children's display of attachment behaviors. Studies like this are valuable in developing a theory of attachment that is culturally informed. They connect young children's social experiences with efforts to denote their effects on the child's responses to putative attachment figures and other partners, with the goal of developing understanding of the *meaning* of these relationships to the developing child in the contexts in which they develop.

How Should Differences in Attachment Relationships Be Characterized?

The quality of adult-child relationships can be characterized in many ways: warmth, regulation, acceptance, autonomy support, communication, connectedness, guidance, and mutuality are some of the terms currently used in contemporary developmental science. Attachment theorists' early and sustained emphasis on the concept of *security* reflects several assumptions about attachment relationships.

The first is a focus not on the caregiver's behavior per se but primarily how it psychologically impacts the child. This recognizes that parental conduct is diverse but that its consequences for the child are most important, especially as they are moderated by characteristics such as child age, the child's representations of the parent's behavior, beliefs and expectations, and other factors. Indexing attachment in terms of security focuses on the effects that a caregiver's behavior has on the child.

Second, the concept of security underscores the significance of the child's trust or confidence in the caregiver, particularly in conditions of threat, distress, or danger when the solicitude of others is most needed. In Bowlby's portrayal of the environmental conditions of human evolution, such trust was necessary for human young to engage in exploratory forays to learn from the environment while maintaining access to the protection and nurturance of care-givers (i.e., the attachment-exploration balance). This may be true in many, but not necessarily all, developmental contexts. When infants are being continu-ously carried by one or more caregivers and do not engage in exploratory for-ays, for example, the functions of security might well be reconceptualized in relation to developing attachment. The general argument of attachment theory, however, is that young children's learning and healthy growth are facilitated when threat vigilance is reduced by the child's reliance and trust in caregivers during a developmental period when dependency on the others' solicitude and protection is high.

Third, in the concept of security, early attachment theorists wedded multiple meanings of *adaptation* (Lamb et al. 1985). Secure attachments were consid-ered to be:

- *evolutionary* adaptive by promoting survival to reproductive maturity in the ecological conditions in which humans evolved, especially in light of complementary behavioral systems in adults promoting solici-tude toward the young,
- more *developmentally* adaptive than insecure attachments, because they promoted learning and sociability in the context of strong human connections to one or more caregivers, and
- more adaptive in the sense of *psychological health*, because of how the characteristics promoted by a secure attachment foster stronger person-ality characteristics and better coping skills when children encounter difficulty.

There have been significant theoretical and empirical advances in evolutionary biology and behavioral ecology since Bowlby's time that call into question some of these formulations related to the importance of infant security in the ecological contexts of human evolution. These include:

- reconceptualizing the ecological conditions of the environment (or, more properly, environments) of human evolution as complex and changing over time,
- recognizing that there are different kinds of evolutionary adaptations in the juvenile years and they do not necessarily all have implications for psychological health,
- understanding that there are conditions in which parental solicitude to-ward the young can and cannot normatively be expected, and
- recognizing that evolutionary adaptations likely involve multiple con-ditional behavioral strategies applied situationally rather than a single fixed species-typical strategy (i.e., the concept of adaptive phenotypic plasticity; Thompson 2013a; Simpson and Belsky 2016).

Many of these problems for Bowlby's theory were raised early on by attach-ment researchers (e.g., Lamb et al. 1984b) as well as by researchers external to attachment theory (e.g., Trivers 1974; Hinde 1982). Although these advances in evolutionary thinking have not been fully integrated into attachment theory, a general agreement among many twenty-first century attachment theorists is emerging that Bowlby's view of secure attachments as a species-typical norm, adapted to a single, species-typical ecological niche, and insecure attachments as deviations from this norm is inadequate. Instead, it is increasingly recognized that different forms of attachment may be adaptations to different conditions of care which are themselves adapted to different ecological niches.

Guided by life history theory, for example, some views regard different forms of attachment as conditional adaptations to different developmental contexts (Simpson and Belsky 2016). Ecological conditions—including resources that are plentiful, scarce, or unpredictable, and environmental conditions that are benign or threatening—contribute to differences in parental investment which foster different forms of attachment in offspring that are, in this sense, preparations for living in those ecological conditions (e.g., Belsky et al. 1991; Chisholm 1996). Thus, for example, avoidant attachment to caregivers who are indifferent or rejecting—perhaps because they have few resources to provide the infant—motivates young children to look elsewhere for support and to develop other behaviors suited for competing with others in an environment of scarcity and potential deprivation. From this perspective, each form of attachment is biologically adaptive in the ecological contexts leading to its development, and this is consistent with the value of viewing parent-child attachment within a social network perspective.

These formulations also reflect how current evolutionary theory underscores the context sensitivity of biologically adaptive behavioral patterns. They highlight the interaction of what Gaskins (2013) describes as the universal and cultural contributions to attachment. Cultural contributions are manifested in the diversity of beliefs, goals, and practices that influence the development and functioning of attachment relationships. But these cultural contributions must, in some way, address a central issue faced by all cultures: how to ensure that the young survive to reproductive maturity (and that their offspring do as well). Modern evolutionary theory is pushing attachment theory to substitute for Bowlby's portrayal of species-typical secure attachments an alternative model of different forms of attachment adapted to different ecological conditions, where each enables infants to survive to reproductive maturity in those conditions. Within and outside of those conditions, moreover, different forms of attachment have different implications for the child's behavior, integration into the social context, and well-being.

Even when they recognize that different attachment patterns may be conditional adaptations to different developmental contexts, attachment theorists do not conclude that young children's developmental outcomes are equivalent or comparably constructive even in the contexts in which they develop. Not all cultural adaptations are psychologically constructive for children. Seymour (2013), for example, drawing on the ethnography by Du

Bois (1944) of the Alor (a small horticulture community living in an island in Indonesia), argues that the dispersed caregiving practices Du Bois documented are consistent with Alorese values of self-reliance, open expression of anger and hostility, low interpersonal trust, and high aggression of this community that was just emerging from a period of continuous warfare. But the attachment patterns derived from these conditions—in which infants experienced varying degrees of hunger and unpredictable care, the withdrawal of social support with the onset of walking, and adults provoking young children with teasing, threats, ridicule, and intentional scare tactics—yields a portrayal of young children whose working models of social relationships were characterized by "fear, anger, and distrust," according to Seymour (2013:123). These developmental outcomes may reflect an adaptation to a particular ecological context but also derive from what Carlson and Harwood (2014) call a "disabled caregiving system."

The conditions of the Alor children, and equally disturbing contemporary observations of children in conditions of war and social upheaval, establish a boundary of conditions that have failed children, no matter how much they reflect adaptations to ecological contexts (Carlson and Harwood 2014). On the other side of the continuum of care, in which caregiving conditions seem more constructive for supporting young children's well-being, cultural informants differ on what behaviors in young children denote positive attachments. Research in a variety of cultural contexts has revealed a range of parental portrayals of desirable child conduct, whether it involves interdependency (rather than autonomy), minimization of emotionality (rather than emotional expressivity), assertiveness, or other characteristics (Rothbaum et al. 2000b). Whether these are relevant to attachment theorists' emphasis on security, whether typical conceptions of security need to be broadened, or whether alternative ways of conceptualizing attachment are needed, as Crittenden (2000) has done, remains to be seen, and thoughtful ethnographies will make important contributions to thinking through the intersection of parental values, children's needs, and the broader ecological conditions they share (see, e.g., the portrayal of "concentric circles of attachment" in the Pirahã of Amazonia, Brazil, by Everett 2014).

It is important to note, however, that attachment theorists' fidelity to Bowlby's concept of security is not because it was or is a preeminent child-rearing value of British or American parents. Rather, viewing security as the core of attachment relationships was consistent with Bowlby's portrayal of the evolutionary adaptations that enabled human young to survive to reproductive maturity. In addition, in studies conducted in the West, secure attachment has been consistently associated with a wide range of positive developmental outcomes. Alternative portrayals of the functions of attachment relationships in other cultures must, therefore, move beyond descriptions of parental beliefs alone to offer comparable evidence that they are also associated with important developmental outcomes in children, preferably in longitudinal research, in order to provide a strong alternative to the concept of security. This is consistent with the view that attachment embraces both cultural and universal dimensions.

Finally, the Strange Situation Procedure was developed as a laboratory assessment of differences in the security of attachment in middle-class American families, and misapplications of this procedure to cultural contexts involving different normative early experiences were recognized quickly (Lamb et al. 1984b). Although this did not eliminate inappropriate applications of the Strange Situation to contexts very different from those for which it was developed, it has cautioned attachment researchers in their interpretation of children's behavior in this procedure. The Strange Situation has been a very useful tool for assessing security-oriented differences in attachment in Western contexts. Determining whether the Strange Situation or alternative validated measures, or their adaptation, are appropriate for assessing attachment in different cultural contexts, however, hinges on a prior determination of the place of security in the infant's experience of caregiving relationships and the suitability of the prevalent attachment classifications for characterizing individual differences in alternative contexts. This remains an important task.

What Influences Lead to Differences in Attachment Relationships?

Attachment theorists, beginning with Bowlby, understand differences in caregiver sensitivity to be a major influence on the development of secure or insecure attachments. Sensitive responsiveness is believed to provide confidence in the reliability and helpfulness of the adult's assistance, especially in circumstances when infants are distressed or alarmed, and thus it is believed to contribute to a secure attachment. Based in part on very strong associations between home observations of maternal sensitivity and infant Strange Situation behavior in Ainsworth's original Baltimore sample, attachment researchers, from the beginning, concertedly sought to replicate and extend this predictive association.

Results, however, have been a bit disappointing. According to a meta-analysis by de Wolff and van IJzendoorn (1997), differences in maternal sensitivity are reliably but very modestly associated with the security of attachment in studies conducted in Western industrialized countries. Ainsworth's strong findings have not been replicated, and subsequent research indicates a much weaker association than she found between maternal sensitivity and infant security. Importantly, de Wolff and van IJzendoorn (1997) included studies using a range of measures of maternal sensitivity to ensure that their conclusions were inclusive of diverse conceptualizations of sensitivity. Although experimental studies in multiple countries show that improving maternal sensitivity increases the likelihood of infant secure attachment, which provides causal evidence of their association, research since the 1997 meta-analysis has been

unable to document a stronger association between them in correlational studies (for a review, see Fearon and Belsky 2016). This conclusion has led attachment researchers to consider why the association of sensitivity and security is
not stronger. Belsky (1997b) has proposed, for example, that because children
are differentially sensitive to some environmental influences, it is possible that
some children are more strongly affected by maternal sensitivity than others,
leading to a modest overall effect size when the two groups are combined.

Twenty-first century attachment researchers have moved beyond differences
in sensitivity alone to consider the influence of other variables, including characteristics of maternal personality, the quality of the marital relationship, social
support from outside the family, and infant temperament (Fearon and Belsky
2016). Because the association of sensitivity and security is lower in economically stressed families, for example, family stress may attenuate the strength
of their association and influence directly the security of attachment. Raikes
and Thompson (2005) found that in a sample of lower-income families in the
United States, economic stresses (such as low income) were associated with
insecure attachment because of their effects on maternal sensitivity, whereas
emotional stresses (such as domestic violence or substance abuse problems in
the family) were directly associated with insecure attachment independently
of maternal sensitivity. Cowan (1997) has proposed that a family system approach is required for better understanding. Clearly, more work is needed. The
authors of the 1997 meta-analysis stated the clearest conclusion succinctly:
"Sensitivity is an important but not exclusive condition of attachment" (de
Wolff and van IJzendoorn 1997:571). Most attachment researchers would con-
cur that it is important to look beyond sensitivity alone in understanding the
origins of differences in attachment relationships.

Research that incorporates a greater focus on culture can help identify other
ways of thinking about influences that guide the development of differences
in early child-caregiver attachment. This might involve developing measures
of sensitivity that are adapted to cultural practices, values, and goals for children (see Keller 2007). It is important to understand that how sensitivity is
expressed in situations, for example, where mother and child are in nearly
continuous physical contact, or when multiple figures in the community provide care, is likely to be different than in a context in which child and parent
are often physically separated and responsiveness to signals is central. One
illustration of relevant cultural research is when Hewlett et al. (2000) conducted extended, time-sampled observations of mothers and infants among the
Aka foragers and Ngandu farmers from Central Africa and upper middle-class
Americans in Washington, D.C. Their findings confirmed differences in infant behavior and maternal responsiveness relevant to each ecological context.
Almost always held, Aka infants cried least, but when they did, their mothers responded immediately with rocking, feeding, and other kinds of soothing behavior. American infants were more likely to be picked up when they
fussed, with American mothers engaging in more vocalizing, distraction, and

stimulation of the baby. Ngandu mothers, whose infants fussed most, engaged in other kinds of soothing. Hewlett et al. (2000) argue that mothers responded sensitively in each context in a manner consistent with other caregiving conditions (e.g., carrying, foraging, proximity to the infant).

Shared caregiving, which is observed in many cultural contexts, may also require the consideration of sensitivity in a collective manner. This was the conclusion of a study reporting that children in nonparental care in Western countries were more likely to be securely attached to care providers when these adults manifested *group*-based sensitivity (e.g., such as interacting positively with a child while supervising the other children) rather than sensitivity expressed dyadically to *individual* children (Ahnert et al. 2006). Not just multiple children but multiple caregivers also warrant consideration of collective sensitivity (Keller and Chaudhary, this volume). In many non-Western contexts, observers must take into account the sensitivity of several caregivers and the security this confers on the child's experience, in which children learn about the trustworthiness of multiple people. Consideration of what is locally defined as good parenting is also important.

It is also likely to be necessary to look beyond variability in sensitivity to understand the origins of differences in early attachment relationships. Attachment researchers have not examined variability in how parents regulate the child's behavior, for example, even though the period during which attachment security takes shape is also a period when young children become increasingly mobile and goal-directed, assert independent intentions, and become more capable of acting in a more dangerous or disapproved manner (at least in Western contexts). Yet differences in parental regulatory behavior may be important to the security of attachment in ways that do not fully overlap with differences in sensitivity. Bowlby (1969) himself recognized that even in infancy, attachment is only one of several components of the parent-child relationship: parental roles as attachment figures are complemented by their roles as play partners, teachers, and behavioral managers. In other cultural contexts, other parental roles can be observed, and this is certainly noted in ethnographies in non-Western contexts; see, for example, Barlow's study of the influence of feeding in the development of attachment among the Murik of Papua New Guinea (Barlow 2013). It would be premature, in a Western or non-Western context, to expect that these alternative parental roles do not intersect in shaping the infant's experience of the parent-child relationship. Thus further exploration of their contribution to the development of differences in attachment relationships seems warranted.

What Are the Outcomes of Differences in Attachment?

Developmental psychologists were drawn to attachment theory as the result of two sets of research findings that appeared in the late 1970s. First, Waters

(1978) reported that the security of attachment was remarkably stable when infants were observed in the Strange Situation at 12 and 18 months, with 96% of infants classified the same each time. Second, a number of researchers began reporting longitudinal findings indicating that differences in the security of attachment were associated, in ways predicted by attachment theory, with later measures of social-emotional functioning such as peer sociability, positive affect, and cooperativeness. These findings were important in light of the failure of previous measures of parent-child relationships to show any kind of consistency over time or to predict important aspects of the child's subsequent development. They were also consistent with the claims of attachment theory and also with other developmental perspectives, such as Eriksonian theory. However, both conclusions—that early attachments are necessarily stable and that they predict later social-emotional competencies—have been modified over time in the face of accumulating research evidence.

Concerning the first, it is now clear that early attachments are not necessarily stable over time. Subsequent longitudinal studies have failed to replicate the findings by Waters; instead, researchers have reported a broad range of stability estimates when attachment assessments have been separated by a few months to a few decades (Thompson 2000).[1] There is some evidence that changes in the security of attachment over time are associated with concurrent changes in family stresses and/or circumstances of care, which may account for the wide range of estimates of stability in different samples. The most confident conclusion that can be derived from this research is simply that "sometimes early attachment relationships remain consistent over time, and sometimes they change" (Thompson 2000:146).

Concerning the second, several decades of research in Western settings have confirmed that early mother-child attachment is associated with the later social-emotional competencies identified by Bowlby. Attachment researchers have found associations between early security and later relations with parents, peers, and other social partners, as well as with self-concept, aspects of developing personality, #social cognition, emotion regulation, and behavior problems (reviewed by Thompson 2016). By and large, most of these predictive associations are consistent with the expectations of attachment theory, although the proportion of variance explained in these outcomes is small. Nonetheless, attachment researchers have also tested, and confirmed, that attachment is associated with a remarkable range of other outcomes that are well beyond those predicted by Bowlby and with the attachment theory he formulated. Guided by a general expectation that a secure attachment would predict better later functioning, researchers have broadened their inquiry to explore how security predicted later cognitive and language development, exploration and play,

---

[1] A meta-analysis by Fraley (2002) concludes that there is modest stability in the security of attachment, although this analysis used a wide variety of attachment measures and only indexed stability over time in whether individuals were secure or insecure.

curiosity, math achievement, and even political ideology, extending the range of predictive correlates far beyond what Bowlby originally envisioned. As Belsky and Cassidy (1994) mused, one might wonder if there is anything to which attachment security is *not* related. The broadening range of later correlates derived, in part, from the availability of large longitudinal data sets with early measures of attachment and a wide range of later measures that could be studied as developmental outcomes (whether theoretically expected or not), together with the flexibility of the unmeasured internal working models concept to "explain" the significant associations that emerged.

What does it mean when attachment researchers find associations between attachment security and outcomes (such as math achievement) that are not really consistent with the theory? One response is to modify the theory to harmonize with the findings, which helps to account for the current diversity of theoretical perspectives about the developmental outcomes associated with secure or insecure attachments. Another is to dig more deeply into (unmeasured) mediating influences that might explain the association. For example, it might not be true that a secure attachment makes children more mathematically competent. Rather, the security of attachment might be associated with stronger teacher-child relationships, parental support at home (and with homework), and other social influences that contribute to math achievement (see Teo et al. 1996). Consistent with the latter approach, attachment researchers have begun to enlist more sophisticated methodologies, beyond simple test-retest longitudinal designs, to consider more complex associations between early attachment and later outcomes. These include the use of growth curve modeling, mediational analyses, and biologically informed designs to examine, for example, whether a secure attachment moderates the effects on children of other parental practices (such as disciplinary styles), or how the social cognitive correlates of a secure attachment facilitate relationships with others, such as peers (for a review, see Thompson 2016). Although there remain many problems to resolve (and cautions in the overstatement of correlational research conclusions), the association between the security of attachment and later competencies in Western contexts is important, even though attachment researchers need to be more theoretically self-disciplined in this work.

One implication of this research, however, is that twenty-first century attachment researchers do not embrace the view that early attachments are rigidly stable or an unduly narrow interpretation of the competencies that should derive from an early secure attachment. Indeed, the dizzying variety of outcomes which have been documented in Western contexts seems to invite an open regard for the kinds of competencies that might be associated with attachment in other cultural contexts, especially in light of current interest in indirect and mediated associations. Consider, for example, the association between secure mother-child attachment and the child's nutritional status identified by Kermoian and Leiderman (1986) among the Gusii in East Africa. Western researchers would be unlikely to consider nutritional status as a predictable outcome of secure attachment, and thus it broadens the range of potential benefits of attachment security, especially if such associations are replicated in other contexts. At the same time, it invites the same consideration of direct and indirect influences (e.g., through duration and social interaction during feeding) by which attachment and nutritional status are related.

Viewed more broadly, perhaps a good starting point for thinking about how early attachment influences development is in terms of the various ways it contributes to the integration of the child into the social context, initially through the development of social trust and the growth of behavioral competencies relevant to becoming a well-functioning member of the cultural group (Weisner 2016b). This would be a way of characterizing some of the developmental outcomes proposed by attachment theory for children growing up in Western contexts, and it might offer a useful general heuristic for thinking about developmental outcomes in other cultures also. Young children who are learning, in the context of their own culture's values, to coordinate their needs with the needs of others in the interests of interdependent social harmony, to manage the expression of strong emotions in the interests of maintaining respectful relationships, or of maintaining appropriate ingroup-outgroup distinctions are each developing cultural competencies in ways that could be consistent with cultural values, desirable parenting practices and, one might suggest, the claims of attachment theory.

Attachment and Culture

It should be apparent to the reader that there is considerable diversity of perspectives among attachment researchers. In a community of scholars as conceptually diverse as this, it is probable that some would disagree with the portrayal of twenty-first century attachment theory presented here. The characterization of contemporary attachment theory presented here is based on research and conceptual advances during the past half century, however, and it is likely to be close to a consensual view, although I claim sole responsibility for this portrayal.

The portrayal of attachment presented in this chapter underscores the continuing challenges that derive, in part, from the theory's breadth and longevity. The tension between a monotropic view of attachment and recognition of the importance of multiple attachments that characterized Bowlby's theory remains true in contemporary research—the tension still exists—even though contemporary attachment researchers recognize that a much wider range of normative attachments develop in the early years. Bowlby's evolutionary model of secure attachment as biologically normative is being superseded by more current, complex evolutionary models that portray attachment patterns as different behavioral strategies adapted for different forms of parental investment under different ecological conditions. But attachment researchers are still working out the implications of this view and, in particular, how sensitivity to context is incorporated into a species-typical developmental formulation. The realization that sensitivity is a reliable but not especially strong predictor of the security of attachment has forced researchers to consider other contributors to developing parent-child relationships, including those from multiple caregivers. Finally, the remarkably broad range of developmental outcomes to which attachment security is linked is requiring attachment researchers to examine more carefully how and why these outcomes should be associated with the security of attachment in order for the theory to remain coherent.

What does this mean for better understanding the cultural context of attachment? For many years, critics have argued that culture has been ignored by mainstream attachment researchers in two ways: First, attachment researchers have failed to adequately qualify the generalizability of their conclusions to the cultural contexts (primarily Western industrialized nations) from which they were derived. As a consequence, Western conceptions of infants' needs and care have become a universal standard against which others are evaluated under the umbrella of Bowlby's evolutionary model. Although attachment researchers have made some progress in recognizing the limitations of their research literatures (see, e.g., Mesman et al. 2016b), this has not been satisfactory to many critics (see, e.g., Morelli and Henry 2013:17), and more progress is certainly needed. Second, critics argue that mainstream attachment researchers perpetuate this problem by continuing to focus their attention on developmental processes in Western industrialized countries and thus fail to build the database necessary for a truly culturally informed attachment theory. Quinn and Mageo (2013:3–32) describe what that research enterprise would look like:

> [I]f we ever hope to derive culturally meaningful patterns of variation in attachment, we must deduce that variation from a large set of such cross-cultural studies, representative of the full range of human societies and human caregiving practices.

Although the lack of cultural diversity could be regarded as a limitation of virtually every research literature concerning normative development, this criticism certainly applies to attachment theory. Addressing it as Quinn and Mageo propose is a daunting challenge, as it would be to any developmental formulation that claims to address broadly generalizable, species-typical developmental processes.

It is, however, possible to take a somewhat more optimistic view of the opportunities for a constructive integration of culture with attachment theory. Many of the current challenges facing attachment theory can be addressed, at least in part, through greater consideration of findings from different cultural contexts. Greater understanding of how multiple attachment relationships develop and function for young children requires studying children in contexts where these networks are normative. Understanding how species-typical processes underlying attachment incorporate sensitivity to the physical and social ecology requires studying attachment in diverse ecologies. Deeper consideration of the utility of concepts such as attachment "security" and parental "sensitivity" can benefit not just from conceptual critiques of these terms but also from focused studies of the interactions between young children and their caregivers in different contexts and, most importantly, the meaning of these interactions to the child. Likewise, if attachment contributes to the development of social trust and the behavioral competencies necessary to function effectively in the social world, then better understanding of how this occurs in diverse cultural contexts could contribute to clarifying the developmental outcomes that attachment should—and should not—predict.

Taken together, it is reasonable to conclude that twenty-first century attachment theory offers today a more open field for integrating cultural perspectives than has previously existed or been understood. Evidence from carefully designed, culturally informed studies of attachment has broadened perspectives that have emerged from research conducted primarily in Western contexts and can continue to do so in the future. Seizing this opportunity is a challenge for twenty-first century attachment research.

It is also a challenge for developmental researchers who focus on culture. They need to appreciate one further reason that cultural criticism of attachment theory has tended to fall on deaf ears. While culturally oriented researchers ask for greater *culturally informed attachment research,* attachment researchers sometimes wonder where they can find greater *attachment-informed cultural studies.* When they survey the research literature on culture and attachment, attachment researchers find relatively few studies that address the central claims of attachment theory in an informative way: as indicated above, research that might be relevant is often not focused on the developmental experience of young children. The perplexity of attachment researchers finds resonance in Alma Gottlieb's (2004) remarkable study on the culture of infancy among the Beng of West Africa, which opens with the question: "Where have all the babies gone?" In posing this question, Gottlieb reflects on the absence of attention to infancy by contemporary cultural anthropologists. Her question thus helps to explain why so many of the questions posed in this chapter concerning the intersection of culture with attachment theory still do not have useful answers. Although critics of attachment theory often point to shared caregiving as a cultural norm inconsistent with Bowlby's theory, for example, cultural research described earlier suggests that all alloparents are not necessarily attachment figures; thus, the significance of shared caregiving to contemporary attachment theory remains unclear until the meaning of different caregivers to infants in these contexts are better studied. When LeVine (2014) draws on his work with the Hausa of northern Nigeria and notes that Hausa mothers practice a custom of avoidance with their infants, he posed a question that would interest an attachment researcher: What is it like to be raised by a mother who avoids you in public? Unfortunately, he offers no answer. Attachment researchers have failed to properly incorporate culturally informed studies into their theoretical conceptions of attachment relationships, but they have not been aided by cultural critics whose (sometimes strident) criticisms have often failed to be substantiated by informative data relevant to attachment concerns.

Fortunately, this situation is beginning to change as Gottlieb's question is being addressed by a growing research literature focused on infancy and cultural conditions of early care (e.g., Hewlett and Lamb 2005; Quinn and Mageo 2013; Otto and Keller 2014). Attachment theory would be stronger with the thoughtful inclusion of culturally diverse studies. This is a goal to which researchers with a variety of perspectives should contribute.

Acknowledgments

I am grateful to Michael Lamb, Gilda Morelli, Doug Teti, Julia Lupp, and the editors of this volume for their thoughtful reading and helpful comments on an earlier version of this paper.

13

Implications for Policy
and Practice

Suzanne Gaskins, Marjorie Beeghly, Kim A. Bard,
Ariane Gernhardt, Cindy H. Liu, Douglas M. Teti,
Ross A. Thompson, Thomas S. Weisner, and Relindis D. Yovsi

Abstract

Ideas and claims about children's development (e.g., concerning attachment relationships) that have found broad acceptance in the academic community have impacted the development of policy in governmental and international organizations. These accepted ideas and claims, in turn, have been incorporated into practice and services provided to families in various forms (e.g., social work, child care). The reconceptualization of attachment systems proposed in this volume—in particular, the explicit evaluation of the influence of multiple attachment figures on children that is normative in many societies—should have profound effects on both policy and practice. This chapter addresses issues that need to be considered if society is to integrate current understanding of the cultural nature of attachment into policy and practice.

Policy: International Organizations, Governments, and Professional Organizations

International bodies and national governments set policies that affect families the world over, based on a limited conceptualization of children, caregivers, and their relationships that are incongruent with the actual values and practices of many societies and cultures (Serpell and Nsamenang 2014). Such policies often conflict with the meaning systems and goals for parenting in particular settings, as well as people's everyday experiences, and are thus likely to be ineffective. This situation has emerged, in part, because academic experts (on whom organizations rely for theories and models) have reached "scientific" conclusions about "universal" behavior by studying a narrow range of human behavior that primarily reflects Western thought and practice, and then over-generalizing those conclusions to the rest of the world. Governing bodies, like scientists themselves, are often unaware of the cultural assumptions that are embedded in the very foundations of their policies. This bias not only threatens the validity of a policy, it reduces the likelihood that it will be embraced by the people it is intended to support.

The UN Convention on the Rights of the Child (UNCRC) delineates a broad range of rights applicable to children around the world (United Nations 1989). It has been ratified by the vast majority of nation-states, with the United States a notable exception. The range of rights covered under the UNCRC is extraordinary: Article 27 recognizes the right of children to a standard of living adequate to their physical and mental needs. It also stresses freedom of expression, the right to have their opinions respected, freedom of association, the right of privacy, access to information, and freedom of thought and conscience. UNCRC provides an expansive interpretation of children's rights intended to guarantee minimal requirements of health, safety, and well-being. It also urges nation-states to adopt a much broader view of children's rights than most currently embrace. In a subsequent document (United Nations 2005), the UNCRC clarifies that these rights apply to young as well as older children, emphasizing that both parents and extrafamilial stakeholders play important roles in their implementation.

More recently, the World Association for Infant Mental Health (WAIMH), an international organization that works with infants and parents from diverse societies and cultural groups, published a position paper which argued that infants require additional rights beyond those listed by the UNCRC, because immature infants, unlike older children, are totally dependent on caregiving for survival (World Association for Infant Mental Health 2016). From animal and human research, it is clear that early experience contributes significantly to brain development during the first three years of life, as well as to a child's positive adaptation and well-being later. Thus, there is an urgency to provide developmentally informed care and appropriate protection for infants. WAIMH seeks to inform and guide policies that provide support for parents and other caregivers. It also aims to raise awareness of the special needs of infants, particularly those reared in "high-risk" environments (e.g., poverty, violence).

Without doubt, these types of international initiatives, which are designed to support the health and well-being of all infants and their families, are important. However, significant barriers exist that limit their efficacy and ease of implementation across diverse cultural settings. With respect to parent-child relationships, for example, some of the rights guaranteed to children reduce, as a consequence, the authority and autonomy of parents to regulate their children's care and conduct. This sets up a conflict in cultures where a strong commitment to parental authority (and even physical punishment) is used to instruct children in what they need to learn. In such cultural settings, some might view these provisions as inconsistent, or at least in tension with the UNCRC's guarantee of the rights of parents to direct and guide their children

(Article 4). The UNCRC initiatives concerning children's rights often challenge existing values and beliefs of indigenous cultures (see Rosabal-Coto et al., this volume). This may explain why the global adoption of UNCRC and its implementation has been inconsistent.

To maximize acceptance and compliance with these international initiatives, policy-makers and practitioners need to recognize and respect the distinct meaning systems of families in diverse settings. When policy and practice procedures are developed, local stakeholders need to be included in the process, a caveat that is in line with the recommendations delineated by Serpell and Nsamenang (2014) for childcare initiatives.

Similarly, individual national governments often produce top-down policies that affect families and young children, based on guidance received from international agencies and Western academic research. Still, because national governments have access to local knowledge and meaning systems, they are in a position to develop policies that are culturally informed and sensitive to the very people they are designed to serve. This process will be enhanced if bias (racial or cultural) is recognized and mitigated.

Agencies (national and regional) charged with carrying out government policy must be able to interpret and apply national policies to the "pressing" needs of local communities being served. For instance, a policy designed to support the physical, emotional, social, and cognitive aspects of child development may well respond to different needs: In one community, a program may be needed to reduce child mortality and improve children's physical development, as in the face of dietary deficiencies (Abubakar et al. 2011). In another, a program might be used to optimize social development, as when children have been separated from their families because of war (Macksoud and Aber 1996; Hasanović et al. 2006). In still another community, a program may be needed to support cognitive development, when schooling is limited (Koller et al. 2012). The particular programs that government or other agencies might prioritize in each of these communities may look quite different, yet they all address the common goal of supporting children's development and well-being.

At times, national governments may resist the recommendations of international organizations. For example, the UNCRC articulates the importance of parent-child relationships and the need for governments to support them. This responsibility, however, may not be accepted or considered a priority in certain areas of the world. Some governments, for example, may believe that a child's well-being is the primary responsibility of individual parents, and thus government policies should not intrude. In such cases, it is important to persuade governments that supporting families does fall within their area of responsibility, but that this support needs to be meaningful to local communities.

When there is a significant mismatch between the articulated policies of an international agency or government and the existing ways of raising children inherent to the cultural setting, programs may not be fully embraced or implemented. Two examples of issues (as locally interpreted) raised by the UNCRC that may be seen to be in conflict with local understanding and practice are (a) child labor, which may be viewed as a valuable learning environment as well as an important contribution to a marginal economic system (Gaskins 2014), and (b) corporal punishment, which (when differentiated from child abuse) may be viewed as an effective teaching technique that also supports a valued hierarchical social order (Nutter-El-Ouardani 2014).

To illustrate the complexities involved in developing and implementing a policy at odds with parental commitments, let us consider the example of spanking, as it has been discussed so thoroughly in the literature. Many nations have adopted the position held by the United Nations that spanking is a form of "legalized violence against children" and that it should be banned altogether. The UN has banned the use of corporal punishment in member countries, without considering that it may be viewed as an effective tool for teaching by parents in some cultural groups. This international declaration, by itself, does little to change parents' behavior or attitudes toward spanking, and for that reason it is often ineffective.

The United States, for example, has no national governmental policy on spanking. However, a growing empirical literature suggests that spanking and other forms of corporal punishment are ineffective disciplinary techniques and may have detrimental effects on children (Gershoff 2013). Many professional groups (e.g., the American Academy of Pediatrics, the American Psychological Association) recommend that parents refrain from spanking young children as a disciplinary strategy (American Academy of Pediatrics 1998; Hagan et al. 2008; Smith 2012). Local government agencies, such as child protection services, rely on these professional recommendations when they establish their own guidelines about what is acceptable parental behavior. However, as in other countries, many parents of young children in the United States—depending in part on their ethnicity, religion, and class (e.g., Berlin et al. 2009; MacKenzie et al. 2011)—ignore these recommendations. The majority of parents in some groups report that they have spanked their children at least once during the past year, and many endorse "a good hard spanking" as an effective disciplinary strategy in certain circumstances (Child Trends 2015a).

Official policies on spanking are adopted without consideration of the diversity of parental beliefs about corporal punishment, the consequences that may ensue once this socializing technique is eliminated, or the broader parental ethnotheories that motivate and validate them. Such policies may not only be insensitive to different systems of parenting, they may also be inappropriate. Some research shows that spanking may or may not have long-term negative consequences for children, depending on the cultural environment in which it is administered (e.g., Deater-Deckard et al. 1996; Lansford 2010). Other findings suggest that spanking has negative effects on children even when it is in accord with cultural traditions (Gershoff 2013). Of concern here is that government and institutions have formed policy based on (a) inadequate knowledge of what families do and, especially, why they do it, (b) with little consideration of how the local meaning of the targeted parental behaviors may affect the implementation of the policy, and (c) incomplete scientific evidence.

For international, national, and professional organizations that are focused on young children, the overall goal is to support and increase their well-being. This goal is compromised if an organization's expectations and recommendations conflict with reality concerning the role of children, their social support systems, and the powerful forces of socialization that mold them through their everyday experiences to become members of particular cultural groups. Policy-makers need to recognize that parents and caregivers share the goal of supporting the well-being of children, and policies need to be designed to enable caregivers to be more successful in raising healthy and well-adapted children. Unfortunately, policies can actually achieve the opposite effect. When they deny resources or impose penalties on caregivers whose perspectives do not match the policy more harm than good may result. Such policies can also sow doubt in the minds of caregivers about the appropriateness of their cultural commitments to child-rearing. Thus, it is imperative for academic researchers and scientific advisors to incorporate a cultural view into their claims, and to emphasize the importance of cultural considerations when translating recommendations about young children and their families to policy-makers. This will help governing bodies, reliant on their expert advice, to develop policies that support the full range of cultural contexts of development and, in the process, better serve children's well-being.

Practice

When agencies develop goals based on assumptions that do not match the groups being served, the methods selected to realize such goals may prove ineffective or even harmful to families. This is particularly relevant for services developed by international nongovernmental organizations (NGOs) for implementation in developing countries. It is also an issue in Western nations, where migration is increasing in prevalence, especially from African and the Middle Eastern countries. Uprooted immigrant or refugee families often depend on social workers and educators, who are trained in Western "best practice" principles but often lack culturally sensitive information about their clients' specific parenting beliefs and practices. This situation does not just go away. Instead, the "native" population of the society becomes increasingly diverse as second- and third-generation immigrants make up a larger percentage of the population. Generally, the parenting beliefs and practices of immigrant families are conservative, in comparison to those of native families in the host country. Many immigrant parents retain the child-rearing beliefs and practices that they learned from their own parents and grandparents, often without even realizing it, even as they assimilate into the dominant culture in other ways.

It is not difficult to find examples where the application of current pub-lic policy relevant to attachment theory is interpreted in culturally insensitive ways, potentially resulting in inappropriate negative judgments about child-care practices because they differ from normative Western practices:

- Leaving a child home alone or in the care of another preadolescent child.
- Parent-child co-sleeping arrangements.
- Leaving children in the home country or sending them back to the home country to be cared for by relatives (see Liu et al., this volume).
- Fathers being denied adequate visitation with their children to ensure that children maintain a close relationship with their mothers.

In each of these examples, as in many others, the issues at stake are com-plex, nuanced, and potentially significant in terms of negatively influencing the well-being of families. Parents and caregivers who are immigrants or who come from minority communities are often evaluated negatively for engaging in practices that have been handed down to them over generations—practices judged positively by their communities. In addition, significant legal and finan-cial consequences may result when such practices violate the host country's laws and practices (e.g., use of physical punishment). We are not suggesting that there is no room to interpret the concept of harm by a caregiver or danger for the child. Instead, we suggest that all social service interpretations of at-tachment behavior should begin by asking the following questions:

- Why is the parent doing that particular practice?
- Is the intent to hurt or harm the child?
- Is there a reason that motivates the behavior, perhaps stemming from a belief in particular socialization practices or from difficult circum-stances outside the child-rearing domain?
- Is there actual evidence that the practice causes harm to the child being raised in this environment?

The effectiveness of professionals who work with children and families from diverse groups would be greatly improved if they had increased knowledge of cultural variation in parental beliefs and traditions, including those related to children's attachment systems and their associations with variations in parent-ing practices—the same range of information we raised earlier for researchers to consider (see Chapter 8, this volume). Practitioners need to explore and reflect on the meaning of parenting practices (and their attitudes toward them) when those practices vary from those considered "optimal" in Western society. They also need to recognize that different parenting styles do not automatically reflect "poor" parenting or a lack of care or concern about the child. When practice is culturally insensitive, parents are likely to ignore the guidance of-fered to them. This puts them at risk of being misunderstood and judged nega-tively, or even of being accused of abuse or neglect.

Many practitioners in fields such as medicine, mental health, law, education, social services, and economic development are fully aware of this problem (e.g., Forehand and Kotchick 1996; Lillas and Marchel 2015). There are, however, significant barriers to more culturally sensitive practice. The sheer number and diversity of cultures that need to be understood is daunting. In addition, individuals in any society have difficulty recognizing and accepting that one's own beliefs are culturally motivated, yet without this awareness, the beliefs of others cannot be recognized as legitimate. In her account of a Hmong immigrant family's struggle with the Western medical system during a serious illness of their infant, Fadiman (1997) captured the complexities involved in trying to help a tiny child when two cultural systems collide.

When evaluating services and interventions, it is important to integrate "culturally competent," evidence-based perspectives into the process over time. The initial experiences of the practitioner, the researcher, and the client trying to work together may not be one of ease: significant disorientation may leave all parties feeling less assured about how to interpret behaviors and experiences, much let alone what should be said or done. These experiences, however, provide an opportunity to learn about the other person as well as the impetus to get the interaction and support "right" for all participants. This process is crucial if "culturally pluralistic" situations are to be addressed competently (Weisner and Hay 2015:2–3) and needs to occur both in the application of attachment theory as well as in the study of attachment. It is important to observe the process, evaluate it, and reflect on it as part of taking up a commitment to embrace a cultural perspective.

We encourage policy-makers and practitioners to use information to support parents from all cultural groups in the care of their children. All parties need to be more aware of their own cultural commitments and more open to recognizing those of others. To illustrate this problem further, we discuss three types of agencies and institutions that illustrate the perils of cultural mismatch: NGOs, social work, and child care.

International Nongovernmental Organizations

How do NGOs develop programs to help communities and families achieve the best outcomes for their children? Unfortunately, all too often, NGO programs fail in this endeavor. To improve effectiveness, programs need to be designed and established through collaborative partnerships with the communities and individuals they are intended to serve, and the resulting goals need to be effectively communicated to the target group(s).

Although many organizations impose standards and make judgments without local community engagement and collaboration, and without understanding the negative consequences of their good intentions, others are committed to working with local partnerships (Pence 2013). Some NGOs do not provide direct services from a national or international office but rather partner with local community organizations to provide programs (e.g., preschools, women's health, youth programs, child protection) that are linked to ongoing local projects (e.g., connecting preschools to local primary schools or women's organizations, connecting youth programs to schools or work programs). By partnering with local organizations, programs can be adapted to local goals and may even be jointly funded through local or regional sources. No matter how desirable and empirically based, intervention proposals will only be successful if they are integrated meaningfully into the daily lives of the families and communities being served. Since many NGOs are internationally based, precautions need to be taken to ensure that NGO workers understand the cultural landscape of the local setting and develop the programs in partnership with that community.

Social Work in Support of Children's Health and Well-Being

Social workers assigned to help families often analyze the social relationships within a family to see if they are "healthy" and capable of promoting a child's "well-being." Yet the mode of evaluation often depends on the current interpretation of academic research and the recommended best practices derived from that research.

Attachment theorists and researchers posit that the establishment of secure attachment relationships with caregivers in early childhood is foundational for children's later successful functioning, but not all have incorporated cultural variation into their model of attachment. Instead, based on research conducted mainly in Western societies, they hold that a primary antecedent of secure attachment is, for example, an adult caregiver's "sensitivity." Further, Western professionals widely accept these findings as defining "universal" behavior and use them to develop and implement attachment-based therapeutic interventions for at-risk families. For instance, a goal of many Western attachment-based interventions is to support parents in providing sensitive, nurturing care in the context of child distress (Dozier et al. 2001; Cyr et al. 2010) and in engaging in contingently responsive ("serve-and-return") interactions with children (Bernard and Dozier 2010) (see also Chapters 5 and 8, this volume).This, however, overlooks the fact that even though a significant association between "sensitivity" and child's "attachment security" was found in the Western populations studied, only a small amount of the variance in the children's behavior has actually been explained.

Given the conceptualization outlined in Chapter 8 (this volume) and developed throughout this volume, it would follow that the concept of sensitive parenting and attachment-based interventions derived from it are not necessarily appropriate for use with all families in non-Western societies or from minority communities within Western societies. Best practices need to be culturally specific. We believe that the effectiveness of parenting interventions will be maximized when they are informed by culturally specific ethnographic information (e.g., local parental/community socialization goals for children) and developed and implemented in direct collaboration with local "stakeholders" in community organizations. When the goals and content of interventions clash with local parenting beliefs and practices, the extent to which they will be adopted will be reduced and inappropriate moral judgments about parents who use parenting practices that vary from Western norms will increase.

Child Care

In the United States, changes in attitudes toward child care over the last fifty years illustrate how perceptions in child care (and its developmental consequences) are both empirical and culturally constructed, and how attachment theory has played a role in this evolution. Since the 1950s, public attitudes have evolved past the initial deep concerns about out-of-home care and its effects on young children. In many cases, resistance to out-of-home care was originally "justified" by reference to attachment theory and the view that very young children required continuous access to their primary attachment figure in order to develop security. (The same view, incidentally, was and still is used to justify sole child custody to mothers, as primary attachment figures, during divorce proceedings.) These attitudes affected public acceptance of preschool and nursery school care, combined with strong doubts about the developmental consequences of infant/toddler care, which again were justified based on the claims of attachment theory—this time with reference to the stresses on the infant of developing and maintaining a secure attachment when early and extended child care begins. In the current era, both infant care and preschool-age care are normatively enlisted by the majority of families, and it is recognized that young children form significant attachments to their regular childcare providers as well as to their parents. Families who live in different cultures throughout the world, however, would regard the debates in the United States as inexplicable, since children experience normative shared care virtually from birth without any concerns about young children becoming insecure or overwhelmed by the experience. Although the concerted empirical study of child care and its effects on children in the United States is warranted (NICHD Early Child Care Research Network 2005), public attitudes have changed—and are varied—independently of, or in transaction with, expanding research knowledge.

Beyond societal attitudes about child care in general, there has been much debate about what constitutes a quality childcare program, and much of that discussion has focused on what kind of attachment relationships should be nurtured between caregivers and children. We suggest that achieving social trust in the classroom and school community should be the primary goal and quality to measure—not solely warm teacher-child attachment relationships, which vary in how they are understood and recognized in diverse cultural communities. Howes (2009) provides a model of early child care and childhood education programs that represents a contextual theory of relationship quality and security. The development of positive relationships integrates antecedent factors (e.g., child and family circumstances, caregiver internal processes, caregiver practices and beliefs, the peer group), which then influence the relationship quality of the child-caregiver dyad and peer relations. These relationships, in turn, shape the social and emotional "climate" of the early childhood education program itself: from childcare quality indicators to responsive teaching. There are many indirect pathways as well (Weisner and Hay 2015).

Providing a quality relationship between providers and children in an early childhood education setting is among the standard measures of process quality in early childhood programs (Howes 2009). Good teachers are those who understand that they must "balance the needs of the child, the group, and the child within the group" (Howes 2009:34). This balance is crucial to a reconsideration of the idea of a secure base and conventional measures of attachment security. The importance of social trust (not only a secure base from a single caregiver in a dyadic relationship) is a construct that fits with the early childhood education experience for children from a broad range of backgrounds. Social trust emphasizes the distributed relationship across a *community*. It does not focus on an individual child's internalized sense of security from the exclusive dyadic relationship with the mother. This model also makes clear that quality childcare practices may indeed vary for programs that serve ethnically diverse clientele. This enables the needs that children bring from home to be adequately addressed, especially in regards to relationship formation, security, and attachment.

Howe's model is particularly sensitive to cultural differences, yet many childcare programs in Western countries do not use such a model. As in social services, the primary focus of some curricula is often on caregiver sensitivity, drawing directly from the work of Ainsworth and her colleagues (e.g., Gutknecht et al. 2012). Some programs, for instance, use a formal transitional period to introduce children to the daycare setting. Based on attachment theory, it is assumed that this adjustment period is necessary to ensure a child's well-being. In Germany, for example, this period lasts up to four weeks: parents, children, and caregivers experience different phases designed to enable the caregiver (who is initially perceived as a stranger) to become an attachment person to the child. Parents are required to stay with their child during the first days in the daycare center so that the child can get used to the new caregivers while their primary attachment figure is present. Gradually, as the child adjusts to the new surroundings, parents spend less time at the daycare center (Laewen et al. 2011; Keller and Chaudhary, this volume).

Western middle-class parents usually perceive this procedure to be a sensitive and appropriate solution for an emotionally difficult transition. However, many immigrant parents view the procedure as inappropriate and, in fact, inexplicable. For parents whose children already have extensive experience with different caregivers and who show no stranger anxiety, such a transition period may be perceived as incompetence on the part of the childcare center. These families may, as a result, choose not to send their children to day care because they are uncomfortable with the policies and/or are skeptical of the care their children would receive (Keller and Bossong, unpublished). In Germany, where every third child has some form of migration background, cultural mismatch in child care is an issue of great significance.

Attachment and Cultural Understanding
of Parent-Child Relationships

In the United States, *attachment parenting* has become a popular, albeit controversial approach to infant and early child care in the lay population, because it is perceived to be based on caregiving practices that are deeply rooted in human evolution. Based on the work of Sears and Sears (1993), attachment parenting urges the adoption of practices such as immediate postpartum skin-to-skin contact between mother and baby, breastfeeding on demand throughout the child's early years, continuous contact between mother and baby ("baby-wearing"), co-sleeping throughout the early years, and other related practices. These practices are justified, in part, as being "natural" and "instinctive" to humans because they existed throughout evolution and can be seen in many indigenous cultures around the world. Thus they are deemed to be "best practice" for raising infants and young children.

Although attachment parenting shares with attachment theory an emphasis on parental sensitivity, nurturant care, and warmth toward young children, attachment researchers do not endorse attachment parenting practices. Attachment researchers criticize the emphasis on maternal care (atypical in most indigenous cultures) as undermining fathers as significant attachment figures and feel that it may upset the balance between attachment and exploration, which attachment researchers underscore in early childhood development. More generally, attachment parenting presents a narrow portrayal of which caregiving practices are "natural" to humans as a species, despite scholarly disagreement about which practices actually characterize care of the young (Fuentes 2009) and the extraordinary range of actual caregiving practices throughout the world in which infants thrive (Morelli et al. and Keller and Chaudhary, this volume). Reifying one pattern of child care plucked from ethnographic accounts as species typical, and therefore the norm against which other patterns are evaluated, seems as culturally uninformed as is reifying childcare practices of the United States middle class as the norm. In the end, because much more research is needed to identify the effects of attachment parenting on young children's development in contemporary conditions, it seems highly premature to identify these practices as ones for which adults and children are naturally best suited.

This is an interesting example of traditional attachment theorists making an argument that criticizes policy reliant on naïve and uninformed ideas about attachment. It is particularly instructive here, because it represents the same kind of critique that we are making more generally about the dangers of applying Western-centric theory to child and family policies in other cultures.

Supporting Research, Policy, and Practice through Dialogue

Translating research from any scientific field into policy and practice is difficult. The multifaceted aspects of research are difficult to distill into a form that is accessible to policy-makers, who often seek direct, concrete information communicated in statements that resemble black-and-white dichotomies rather than shades of gray. Many researchers resist giving up the nuances inherent in their research or arguments. Moreover, data may only be available in basic research journals; policy implications may not have been fully developed or communicated, thus severely impacting the translation of findings into policy and practice.

In addition, translating research into policy involves bias that must be recognized. For example, although policy-makers often assert that designing and implementing culturally sensitive practice is an important goal, the successful implementation of this goal remains elusive. Part of the problem is that attachment-informed research still does not make the study of cultural differences a priority and, as such, it is reasonable to expect that recommendations from such research would not emphasize the creation of culturally appropriate practices. Simply put, unless culture is incorporated into the empirical database, culture will never finds its way into policy. To minimize bias and retain scientific integrity in the translation of scientific research, what would an ideal process look like? We argue that greater effort needs to be given to dialogue and debate, and that the construction of this process should be carried out with the relevant stakeholders.

In support of an inclusive translation process, we recommend that policy-making agencies which serve families be directly involved in the development of new research projects intended to address cultural differences in beliefs and practices surrounding young children's everyday lives, including attachment. This research needs to go beyond a basic description of attachment systems in different cultures and examine how attachment affects children's development and engagement in their social worlds.

We also recommend that greater applied research be conducted to help articulate how agencies and institutions can best support families in a variety of communities. This includes how to develop clearer and more effective recommendations for evidence-based best practices, which may vary across cultures. This research requires a multi-method, multidisciplinary approach (see discussion in Chapter 8, this volume, on a methodological tool kit), and needs to include the means to disseminate findings to policy-makers and practitioners as well as to strengthen capacity to design supportive policies for local cultures.

In addition, effort should be given to increase ecocultural training for professionals who work with families with young children: psychologists, psychiatrists, teachers, pediatricians, nurses, midwives, aid workers, social workers, and other community leaders.

Such an integrated approach to research would bring a clear advantage. Practitioners may be better able to appreciate the amount of diversity in attachment systems than researchers. Thus, by working together, researchers may come to see more clearly the ways in which infants and their attachment figures interact within a framework of cultural values and practices. In turn, this insight would enable researchers to correct theoretical blind spots (e.g., regarding the role of culture in forming healthy attachment systems).

As our world shrinks, through immigration and better communication, the need for a more culturally informed understanding of children's worlds increases (Jensen 2011). Those who set policy and provide services need to improve their understanding of how culture organizes and informs caregiving practices and children's everyday lives, so that they can respond effectively to the increasing diversity in the communities they serve. The academic community shares a responsibility in increasing this knowledge. Constructive dialogue, such as was accomplished at this Forum, is of paramount importance: between academics who study attachment in traditional ways and those who have critiqued that approach based on their study of families in cultural contexts, as well as between policy-makers and academics. None of this will be easy and considerable challenges exist. Yet with concerted effort and persistence, cultural diversity can be understood and incorporated into the most fundamental models of children's development and well-being. We hope this discussion will inspire you, the reader, to continue this much-needed discourse on culture and early attachment systems.

14

Real-World Applications of Attachment Theory

Mariano Rosabal-Coto, Naomi Quinn, Heidi Keller,
Marga Vicedo, Nandita Chaudhary,
Alma Gottlieb, Gabriel Scheidecker,
Marjorie Murray, Akira Takada, and Gilda A. Morelli

Abstract

Attachment theory has its roots in an ethnocentric complex of ideas, longstanding in the United States, under the rubric of "intensive mothering." Among these various approaches and programs, attachment theory has had an inordinate influence on a wide range of professions concerned with children (family therapy, education, the legal system, and public policy, the medical profession, etc.) inside and outside the United States. This chapter looks critically at how attachment theory has been applied in a variety of contexts and discusses its influence on parenting. It examines the distortion that often results when research findings are translated into actual applications or programs, ignoring any particularities of cultural context. It describes how attachment theory has been used as the basis for child-rearing manuals and has influenced programs and policies more directly, to form legal decisions that affect families, as well as to develop public policy and programs—all without requisite evidence to support such application and, more importantly, without regard to cultural context. Because child-rearing practices vary among cultures, the value systems that motivate these different practices must be recognized and accounted for when applications are developed and implemented. It concludes with a call for researchers to become proactive in rectifying misuses of attachment theory and holds that doing so is a matter of social responsibility.

A Critical Appraisal

Western societies, and especially their middle classes, have moved away from traditional parenting practices handed down across generations, toward validation of these practices by designated experts (Arendt 1958; Nolan 1998). Prominent among such expert theories today is attachment theory. The extreme influence that attachment theory has on contemporary parents was recently captured by Bethany Saltman (2016) in the popular press. In this narrative, a single mother recounts her preoccupation with whether or not her daughter may be insecurely attached to her, due to her own inadequate parenting. Her anxiety leads her to undertake training in the Strange Situation Procedure and then undergo a clinically administered Adult Attachment Interview. Her anxiety and guilt are only put to rest when the interview results are interpreted by the clinician to mean that her daughter is indeed "securely attached."

A striking feature of attachment theory and its appeal is how widely its fundamental tenets have been disseminated to a range of lay audiences, all targeting parents such as the one in the example above. Books, brochures, and videotapes proliferate, with the Internet now serving as a platform to facilitate parental information searches, counseling, and peer support in every major language (Niela-Vilén et al. 2014; Shah 2014; Montesi 2015). The Internet has thus become a supplement to the advice that might once have been sought exclusively from pediatricians or other child health-care professionals (Fischer and Landry 2007; Gottlieb and DeLoache 2017). This new technology may well account for the rapid, global spread of attachment theoretical approaches to parenting as well as a rash of other fashionable approaches. Although many of these are independent of (and may even predate) attachment theory itself, they may bear some resemblance to the latter.

In their U.S. versions, these approaches have been classed together as advocating "intensive mothering" (Hays 1996). They may be variously labeled, in both the United States and other countries, as "child-centered parenting," "natural parenting," or "evolutionary parenting." These concepts project a cultural model of mothering that has been described as being "so sacred, so deeply held, and so taken for granted" in U.S. society "as to remain generally unquestioned and regularly treated as common sense" (Hays 1996:13). The U.S. cultural model holds that:

1. Child-rearing is the responsibility of individual mothers.
2. Child-rearing entails constant nurture centered on the child, nurture that is labor-intensive, emotionally absorbing, and financially expensive, "even if this means that the mother must temporarily put her own life on hold" (Hays 1996:111).
3. Children themselves are innocent and pure, and hence worthy of a mother's love, care, and sacrifice; that is, children, and mothering them, are "sacred," in opposition to the expectations of self-interest and personal gain in the outside world.[1]

---

[1]   While attachment theory can certainly be classified as an example of "intensive parenting," the story that Hays tells is much broader historically. Indeed, Bowlby, the only attachment theorist to be mentioned, appears only twice in her book, quoted once for his stance on maternal deprivation (Hays 1996:47), and then again as one of three "maternal-attachment theorists" from diverse disciplines (Hays 1996:155).

This cultural model of mother-centered child-rearing has its roots in the post-World War II white, middle-class United States (Ehrenreich and English 1978). In our view, the rise of attachment theory helped validate this general approach to child-rearing, along with specific assumptions such as that children must be taken care of by their mothers. Attachment theory and other versions of intensive mothering are now widely disseminated to parents outside the United States, particularly in other Western countries, such as France, the United Kingdom, and Spain (Montesi 2015), as well as within the Westernized middle class in the Global South in countries such as Chile (Faircloth 2013; Murray 2014). In Latin America, this cultural model of mothering finds ready validation from the preexisting model of *Marianismo*, with its emphasis on the "devotional and self-sacrificial mother" (Murray 2014:5). Whatever their cultural resonance, appeals to intensive mothering are bound to pose an undue opposition between a woman's own interests and those of her child, with an inevitable accompaniment of maternal guilt (Rippeyoung 2013). Everywhere it has penetrated, this cluster of approaches has had profound effects on views and practice of parenting, particularly of mothering.

As Hays (1996:52) recounts, the cultural model came to dominate the advice given in the best-selling child-rearing manuals, spanning decades, that middle-class U.S. mothers (among others) were (and still are) encouraged to consult. The three best sellers that Hays identified were (a) pediatrician Benjamin Spock's *Baby and Childcare* (Spock 1968), with its multiple updated editions; (b) pediatrician T. Berry Brazelton's *What Every Baby Knows* (Brazelton 1987) and his various other books on this topic; and (c) British social psychologist Penelope Leach's *Your Baby and Child* (Leach 1986) and its successive editions. Notably, these authors' approaches to parenting presage that of attachment theorists. In the words of Hays (1996:57), "[t]he mother's day-to-day job is, above all, to respond to a child's needs and wants." In addition, Brazelton calls for "a sensitive parent" and argues that "[g]ood parenting follows from attention to a child's cues and requests" (Hays 1996:57).

Ideologies that call for intensive parenting and put this burden on mothers are hardly new. Hays (1996:152–178) attributes this cultural preoccupation in the United States, where it originated, to several factors. Perhaps the most interesting of these is that motherhood is but one field in which a struggle is waged between the logic of self-interested gain (which characterizes U.S. society at large) and the oppositional pull of human social ties. The current upwelling of such views with regard to mothers, in particular, may represent a backlash against the upsurge of women, especially white, middle-class women, entering or reentering the workplace in unprecedented numbers, coupled with the ascendancy of a neoliberalism that promotes individual over governmental solutions to social problems, including those that emerge when mothers work outside the home. However, these ideas and other components of attachment theory were evident earlier from other influential sources in Western psychology. One example of this is the influential book, *Beyond the Best Interests of the Child*, by Goldstein et al. (1973), which was published before attachment theory gained prominence and addressed child placement within the court system. The coauthors (a law professor, a child development researcher, and a clinician) coined the phrase "psychological parent" (Goldstein et al. 1973:17–20) and advocated that placement in child custody cases be dictated by "the need of every child for unbroken continuity of affection and stimulating relationships with an adult" (Goldstein et al. 1973:8). This early psychoanalytically grounded advice is an obvious precursor to the attachment theory notion of sensitive care described by Morelli et al. (this volume).[2]

Some attachment theorists reject contemporary approaches to intensive mothering that do not derive from their own theory. Main et al. (2011:438–439) discuss a number of misconceptions, unsupported by attachment theory, which these more popular approaches to parenting perpetuate, including:

- An adult needs to have been present from the infant's birth in order for the infant to form a secure attachment to that adult.
- The window of opportunity for formation of a secure attachment endures only throughout the first three years of life.
- The amount of time spent with a child is the most important element in forming an enduring attachment relationship.

However, such disavowal does not prevent other schools of thought—such as the programs *attachment therapy* or *attachment parenting*—from sharing their name or, notably in the case of attachment parenting, from leaning on the tenets of attachment theory for scientific authority. For evidence of this, see the website of Attachment Parenting International, which (a) promotes attachment theorist Mary Ainsworth's research and her idea of "maternal sensitivity," (b) lists the four categories (secure, insecure-ambivalent, insecure-avoidant, and insecure-disorganized) that emerged from her research, and (c) describes the Strange Situation as the instrument used to reveal these categories. Notably, the attachment parenting literature also advocates practices such as mother-child bonding immediately at birth, continuous bodily contact with the infant (e.g., wearing the child in a sling), co-sleeping, and breastfeeding up to four years of age (Sears 2011, 2016). Adoptive parents of older children are cautioned that since these children were unable to bond with their attachment figure at birth, they may exhibit attachment disorders or insecure attachment if the parents are not trained to use the other parenting techniques that attachment parenting recommends.

The validity of such practices should be seriously questioned. The approach is introduced as an "evolutionary and natural-based" one, but this assertion relies on pseudo-ethnographic observations of child-rearing practices

---

[2] Bowlby and Ainsworth are each mentioned only once, together in a footnote (Goldstein et al. 1973:115) which cites their earliest work on maternal deprivation, along with half a dozen other citations.

in selected indigenous communities (Liedloff 1997)—practices that are then simply declared to be more "natural" (Sears 1983). Unfortunately, despite being questioned by attachment theorists and others, attachment therapy and similar movements are legitimized when acclaimed by even a few scholars (e.g., Miller and Commons 2010), thus paving the way for these movements to be adopted uncritically by practitioners.

It is not enough for attachment theorists to merely disassociate themselves from an extreme approach, such as attachment parenting, that draws on common cultural preoccupations, and promotes parental guilt in doing so, as well as pirating key components of the academic theory to make its tenets seem more scientific. Academic practitioners of attachment theory should mount a full-fledged critique of such a rival school, with its extreme, often economically unaffordable, and even nonsensical proposals for parenting. Moreover, the academic community (e.g., anthropologists, cultural and developmental psychologists, attachment scholars) has a responsibility to become directly involved with the topic of child-rearing and to make clear, in both academic and public forums, that child-rearing practices in non-Western communities are not more "natural" than others, and thus they cannot be used to legitimate the tenets of a dogmatic movement, such as attachment parenting. At the same time, investigations of child-rearing practices in these communities do provide cross-culturally diverse correctives to academic thinking that has been, up to now, strikingly ethnocentric.

Recommendations regarding breastfeeding deserve special mention. We preface this discussion with the comment that breastfeeding arrangements vary widely across groups. For example, among the Pirahã people in the Amazon, consonant with a cultural emphasis on kinship, other women related to the mother may breastfeed an infant. Depending on the food supply, the mother's health, and other contextual circumstances, children may be nursed by a mother's sister; Pirahã women also occasionally nurse nonhuman mammals as well (Everett 2014:176–177). This case of breastfeeding by nonmothers is not singular, as we discuss further below (see section on Public Policy). Nonetheless, attachment theorists and policy-makers alike consider biological mothers to be the only ones to breastfeed.

The American Academy of Pediatrics and the World Health Organization (WHO) both subscribe to some version of the standard advice regarding maternal breastfeeding, routinely promoting it as a best practice until the child is one or two years of age. A large impetus behind the WHO's recommendation is so that mothers from countries in the Global South avoid feeding their infants and young children dangerous alternatives to breast milk. More ominously, the U.S. Department of Health and Human Services once held a national campaign in the United States, teaching that if a mother did not breastfeed, she was putting her child's life in danger (Rippeyoung 2013:10).

Depending on local labor practices, maternity leave structures, transportation networks, and childcare options, breastfeeding of any length may have negative consequences for mothers who work outside the home. The Canadian National Survey of Youth found that mothers who breastfeed for six months or longer suffer more severe and prolonged loss of earnings than do mothers who breastfeed for shorter durations or not at all (Rippeyoung and Noonan 2012). The report observes that to facilitate this practice, La Leche League, one of the largest organizations in the world to promote breastfeeding, tends to encourage mothers to work part-time rather than full-time, to this end. In addition, the attachment parenting literature recommends strongly that mothers not work outside the home, not only so that they may breastfeed but to ensure that their children receive the whole package of recommended care, of which lengthy breastfeeding is just one component (see, e.g., Sears and Sears 2001; Bialik and Gordon 2012). For impoverished mothers living in countries without extended paid maternity leave, recommendations to stay at home or to work part-time only represent an unaffordable luxury. By contrast, women living in northern European and other highly developed nations may be eligible for up to a year of paid maternity leave, thus facilitating breastfeeding and other "intensive parenting" practices (Golden 2017; Schug 2017). Yet even these more fortunate mothers in Western countries are targets of advice designed to counter the effects of their employment. In the United States, mothers who work outside the home are counseled, in the attachment parenting literature, to "wear" the child for at least four to five hours every night to compensate for their absence during the day (for critiques, see Schön and Silvén 2007; Faircloth 2013). While breastfeeding is not a prerequisite in attachment theory, it fits well with the theory's ideas of sensitive parenting; in addition to its value as the most beneficial source of food for small children, this practice is readily responsive to children's indications of hunger, chief among their signaled needs.

Academic reports of attachment findings are themselves likely to be replete with cautionary notes about the inconclusive or only suggestive nature of this work. For example, in the *Handbook of Attachment*, Slade (2008) describes research on the links between attachment and psychotherapy and states that "this literature raises more questions than it answers and provides few clear guidelines for practitioners." Still, the application of attachment theory to actual parenting and the mis-education of professionals who oversee such parenting will most likely not benefit from these cautions. Practitioners often present attachment theory findings as proven scientific results. In sum, just as in the case of attachment parenting, suggestions for real-world applications that emanate from attachment theory can evolve into caricatures of the theory itself. Such oversimplified readings of research findings are certainly shaped by ethnocentric ideological and moral standards held unwittingly by those who apply them. Another driver of such distortions is funding, as Berlin et al. (2016:753) note:

> In our experience, whereas university-based researchers pursue numerous and often nuanced program outcomes, community agency personnel and local funders are most interested in basic public health and child welfare indicators,

such as the rate of children in foster placements, or rates of children referred for special needs services, both of which have large financial implications for local and state governments.

When even a minority of attachment theorists advocate extreme positions regarding parenting, outliers can have an outsized influence on both audiences, given the tendency of lay practitioners to oversimplify academic views and the readiness of parents to accept any expert opinion. To offer just one example, some attachment theorists find that adopted children are less secure than nonadopted children. Using a meta-analysis of unpublished data, van IJzendoorn and Juffer (2006:1234) found that 47% of more than 400 adopted children were insecurely attached (and 53% securely attached), as measured by the Strange Situation Procedure. This rate compares, they note, with 67% in normal, nonadopted samples. While researchers, as they most often do, target later-adopted children in such studies, this subtlety may be lost in translation. Adoptive parents are especially susceptible to cautionary messages about their parenting that may originate from reports of such findings, but then get exaggerated in practitioners' retelling. And they do receive such messages: In two separate cases, one from Durham, England, and the other from Durham, North Carolina, adoptive parents reported to Quinn (pers. comm., April, 2017) that they had been told by their social workers not to expect their adopted children to "attach" to them at all. The child in the second case was only a month old at adoption.

Academic attachment theory itself has its own stringent recommendations regarding the kind of care that children need. Chief among attachment theoretic assumptions, already alluded to within the broader context of "intensive mothering," is an exclusive attention to mothers as child caregivers. When fathers get any mention, they are likely to be afterthoughts, and paid mere lip service. However, as becomes obvious in the description of the actual protocol that follows, fathers are not actually incorporated into the intervention. Even worse, they may be seen as impediments to mothering—as illustrated by the following example, drawn from the closing chapter (Video-Feedback Intervention to Promote Positive Parenting) of a book (*Promoting Positive Parenting: An Attachment-Based Intervention*) on a popular attachment theoretic approach from The Netherlands (van IJzendoorn et al. 2008). After acknowledging that fathers "do take part in rearing their children, and may benefit from interventions as much as mothers do," the authors leave the impression that the involvement of fathers is only ever secondary to that of mothers. They point out that the involvement of fathers "may motivate their partners to continue participation and to practice new behaviors at home." Thus, the "presence of the father may enhance the effectiveness of the intervention as well as the permanence of the changes in maternal behavior. It should be noted, however, that paternal involvement may be counterproductive as far as the mothers are concerned." They go on to cite two studies involving fathers in which "the effects on paternal sensitivity were large, but similar effects on maternal sensitivity were absent" and conclude (van IJzendoorn et al. 2008:199):

> Several explanations for these disappointing findings may be considered. First, if fathers are included in the intervention efforts, less attention might be paid to the mothers' needs and abilities. Second, when fathers are also involved in the intervention, mothers may underestimate the importance of their practicing new child-rearing insights and skills.

This is one concrete example of how intervention programs can end up caricaturing the theory upon which they draw.

It is worth noting that van IJzendoorn et al. only consider siblings in reference to how sensitively they are treated by comparison to their siblings—not as potential givers of child care or attachment figures to these siblings. Grandparents are not considered at all, despite the fact that grandmothers and older siblings are key child caregivers or "allomothers" all over the world. This is but one instance of how intervention programs ignore cross-cultural variation, including ethnic variation that certainly occurs within the United States and other Western countries. It also clearly illustrates Morelli et al.'s (this volume) argument that attachment theory and research have been one-sided, overlooking what children not only receive from others but also what they provide to others—in this case to younger children.

Current attachment theory emphasizes the child's need for sensitive mothering. Intervention programs follow Ainsworth's original definition of sensitive mothers as those "who accurately perceived their child's signals of distress and responded to these signals in a prompt and adequate way" (Juffer et al. 2008:3). This may be referred to as sensitive parenting, but the ease with which discussion of suggested parental interventions then turns to mothers and maternal sensitivity, the commonplace use of the feminine pronoun "her" to describe this parent, and the paucity of research to assess interventions that include fathers all betray the assumption, whether explicit or unexamined, that mother is the one who will be providing the sensitive parenting (see Morelli et al., this volume). Today, attachment theorists, unlike attachment parenting advocates, do not explicitly argue that mothers should stay at home to be able to provide appropriate care for their children.[3] Nevertheless, attachment theory does pose a tension between the kind of mother one is supposed to be and the pursuit of employment outside the home. One practical arena in which this tension plays out is that of day care, especially for infants and younger children (discussed below in the section on early education programs).

When parents themselves consult popular parenting manuals, the effects of attachment theory and related approaches are brought to bear directly on them.

---

[3] Bowlby, however, did. He called daycare centers "a dangerous waste of time and money" and argued that, other than the communists, the only ones who opposed his views were "professional women." He continued: "They have, in fact, neglected their families. But it's the last thing they want to admit" (quoted in Vicedo 2013:225–226).

In addition to this direct appeal to parents, attachment theory often infiltrates parenting more indirectly from four areas that will be discussed in turn:

1. in family therapy programs designed around attachment theory,
2. in early childhood education programs similarly designed,
3. in jurisprudence in connection with child custody and placement, and
4. in public policy relating to children and child development.

Implications for Family Therapy

In the third edition of the *Handbook of Attachment* (Cassidy and Shaver 2016), Berlin et al. (2016:746) describe four intervention programs derived from attachment theory: Child-Parent Psychotherapy, Attachment and Biobehavioral Catch-up, Video-Feedback Intervention to Promote Positive Parenting, and Circle of Security. The authors note that Circle of Security is the one most directly derived from attachment theory and research, and that all four programs have a strong evidence base. In selecting these programs for the third edition, Berlin et al. dropped two programs previously included in the second edition (Berlin et al. 2008)—Skill-Based Treatment from Leiden and the UCLA Family Development Project—presumably because these had a weaker evidence base. It can be assumed that both programs are still, however, widely practiced. Presumably other family therapy intervention programs are in use as well. Berlin et al.'s treatment in the third edition is briefer and more narrowly focused than in the second edition; they build on the 2008 chapter to consider new community applications of these intervention programs and refer back to it as needed, as will we.

The foundational tenet of contemporary attachment theory—the importance of attachment security for a child's present and future well-being—is unquestioned in all of these programs. Indeed, what such security might look like is wholly unexamined in these attachment theory-based intervention programs, as is the notion of the "secure base," which the parent is thought to provide, and the Strange Situation Procedure through which secure and insecure attachment are assessed. After all, practitioners must assume that these assessments are based on proven scientific theory, tangibly demonstrated in the experimental procedure upon which they rest. Moreover, the programs never address how attachment and these associated constructs might vary across cultural and ethnic groups, in ways that might suggest different interventions.

Another key issue, one characterizing therapy and counseling based on attachment theory, arises from the "internal working model" of parenting, a construct originally proposed by Bowlby. When a child exhibits behavioral problems or relationship conflicts with his or her parents, the chief intervention is to try to "reframe" or "restructure" the working model of the parents, the child, or both. This might involve commonsense interventions, such as counseling parents to discontinue labeling their child as a "bad kid," or teaching the child to be a safe haven and secure base to a younger sibling rather than engaging in competition for a parent's affections (see Johnson 2008). However, although internal working models may be the starting point for therapy derived from attachment theory, Berlin et al. (2008:747) confess that the "mechanisms through which parents' working models affect child-parent attachment are not well understood," an assertion that still seems to hold true in 2016. Specifically, the applications never interrogate the working model as being the product of one among many diverse cultural or ethnic ideologies used to describe what a virtuous adult should look like and how child-rearing is set up to achieve that result, as discussed by Morelli et al. (this volume). One theory now being put forth to account for mixed findings, regarding the efficacy of interventions (called "the transmission gap"), is that some children may be inherently more susceptible to environmental conditions that lead to insecure attachment (Belsky et al. 2005). This, however, has effectively closed off consideration of other obvious explanations for mixed findings; namely, there is cultural diversity in models of child-rearing.

An illustrative anecdote was offered by anthropologist Thomas Weisner (2005:89), who remembers standing alongside a single mother watching her son through a one-way mirror as the child played with toys during a phase of the Strange Situation Procedure when she was absent. The mother comments proudly on her son's independence. The researchers, though, clearly assuming a different working model to interpret the boy's behavior, have classified him as "avoidant." Followed through his adolescence, this boy exhibited none of the symptoms of insecure attachment for which attachment theorists would have predicted him to be at risk. At issue seems to be a cultural difference between the experimenters' middle-class Western notion of what constitutes insecure attachment and the understanding of this working-class mother, who imagines a tough, resilient child, one who will grow into an adult able to make it in the world on his own. Such a cultural difference is well documented in an ethnography of child-rearing practices in three U.S. neighborhoods varied by class (Kusserow 2004). Attachment theory and the therapeutic interventions and recommendations based on it do not recognize such class-based nuances. Weisner (2005:89–90) cites systematic evidence in which mothers prefer behaviors that are coded by researchers as "insecure."

In therapeutic programs designed to reframe the internal working models of children and parents, such as those described in the *Handbook of Attachment* (Cassidy and Shaver 2008, 2016), there are methodological problems, as revealed in this review of studies offered in support of these programs. Interventions aligned with attachment theory principles are often mixed with standard therapeutic ones (Berlin et al. 2008) and are said to be but one component of the overall program (Berlin et al. 2016). This intermingling makes it difficult to discern exactly what accounts for an improvement in a child's behavior or a parent-child relationship. Any supportive intervention is likely to have a positive effect. Moreover, studies of these therapeutic interventions are often performed on children at risk of various sorts (e.g., temperamentally irritable infants, maltreated children, children with depressive mothers), yet Berlin et al. (2016), just as does Bowlby, find that the mechanisms posited behind the proposed interventions apply equally to children and their families not at risk. Even so, mixed and even sometimes null results are often reported (Slade 2016:751–752). Sometimes, the mere presence of a behavior in early infancy is taken as evidence that it is related to attachment, as Slade (2008:771) candidly concludes: "The term 'attachment' became a code word for early experience" in psychoanalytic circles. In general, reviewers' enthusiasm for such therapeutic interventions runs well ahead of hard evidence for the efficacy of the attachment theoretic components of these programs. Researchers in this paradigm continue to pile on citations purporting to prove that therapeutic interventions work. But work to what end? The underlying goal of achieving "secure attachment" remains unquestioned.

Beyond these therapeutic intervention programs, attachment theory has also seeped into standard psychotherapy. Slade, the author who wrote in the second edition of the *Handbook of Attachment*, that among therapists "attachment" had become a code word for any early experience, is even more outspoken in the 2016 edition. There she asserts that "many tests of attachment theory's use in psychotherapeutic research and practice over the last 25 years have been limited in significant ways" (Slade 2016:759–760). Her critique of research bearing on psychotherapy has to do with the lack of "measures sensitive to the dynamically meaningful and theoretically predictable differences" among categories of secure and insecure attachment. As to clinical practice, she observes that there is "a surprising lack of depth in the way attachment constructs are applied to the clinical enterprise" (Slade 2016:260). This observation led Slade (2016:760; italics in the original) to iterate her caution from the second edition, concluding that "the assumption that *attachment* is shorthand for *relationship* is both incorrect and incomplete."

Implications for Early Education

Attachment therapy programs in the United States may contain early education components. Although they are not designed for schools, these components can be considered "educational" in a loose sense because they train parents at home in parenting skills. All of these programs adhere to attachment theory principles: they are designed to train caregivers in improved parenting skills according to these principles, with the objective of either enhancing the security of attachment or preventing risk factors that might lead to insecure attachment. Training typically involves home visits and is augmented in one program, called STEEP (Steps Toward Enjoyable Effective Parenting), by group sessions for parents (Erickson and Egeland 2004).

Sometimes the parenting skills are taught through parent-child play (e.g., in a program called Time Together) and sometimes through direct training, which may be supplemented with modeling (as in STEEP). While STEEP recruits parents during pregnancy, other programs are more likely to be directed to (a) categories of children considered to be at high risk, (b) those identified by means of the Strange Situation Procedure as having insecure or disorganized attachment (Circle of Security; Marvin et al. 2002), or (c) those who appear to be socially isolated or having relationship difficulties (Time Together; Butcher and Gersch 2014). Such attachment theory-based programs have become so popular that as individualized versions suitable for different countries are developed, they have been translated into nine languages. By 2015 more than 6,000 providers have been reportedly trained in these programs (Berlin et al. 2016:751).

Other early education programs are focused on teaching children directly. Some research considers whether educators (early school, nursery school, kindergarten teachers) can potentially serve as attachment figures for the children in their institutions (Riley 2013). In daycare centers and other early-learning settings in Germany, for example, teachers may be taught to incorporate basic tenets of attachment theory into their behavior toward the children by (a) addressing the individual child, thus laying the groundwork for an autonomous self; (b) following the children's initiatives as sensitivity requires; and (c) operating as a secure base for the child. This adult becomes the relational and educational partner for the child. It is thought that the interests of this individual child should take precedence over those of the group (e.g., Infants Program; Laewen et al. 2006). Some German programs target adults other than teachers (e.g., doctors, medical students, social workers), but these instances are scattered. Thus, the United States is not the only country to have embraced early education programs that are extensions of attachment theory-based therapy programs. In German-speaking countries, in particular, a widespread assumption is that children cannot properly learn and be educated when they are insecurely attached. The slogans "*keine Bildung ohne Bindung*" (no education without attachment) and *Bildung geschieht durch Bindung* (education happens through attachment), used in German daycare centers to promote early education programs, show the extent to which attachment theory has influenced the daycare curricula (Julius 2009; Haderthauer and Zehetmair 2013).

A program widely used in Germany is the *Berliner Modell zum Übergang in die Kita* (often referred to simply as the *Berliner Modell*), which aims to enable a child's transition into day care (Laewen et al. 2006). Offering this program contributes to a daycare center's reputation for educational quality. Two additional programs directed at early education were developed by Karl Heinz Brisch, a pediatrician: SAFE (*Sichere Ausbildung für Eltern*: Safe Attachment Family Education), a training program for promoting secure attachment between parents and children, and BASE (Babywatching), a program designed "to counter aggression and fear and to foster sensitivity and empathy" in nurseries and elementary schools.

In Germany, then, it is understood that education should follow the principles of attachment theory, and that attachment on the child's part is necessary for a successful education. Thus school is a locus for the application of this theory (Becker-Stoll 2013). Learning is considered to be based on a child's self-directed exploration. The individual child decides what s/he wants to do, thereby making free play a fundamental activity. The educational partner of the child is an adult who is taught to resemble the sensitive mother. The *Berliner Modell* requires the primary caregiver (usually the mother) to spend decreasing amounts of time, over a period up to four weeks, with the child in the daycare setting, so as to familiarize the child with the *Bezugserzieherin* (the childcare worker who has primary responsibility for the child). Since the introduction of the theory into Germany by attachment theorists Karin and Klaus Grossmann, the challenge of successful learning in schools based on attachment theoretic principles has been strongly promoted and advocated by others (e.g., Claus Koch from the Berlin Pedagogical Institute).

In the United States, day care has not been viewed so benignly. As Howes and Spieker (2016:319) observe:

> A dramatic demographic shift in the rearing experiences of infants in the United States occurred in the closing decades of the 20th century. By the mid-1980s, the number of mothers in the paid labor force with infants under 1 year of age reached 50%. Social scientists began to ask whether the experience of repeated separations from mother, and time away from mother during the development of a child's primary attachments, had adverse consequences for the quality of infant-mother attachment.

Some theorists raised the alarm about the effect of day care on attachment security. However, Howes and Spieker (2016:316) conclude that:

> The formation of toddler-childcare provider attachment relationships appears to be similar to the formation of an infant-mother attachment. When toddlers begin child care, they direct attachment behaviors to the caregivers, and with increased time in the setting, children's interactions with the caregivers become more organized, similar to attachment organizations found in mother-child dyads.

Nevertheless, Howes and Spieker (2016:315) temper their assessment of this similarity:

> While research published in the interval between the second and third editions of the *Handbook of Attachment* has not challenged the assumption that children may establish attachment relationships with their nonparental childcare providers, there is some evidence that asking these caregivers for their perceptions of their relationships with particular children may result in relationship descriptions less aligned with attachment theory than observations of child-caregiver attachment behaviors in childcare settings.

They give the example of "childcare providers as teachers who are responsible for children's school readiness," so that these providers perceive their
relationships with their charges in terms of self-efficacy as well as warmth and
intimacy. This analysis does not consider, however, whether the very definition
of attachment ought to be expanded to include a wider variety of providers or
a combination of functions.

Indeed, their chapter in the third edition of the *Handbook of Attachment*,
though it is entitled "Attachment Relationships in the Context of Multiple
Caregivers," is entirely occupied with the question of whether children can
become attached to their daycare providers (Howes and Spieker 2016). The
range of multiple attachments that can occur cross-culturally, and the various divisions of labor that may apply among these multiple caregivers, is
not even addressed. Instead, as the passages already quoted from this chapter illustrate, its sole concern is how children's attachments to institutional
providers might fare in comparison to those with mother. While the authors
note other attachment relationships in passing (e.g., relationships to fathers,
grandparents, and other relatives), they do not pursue the nature of these relationships. Howes and Spieker (2016:316) do note, however, that "there is
almost no literature on grandparent-child attachment relationships constructed concurrently with child-parent relationships," despite the fact that grandmothers are crucial child caregivers everywhere in the world (see Keller and
Chaudhary, this volume).

Clarke-Stewart (1989) offers quite a different interpretation of the possibility that children who attend daycare programs early in life, and spend many
hours a day there, may end up "insecure-avoidant." She suggests that the autonomy and independence which children gain through their daycare experience is being improperly mistaken for avoidance in the Strange Situation
Procedure. She points out that this experimental procedure may not stress children with daycare experience as much as those without it, because it replicates
the experience to which they have become habituated in key respects: children
with daycare experience expect their mothers to leave them and know that they
will return; they are also used to playing with toys that are not their own and
being cared for by adults other than the mother. Attachment theory researchers
would do well to consider this alternative interpretation.

Another approach to early education different from those reviewed above
has been described by Serpell and Nsamenang (2014). Using the instance
of sub-Saharan communities in Africa, they present the case that policy and
services should be constructed around local knowledge systems, the distinctive cultural practices that have come to surround that knowledge, and the
unique environmental context in which it is set. Unless this is done, they
argue, children will always remain at a disadvantage. Everything that a child
learns in existing centers, which have been designed to provide compensa-
tory education to disadvantaged children, is irrelevant to what they actually
need to know in order to thrive in their worlds. Moreover, these Westernized

centers treat local knowledge, including knowledge of social life, as inadequate and inferior.

Legal Implications

When it comes to the welfare of children, attachment theory influences the law and its practitioners directly, as well as the guidelines adopted by these experts (e.g., court counselors, family therapists, and health professionals including psychologists, pediatricians, and social workers) whose role it is to advise jurists and to testify before the court. Attachment theorists have generally interceded in conversations about child custody by refuting earlier judicial biases toward maternal or, more recently in Western countries, joint or shared custody that typically accompanies dual residence (Maccoby and Mnookin 1993).[4] Instead, these theorists advocate a more child-centered approach, one independent of gender or biological parenthood. In the language of the study discussed earlier (Goldstein et al. 1973:17–20), but fully in line with their own emphasis on sensitive parenting, they ask: Who is a child's "psychological parent"? The common idea that there is one such individual, and that the child should be assigned exclusively to that parent in the case of divorce, has sometimes reintroduced the biased assumption that mothers are the natural child caregivers, with the result that fathers have been needlessly excluded from meaningful engagement in their children's lives (Kelly and Lamb 2000).

Contributors to a special issue of *Family Court Review*, devoted to attachment theory, entertain the possibility that fathers can also be attachment figures. While some consider this strictly a gender-neutral matter (e.g., Siegel and McIntosh 2011:519), others are inclined to posit a division of labor in which "mothers" are more likely to provide "close emotional scaffolding," whereas "fathers" are the ones who encourage autonomy and exploration (Bretherton et al. 2011:542). We support the general observation that the care of children may be divided into different roles. However, these roles can be many and widely varied (see Morelli et al., this volume). We disagree with the ethnocentric notion that these two roles exhaust the cross-cultural possibilities or that one role is more conducive to attachment than the other.

In this special issue of the *Family Court Review*, the most adamant endorsement of a binary difference between a mother's and father's capacity for care, and consequently of the mother as the primary child caregiver in all cases, comes from Allan Schore. As a clinical neuropsychologist, Schore claims that the difference between "females and males" is dichotomous and universal, presumably because it is neurobiologically based (Schore and McIntosh 2011:504, italics added for emphasis):

---

[4]   In Spain, shared custody of children no longer lactating has been federal law since 2000 (García and Otero 2006).

...we know that there is a difference between the father and the mother even in the 1st year and that the father's play is more arousing and energetic, while the mother's is more calming. There are extensive differences between females and males in terms of the ability to process emotional information. Females show an enhanced capacity to more effectively read nonverbal communications and to empathically resonate with emotional states than men. When it comes to reading facial expressions, tone of voice, and gestures, women are generally better than men. This is why, *in all human societies*, the very young and the very old are often attended to by females.[5]

This is a woefully ignorant list of claims, without support from the literature on gender differences. While some gender differences in humans certainly exist, they are not of the categorical kind to which Schore subscribes. Moreover, there is no proof that gender differences that do or may exist would affect caregiving. This is an especially disturbing example of gender bias, because judges have so much leeway to form opinions based on whatever academic positions they encounter and regard favorably.

Arguably, no sector of society today relies more routinely and confidently on attachment theory than do the courts in their consideration of cases involving child custody and child placement. Attachment theorists may demur, cautioning as do Main et al. (2011:428, italics in original) that

> ...*all* present methods of assessing attachment were designed for research purposes...and have yet to be sufficiently tested for their predictive powers with respect to the assessment of individuals.

Previous authors (e.g., Byrne et al. 2005; Emery et al. 2005; Mercer 2009; Symons 2010) as well as other contributors to this same journal issue on attachment theory and family law (Bretherton et al. 2011) warn sharply about the fallibility of the evidence supporting these methods. Still, Main et al. (2011:427–428) recommend, for use in making custody decisions, what they consider to be the "gold standard measures" developed by attachment theorists: the Strange Situation Procedure, the Attachment Q-Sort, and the Adult Attachment Interview. They recommend these instruments (never to be applied singly, they advise, but in combination with one another) because these procedures "come as close as possible to providing scientific evidence" (see Herman 1997; Main et al. 2011:448). Yet, other reports question the reliability of attachment measures like the Strange Situation Procedure and the Attachment Q-Sort, and demonstrate that they produce different results (Ahnert et al. 2006). In addition, parenting is evaluated and decisions about child placement are made on the basis of sensitivity and other attachment tenets, even though the families under evaluation may have profoundly different philosophies about what is best for the child.

---

[5] A similar claim about differences between men and women was made earlier by Bowlby himself when he wrote of the latter being biologically primed to behave in "motherly ways."

Implications for Public Policy

Both governmental agencies and nongovernmental organizations promote attachment theory assumptions in the areas of child development, child care, and breastfeeding. One example of a national policy advocating attachment theory is the Chilean government's Spanish-language Internet program: *Chile Crece Contigo* (Chile Grows with You). In existence since 2005, this program contains a section devoted to "*Apego*" (attachment), which teaches users (a) about sensitive parenting; (b) that infants need parental attention, stimulation, and interaction; and (c) that within this interactional context, infants are already separate "individuals." The program also promotes breastfeeding and offers fully paid six-week maternity leave in support of that practice as well as to foster overall attachment.

When attachment theory is exported from the West to the Global South, ethical problems arise due to the theoretical and methodological bias imposed on children and their caregivers in these non-Western societies. In sub-Saharan Africa, for example, Early Childhood Care and Education programs take a definite colonialist overtone, based on "the progressive appropriation of Western culture in opposition to African traditions" (Serpell and Nsamenang 2014). Their intent is to displace these traditions because they have been deemed to be "deficient and/or outdated." Western standards of nurturing are often held up as being "scientific" and used as the basis for such interventions. This tendency is heightened by the fact that research findings from Western settings have been compiled into readily available databases, whereas there is a paucity of cross-cultural research sensitive to local realities.

There is ample research to demonstrate the strong preference for multiple caregivers (men and women, children and adults) in Indian homes (Trawick 1990; Kurtz 1992; Roland 2005; Seymour 2004; Sharma 2003; Chaudhary 2004). This pattern stands in sharp contrast to the notion of a single sensitive female caregiver, who devotes her time exclusively to raising her children. Breastfeeding, co-sleeping, and physical stimulation of the child are all promoted, but not by a single person. In fact a mother who is alone in bringing up her child is considered to be in a difficult situation because she lacks the support of others. The ideology behind multiple caregiving is founded on the importance of relationships, including the active engagement of elders and other people in the socialization of young children, which is believed to be beneficial for them, for other members of the family, and for overall family cohesiveness.

One telling example of the questionable value of interventions into traditional childcare practices comes from a report by UNICEF, India (UNICEF 2011). This report describes a program, Behaviour Change Communication (BCC), whose proponents argue for transformations in communication patterns in the interests of "better practice." This program seems to have become the latest trend, and its name the latest buzz word, in village-level interventions. However, the program's efficacy is another matter. The report shows, for instance, that there was already a high prevalence of breastfeeding among the families being served prior to introduction of BCC (UNICEF 2011:9). Thus, it is hard to understand how this rate could be attributed to a subsequent intervention resting on a new claim regarding change!

Even as common cultural practices, such as multiple caregiving and breast-feeding, continue today, international NGOs question other of these practices as part of their campaign for what they consider to be universal child rights. Under this banner, welfare programs of all kinds often cast the child as an individual rather than as a member of a collective, within which work and responsibility are routinely shared. Some of the key debates in this domain, in India and elsewhere, relate to (a) care of children by siblings (which counts as child labor by UN standards), (b) children doing household chores and participating in agricultural work in their families (again, conceived of as child labor), (c) care by relatives as an alternative to early childcare centers for young children (which are automatically assumed to be superior), and (d) supposed evidence of child abuse by family members (Aiyar 2015).

Our concern is that international "standards" for child care will make inroads into local practices through various interventions of this kind. When policy and interventions are shaped by such international (Euro-American) standards and imported by NGOs, they may upset the ecological wisdom of community living, all in the name of the millennial development goals that the NGOs pursue. We would guess that social workers as well as legal and medical practitioners, all familiar with village-level practices and their context, may still reasonably accept the local cultural patterns of child care that they encounter, although the absence of corroborating research makes it difficult to be certain.

Another, equally disturbing manifestation of this same cultural conflict between local practices and international expectations has arisen among Indian immigrants to other countries (e.g., in Norway, the United States, and the United Kingdom). Practices such as co-sleeping with parents, feeding by an adult, and discipline by means of physical punishment are still prevalent within Indian families. However, unbeknown to parents, usually recent migrants to the West, these practices come under the scrutiny of childcare services through alerts from daycare facilities, schools, or clinics. Several instances have been reported in which parents who follow these traditions have been treated as being potentially harmful for the children, who were then removed from their families and placed in foster care—the most common solution to such problems (Chaudhary and Valsiner 2015). Long drawn-out legal battles ensued, during which the children are kept separated from parents and not even allowed visits with extended kin.

The same situation faced by these Indian immigrants has transpired elsewhere, for instance among autochthonous Bribri people in Costa Rica, as reported by social welfare staff and local community leaders. Bribri parenting departs sharply from the dominant, Western-oriented caregiving practices that prevail in Costa Rica. For example, it is a Bribri practice to include children in housework. This tends to be interpreted by policy-makers, once again, as forced child labor, and can lead to the removal of children from their households (Keller and Rosabal-Coto, pers. comm., March, 2017).

Conclusion

Attachment theory has had an inordinate influence over a variety of practices by professionals whose province is children: family therapists, educators, jurists, policy-makers, pediatricians, as well as parents themselves. From the start, attachment theorists have noted the practical implications of their views, even if they have distanced themselves from some applications. In addition, some practitioners have borrowed from attachment theory to buttress the scientific credibility of programs for intervention into child care. Moreover, conclusions communicated in academic venues with cautionary notes and conditional qualifications often metamorphose into indisputable, unconditional scientific facts in the hands of those whose responsibility it is to solve real-world problems.

The cultural critique of attachment theory applications that we have raised is far from an idle academic exercise. It is a matter of social responsibility. Attachment theory is being used in a number of areas before sufficient evidence exists to support specific applications. Applications of the theory (e.g., in such venues as child-rearing manuals, the courts, and organizations both governmental and private devoted to public policy) are being made without consideration of cultural context as a fundamental dimension of the practices being addressed. As a result, spurious, often ethnocentric, recommendations for child caregiving are being promoted. We call on the scientific community to undertake the task of uncovering these misuses and to work to rectify the situation.

Morelli et al. (this volume) provide ample evidence that there is no single "best" practice or set of practices for the care of young children. The existence of a single person (the mother), who completely devotes all of her time and energy to the care of her young child, is hardly universal, nor is it realistic or practical. Numerous solutions to child-rearing have been invented by communities in response to their own particular histories, cultures, and ecological settings. Cultural sensitivity in policy, planning, and the delivery of services is a fundamental need for children, families, and communities all over the world. The only way to achieve this is to adopt a pluralistic approach to understanding the variety of attachments that children form as they grow up, and the long-term effects this has on their adult psychological makeup. We argue that practices of child-rearing vary among cultures, and that the variety of values that motivate those different practices need to be recognized (Morelli et al., this volume). To return to our earlier example of how a one-size-fits-all model for child-rearing can lead to profound misfit, we recall the working-class mother who watched her son through a one-way mirror, during a mother-absent phase in the Strange

Situation Procedure, as he played with the toys in the room. In contrast to the researchers running the study, this mother did not observe insecure-avoidant behavior: she saw her son exhibit independence and this made her very proud.

The methodological conclusions delineated by Morelli et al. (this volume) are critical to real-world applications of attachment theory, such as the ones discussed here. It is impossible to fathom what manner of programs, guidelines, and policies are needed by parents and other child caregivers in different societies and ethnic groups other than our own, without accounting for what they want for their children, how they raise them, and how they perceive recommended practices by comparison with their own. To be sure, this intensive method of inquiry and the wide variation it is likely to reveal complicates the practice of therapy, the task of educating children, legal decision making, policy-making, and pediatric advice in all kinds of ways. This poses a fundamental challenge that must be addressed if better social outcomes, of any kind, are to be engineered, especially in complex societies and in societies other than the practitioners' own. To meet this challenge, the goal or goals of such engineering must first be interrogated, understood, and agreed upon by all concerned. Attachment theory does not do a very good job of this.

Bibliography

Note: Numbers in square brackets denote the chapter in which an entry is cited.

Abello, M. T., and M. Colell. 2006. Analysis of Factors That Affect Maternal Behaviour and Breeding Success in Great Apes in Captivity. *Int. Zoo Yearbook* **40**:323–340. [3, 4]

Abubakar, A., P. Holding, M. Mwangome, and K. Maitland. 2011. Maternal Perceptions of Factors Contributing to Severe Undernutrition among Children in a Rural African Setting. *Rural Remote Health* **11**:1423. [13]

Acevedo, B. P., A. Aron, H. E. Fisher, and L. L. Brown. 2012. Neural Correlates of Long-Term Intense Romantic Love. *Soc. Cogn. Affect. Neurosci.* **7**:145–159. [10]

Adams, F. 2010. Embodied Cognition. *Phenom. Cogn.* **9**:619–628. [11]

Adolph, K., L. B. Karasik, and C. Tamis-Lemonda. 2010. Motor Skill. In: Handbook of Cultural Developmental Science, ed. M. H. Bornstein, pp. 61–88. New York: Psychology Press. [8]

Ahnert, L., M. Pinquart, and M. E. Lamb. 2006. Security of Children's Relationships with Nonparental Care Providers: A Meta-Analysis. *Child Dev.* **74**:664–679. [12, 14]

Ainsworth, M. D. S. 1962. The Effects of Maternal Deprivation: A Review of Findings and Controversy in the Context of Research Strategy. In: Deprivation of Maternal Care: A Reassessment of Its Effects, pp. 97–165. Geneva: World Health Organization. [2]

———. 1967. Infancy in Uganda: Infant Care and the Growth of Love. Baltimore: Johns Hopkins Univ. Press. [1, 2, 5, 8]

———. 1969. Object Relations, Dependency, and Attachment: A Theoretical Review of the Infant-Mother Relationship. *Child Dev.* **40**:969–1025. [2]

———. 1972. Attachment and Dependency: A Comparison. In: Attachment and Dependency, ed. J. L. Gewirtz, pp. 97–137. Washington, D.C.: V. H. Winston and Sons. [3]

———. 1976. Systems for Rating Maternal Care Behavior. Princeton: ETS Test Collection. [6]

———. 1979. Attachment as Related to Mother-Infant Interaction. *Adv. Study Behav.* **9**:1–51. [2]

———. 1985. Patterns of Infant-Mother Attachments: Antecedents and Effects on Development. *Bull. NY Acad. Med.* **61**:771–791. [9]

———. 1991. Attachments and Other Affectional Bonds across the Life Cycle. In: Attachment across the Life Cycle, ed. C. M. Parkes et al., pp. 33–51. London: Routledge. [2]

———. 1995. On the Shaping of Attachment Theory and Research: An Interview with Mary D. S. Ainsworth (Fall 1994). In: Care-Giving, Cultural, and Cognitive Perspectives on Secure-Base Behavior and Working Models, ed. E. Waters et al., pp. 3–24. Chicago: Monographs of the Society for Research in Child Development. [5]

———. 1998. Society for Research in Child Development (SRCD) Oral History Interview. Interviewed by Harold Stevenson. www.srcd.org/sites/default/files/documents/ainsworth_mary_interview.pdf (accessed Nov. 18, 2016). [2]

———. 2013/1983. Mary D. Salter Ainsworth: An Autobiographical Sketch. *Attach. Hum. Dev.* **15**:448–459. [2]

Ainsworth, M. D. S., and S. M. Bell. 1970. Attachment, Exploration, and Separation: Illustrated by the Behavior of One-Year-Olds in a Strange Situation. *Child Dev.* **41**:49–67. [2]

Ainsworth, M. D. S., S. M. Bell, and D. J. Stayton. 1974. Infant-Mother Attachment and Social Development: Socialization as a Product of Reciprocal Responsiveness to Signals. In: The Integration of a Child into a Social World, ed. M. P. M. Richards, pp. 99–135. Cambridge: Cambridge Univ. Press. [5, 8]

Ainsworth, M. D. S., M. C. Blehar, E. Waters, and S. Wall. 1978. Patterns of Attachment: A Psychological Study of the Strange Situation. Hillsdale, NJ: Erlbaum. [1, 2, 5, 6, 8]

Ainsworth, M. D. S., and J. Bowlby. 1991. An Ethological Approach to Personality Development. *Am. Psychol.* **46**:333–341. [2, 6, 7]

Aiyar, S. 2015. Indians in Kenya: The Politics of Diaspora. Cambridge, MA: Harvard Univ. Press. [14]

Akesson, B. 2017. A Baby to Tie You to Place: Childrearing Advice from a Palestinian Mother Living under Occupation. In: A World of Babies: Imagined Childcare Guides for Eight Societies, ed. A. Gottlieb and J. DeLoache, pp. 93–122. Cambridge: Cambridge Univ. Press. [6]

Alexander, R. D. 1979. Darwinism and Human Affairs. Seattle: Univ. of Washington Press. [5]

Altmann, J. 1980. Baboon Mothers and Infants. Cambridge, MA: Harvard Univ. Press. [3, 4]

Alvarez, H. P. 2004. Residence Groups among Hunter Gatherers: A Review of the Claims and Evidence for Patrilocal Bands. In: Kinship and Behavior in Primates, ed. B. Chapais and C. M. Berman, pp. 420–442. Oxford: Oxford Univ. Press. [4]

American Academy of Pediatrics. 1998. Erratum. *Pediatrics* **102**:433–433. [13]

APA. 2013. Diagnostic and Statistical Manual of Mental Disorders, 5th edition. New York: American Psychiatric Association. [10]

Apple, R. D. 2006. Perfect Motherhood: Science and Childrearing in America. New Brunswick, NJ: Rutgers Univ. Press. [2]

Apple, R. D., and J. Golden, eds. 1997. Mothers and Motherhood: Readings in American History. Columbus, OH: Ohio State Univ. Press. [2]

Archer, M., M. Steele, J. Lan, et al. 2015. Attachment between Infants and Mothers in China Strange Situation Procedure Findings to Date and a New Sample. *Int. J. Behav. Dev.* **39**:485–491. [7]

Ardrey, R. 1961. African Genesis: A Personal Investigation into the Animal Origins and Nature of Man. New York: Atheneum. [2]

———. 1966. The Territorial Imperative: A Personal Inquiry into the Animal Origins of Property and Nations. New York: Atheneum. [2]

Arendell, T. 1997. Contemporary Parenting: Challenges and Issues. Thousand Oaks, CA: Sage. [5]

Arendt, H. 1958. The Human Condition. London: Univ. of Chicago Press. [14]

Atzil, S., T. Hendler, and R. Feldman. 2011. Specifying the Neurobiological Basis of Human Attachment: Brain, Hormones, and Behavior in Synchronous and Intrusive Mothers. *Neuropsychopharmacology* **36**:2603–2615. [9, 10]

Aviezer, O., A. Sagi, T. Joels, and Y. Ziv. 1999. Emotional Availability and Attachment Representations in Kibbutz Infants and Their Mothers. *Dev. Psychol.* **35**:811–821v. [6]

Axelrod, R. 1984. The Evolution of Cooperation. New York: Basic Books. [11]

———. 1997. The Complexity of Cooperation: Agent-Based Models of Competition and Collaboration. Princeton: Princeton Univ. Press. [11]

Axelrod, R., and W. D. Hamilton. 1981. The Evolution of Cooperation. *Science* **211**:1390–1396. [11]

Bacchiddu, G. 2012. Reticent Mothers, Motherly Grandmothers and Forgotten Fathers: The Making and Unmaking of Kinship in Apiao, Chiloé. *Tellus* **12**:35–58. [5]

Bacharach, M. 2006. Beyond Individual Choice: Teams and Frames in Game Theory, ed. N. Gold and R. Sugden. Princeton: Princeton Univ. Press. [11]

Baimel, A., R. L. Severson, A. S. Baron, and S. Birch. 2015. Enhancing "Theory of Mind" through Behavioral Synchrony. *Front. Psychol.* **6**:870. [6]

Bakan, D. 1966. The Duality of Human Existence: An Essay on Psychology and Religion. Chicago: Rand McNally. [6]

Bakeman, R., L. Adamson, M. Konner, and R. Barr. 1990. !Kung Infancy: The Social Context of Object Exploration. *Child Dev.* **61**:794–809. [6]

Bakeman, R., and V. Quera. 2011. Sequential Analysis and Observational Methods for the Behavioral Sciences. New York: Cambridge Univ. Press. [8]

Bakermans-Kranenburg, M. J., and M. H. van IJzendoorn. 2006. Gene-Environment Interaction of the Dopamine D4 Receptor (DRD4) and Observed Maternal Insensitivity Predicting Externalizing Behavior in Preschoolers. *Dev. Psychobiol.* **48**:406–409. [8]

———. 2007. Genetic Vulnerability or Differential Susceptibility in Child Development: The Case of Attachment. *J. Child Psychol. Psychiatry* **48**:14. [8]

———. 2015. The Hidden Efficacy of Interventions: Gene × Environment Experiments from a Differential Susceptibility Perspective. *Annu. Rev. Psychol.* **66**:381–409. [6]

Bakermans-Kranenburg, M. J., M. H. van IJzendoorn, M. M. Riem, M. Tops, and L. R. Alink. 2012. Oxytocin Decreases Handgrip Force in Reaction to Infant Crying in Females without Harsh Parenting Experiences. *Soc. Cogn. Affect. Neurosci.* **7**:951–957. [10]

Bard, K. A. 1994. Evolutionary Roots of Intuitive Parenting: Maternal Competence in Chimpanzees. *Early Dev. Parent.* **3**:19–28. [1, 3]

———. 1995. Parenting in Primates. In: Handbook of Parenting, vol. 2, Biology and Ecology of Parenting, ed. M. H. Bornstein, pp. 27–58. Hillsdale, NJ: Lawrence Erlbaum, Inc. [4]

———. 2002. Primate Parenting. In: Handbook of Parenting, vol. 2, Biology and Ecology of Parenting, 2nd edition, ed. M. H. Bornstein, pp. 99–140. Mahwah, NJ: Erlbaum. [3, 4]

———. 2007. Neonatal Imitation in Chimpanzees (*Pan troglodytes*) Tested with Two Paradigms. *Anim. Cogn.* **10**:233–242. [3]

———. 2018. Primate Parenting. In: Handbook of Parenting, vol. 2, Biology and Ecology of Parenting, 3rd edition, ed. M. H. Bornstein. Mahwah, NJ: Lawrence Erlbaum, in press. [1]

Bard, K. A., L. Brent, B. Lester, J. Worobey, and S. J. Suomi. 2011. Neurobehavioral Integrity of Chimpanzee Newborns: Comparisons across Groups and across Species Reveal Gene-Environment Interaction Effects. *Infant Child Dev.* **20**:47–93. [4]

Bard, K. A., and D. A. Leavens. 2014. The Importance of Development for Comparative Primatology. *Annu. Rev. Anthropol.* **43**:183–200. [4, 8]

Bard, K. A., M. Myowa-Yamakoshi, M. Tomonaga, et al. 2005. Group Differences in the Mutual Gaze of Chimpanzees (*Pan troglodytes*). *Dev. Psychol.* **41**:616–624. [1, 3, 4]

Barelli, C., K. Matsudaira, T. Wolf, et al. 2013. Extra-Pair Paternity Confirmed in Wild White-Handed Gibbons. *Am. J. Primatol.* **75**:1185–1195. [4]

Barelli, C., U. H. Reichard, M. Heistermann, and C. Boesch. 2008. Female White-Handed Gibbons (*Hylobates lar*) Lead Group Movements and Have Priority of Access to Food Resources. *Behaviour* **145**:965–981. [4]

Barlow, K. 2013. Attachment and Culture in Murik Society: Learning Autonomy and Interdependence through Kinship, Food, and Gender. In: Attachment Reconsidered: Cultural Perspectives on a Western Theory, ed. N. Quinn and J. M. Mageo, pp. 165–188. New York: Palgrave Macmillan. [6, 12]

Baron-Cohen, S. 2002. I Am Loved, Therefore I Think. *Nature* **416**:791–792. [11]

———. 2003. The Essential Difference: Male and Female Brains and the Truth About Autism. New York: Basic Books. [3]

Barr, R. G. 1990. The Early Crying Paradox. *Hum. Nat.* **1**:355–389. [6]

Barr, R. G., M. Konner, R. Bakeman, and L. Adamson. 1991. Crying In !Kung San Infants: A Test of the Cultural Specificity Hypothesis. *Dev. Med. Child Neurol.* **33**:601–610. [6]

Barrickman, N. L., M. L. Bastian, K. Isler, and C. P. van Schaik. 2008. Life History Costs and Benefits of Encephalization: A Comparative Test Using Data from Long-Term Studies of Primates in the Wild. *J. Hum. Evol.* **54**:568–590. [4]

Bartal, I. B. A., J. Decety, and P. Mason. 2011. Empathy and Pro-Social Behavior in Rats. *Science* **334**:1427–1430. [3]

Bartels, A., and S. Zeki. 2000. The Neural Basis of Romantic Love. *Neuroreport* **11**:3829–3834. [10]

———. 2004. The Neural Correlates of Maternal and Romantic Love. *NeuroImage* **21**:1155–1166. [10]

Bartlett, T. Q. 2003. Intragroup and Intergroup Social Interactions in White-Handed Gibbons. *Int. J. Primatol.* **24**:239–259. [4]

Bartz, J. A., J. Zaki, K. N. Ochsner, et al. 2010. Effects of Oxytocin on Recollections of Maternal Care and Closeness. *PNAS* **107**:21371–21375. [10]

Bateson, P. P. G. 1966. The Characteristics and Context of Imprinting. *Biol. Rev.* **41**:177–217. [2]

Batki, A., S. Baron-Cohen, S. Wheelwright, J. Connellan, and J. Ahluwalia. 2000. Is There an Innate Module? Evidence from Human Neonates. *Infant Behav. Dev.* **23**:223–229. [3]

Baumeister, R. F., and M. R. Leary. 1995. The Need to Belong: Desire for Interpersonal Attachments as a Fundamental Human Motivation. *Psychol. Bull.* **117**:497–529. [6]

Beatty, J. 1992. Fitness: Theoretical Contexts. In: Keywords in Evolutionary Biology, ed. E. F. Keller and E. Lloyd, pp. 115–119. Cambridge, MA: Harvard Univ. Press. [6]

Becker-Stoll, F. 2013. Von Der Eltern-Kind-Bindung Zur Erzieherin-Kind-Beziehung. In: Bildung Braucht Bindung, ed. C. Haderthauer and H. Zehetmair. Munich: Hanns Seidel-Stiftung. https://www.hss.de/uploads/tx_ddceventsbrowser/13_AMZ-83_Internetversion_04.pdf (accessed Oct. 27, 2016). [14]

Beckett, C., J. Castle, M. Rutter, and E. J. Sonuga-Barke. 2010. VI. Institutional Deprivation, Specific Cognitive Functions, and Scholastic Achievement: English and Romanian Adoptee (ERA) Study Findings. *Monogr. Soc. Res. Child Dev.* **75**:125–142. [9]

Beebe, B., and F. M. Lachmann. 2014. The Origins of Attachment: Infant Research and Adult Treatment. New York: Routledge. [11]

Behne, T., M. Carpenter, J. Call, and M. Tomasello. 2005. Unwilling versus Unable: Infants' Understanding of Intentional Action. *Dev. Psychol.* **41**:328–337. [11]

Beise, J. 2005. The Helping and the Helpful Grandmother. In: Grandmotherhood: The Evolutionary Significance of the Second Half of Female Life, ed. E. Voland et al., pp. 215–238. New Brunswick, NJ: Rutgers Univ. Press. [4]

Bekoff, M. 2007. The Emotional Lives of Animals. Novato, CA: New World Library. [4]

Belsky, J. 1997a. Attachment, Mating, and Parenting. *Hum. Nat.* **8**:361–381. [2]

————. 1997b. Theory Testing, Effect-Size Evaluation, and Differential Susceptibility to Rearing Influence: The Case of Mothering and Attachment. *Child Dev.* **64**:598–600. [12]

Belsky, J., and J. Cassidy. 1994. Attachment: Theory and Evidence. In: Development through Life, ed. M. Rutter and D. Hay, pp. 373–402. Oxford: Blackwell. [12]

Belsky, J., S. R. Jaffee, J. Sligo, L. Woodward, and P. A. Silva. 2005. Intergenerational Transmission of Warm-Sensitive-Stimulating Parenting: A Prospective Study of Mothers and Fathers of 3-Year-Olds. *Child Dev.* **72**:384–396. [14]

Belsky, J., P. L. Ruttle, W. T. Boyce, J. M. Armstrong, and M. J. Essex. 2015. Early Adversity, Elevated Stress Physiology, Accelerated Sexual Maturation and Poor Health in Females. *Dev. Psychol.* **51**:816–822. [11]

Belsky, J., L. Steinberg, and P. Draper. 1991. Childhood Experience, Interpersonal Development, and Reproductive Strategy: An Evolutionary Theory of Socialization. *Child Dev.* **62**:647–670. [11, 12]

Benedict, R. 1955. Continuities and Discontinuities in Cultural Conditioning. In: Childhood in Contemporary Cultures, ed. M. Mead and M. Wolfenstein, pp. 21–30. Chicago: Univ. of Chicago Press. [6]

Bennett, A. J., and S. Panicker. 2016. Broader Impacts: International Implications and Integrative Ethical Consideration of of Policy Decisions About Us Chimpanzee Research. *Am. J. Primatol.* **78**:1282–1303. [10]

Bennett, A. J., and P. J. Pierre. 2010. Nonhuman Primate Research Contributions to Understanding Genetic and Environmental Influences on Phenotypic Outcomes across Development. In: The Handbook of Developmental Science, Behavior, and Genetics, ed. K. Hood et al. New York: Wiley [10]

Bentley, G., and R. Mace. 2009. Substitute Parents: Biological and Social Perspectives on Alloparenting in Human Societies. New York: Berghahn Books. [6]

Berlin, L. J., J. M. Ispa, M. A. Fine, et al. 2009. Correlates and Consequences of Spanking and Verbal Punishment for Low-Income White, African American, and Mexican American Toddlers. *Child Dev.* **80**:1403–1420. [13]

Berlin, L. J., C. H. Zeanah, and A. F. Lieberman. 2008. Prevention and Intervention Programs for Supporting Early Attachment Security. In: Handbook of Attachment: Theory, Research, and Clinical Applications, ed. J. Cassidy and P. Shaver, pp. 745–761. New York: Guilford. [14]

————. 2016. Prevention and Intervention Programs to Support Early Attachment Security: A Move to the Level of Community. In: Handbook of Attachment: Theory, Research, and Clinical Applications, ed. J. Cassidy and P. Shaver, pp. 761–779. New York: Guilford. [14]

Berman, C. M. 1990. Intergenerational Transmission of Maternal Rejection Rates among Free-Ranging Rhesus Monkeys. *Anim. Behav.* **39**:329–337. [4]

Bernard, K., and M. Dozier. 2010. Examining Infants' Cortisol Responses to Laboratory Tasks among Children Varying in Attachment Disorganization: Stress Reactivity or Return to Baseline? *Dev. Psychol.* **46**:1771. [13]

Bernard, K., M. Dozier, J. Bick, et al. 2012. Enhancing Attachment Organization among Maltreated Children: Results of a Randomized Clinical Trial. *Child Dev.* **83**:623–636. [9]

Bernard, K., E. B. Meade, and M. Dozier. 2013. Parental Synchrony and Nurtuance as Targets in an Attachment-Based Intervention: Building Upon Mary Ainsworth's Insights About Mother-Infant Interaction. *Attach. Hum. Dev.* **15**:507–523. [10]

Bernstein, N. 2009. Difficult Adjustment: Chinese-American Children Sent to Live with Kin Abroad Face Tough Return. *NY Times* **July 24**:A15, A17. [7]

Bernstein, V. J., E. J. Harris, C. W. Long, E. Iida, and S. L. Hans. 2005. Issues in the Multi-Cultural Assessment of Parent–Child Interaction: An Exploratory Study from the Starting Early Starting Smart Collaboration. *J. Appl. Dev. Psychol.* **26**:241–275. [8]

Bialik, M., and J. Gordon. 2012. Beyond the Sling: A Real-Life Guide to Raising Confident, Loving Children the Attachment Parenting Way. New York: Simon & Schuster. [14]

Birch, J., and S. Okasha. 2015. Kin Selection and Its Critics. *Bioscience* **65**:22–32. [11]

Blum, D. 2002. Love at Goon Park: Harry Harlow and the Science of Affection. Cambridge, MA: Perseus Publ. [5]

Blurton Jones, N. G. 2016. Demography and Evolutionary Ecology of Hadza Hunter-Gatherers. Cambridge: Cambridge Univ. Press. [4]

Blurton Jones, N. G., K. Hawkes, and J. F. O'Connell. 2002. Antiquity of Postreproductive Life: Are There Modern Impacts on Hunter-Gatherer Postreproductive Life Spans? *Am. J. Hum. Biol.* **14**:184–205. [4]

Boesch, C. 2012. The Ecology and Evolution of Social Behavior and Cognition in Primates. In: The Oxford Handbook of Comparative Evolutionary Psychology, ed. J. Vonk and T. Shackelford, pp. 486–503. Oxford: Oxford Univ. Press. [5]

Boesch, C., and H. Boesch-Achermann. 2000. The Chimpanzees of the Tai Forest: Behavioural Ecology and Evolution. New York: Oxford Univ. Press. [4]

Boesch, C., C. Bolé, N. Eckhardt, and H. Boesch. 2010. Altruism in Forest Chimpanzees: The Case of Adoption. *PLoS One* **5**:e8901. [3, 4]

Bogart, S. L., A. J. Bennett, S. J. Schapiro, L. A. Reamer, and W. D. Hopkins. 2014. Different Early Rearing Experiences Have Long Term Effects on Cortical Organization in Captive Chimpanzees (*Pan troglodytes*). *Dev. Sci.* **17**:161–174. [1, 10]

Bogin, B. B., J. M. Bragg, and C. Kuzawa. 2014. Humans Are Not Cooperative Breeders but Practic Biocultural Reproduction. *Ann. Human Biol.* **41**:368–380. [6]

Bohr, Y. 2010. Transnational Infancy: A New Context for Attachment and the Need for Better Models. *Child Dev. Perspect.* **4**:189–196. [7]

Bohr, Y., and C. Tse. 2009. Satellite Babies in Transnational Families: A Study of Parents' Decision to Separate from Their Infants. *Infant Ment. Health J.* **30**:265–286. [7]

Bohr, Y., and N. Whitfield. 2011. Transnational Mothering in an Era of Globalization: Chinese-Canadian Immigrant Mothers' Decision-Making When Separating from Their Infants. *J. Motherhood Init.* **2**:161–174. [7]

Bond, M. H., and K. K. Hwang. 1986. The Social Psychology of Chinese People. Oxford Oxford Univ. Press. [7]

Bonner, J. T. 1965. Size and Cycle. Princeton: Princeton Univ. Press. [11]

Bornstein, M., H. Azuma, C. Tamis-LeMonda, and M. Ogino. 1990. Mother and Infant Activity and Interaction in Japan and in the United States: II. A Comparative Macroanalysis of Naturalistic Exchanges Focused on the Organization of Infant Attention. *Int. J. Behav. Dev.* **13**:289–308. [6]

Bos, K. J., N. Fox, C. H. Zeanah, and C. A. Nelson III. 2009. Effects of Early Psychosocial Deprivation on the Development of Memory and Executive Function. *Front. Neurosci.* **3**:16. [9]

Bowlby, J. 1946. Psychology and Democracy. *Political Q.* **17**:61–75. [11]

———. 1951. Maternal Care and Mental Health. *Bull. W.H.O.* **3**:355–534. [2, 11]

———. 1953. Child Care and the Growth of Love. Harmondsworth: Pelican. [2]

———. 1958. The Nature of the Child's Tie to His Mother. *Int. J. Psychoanal.* **39**:350–373. [1, 2, 6, 12]

———. 1969. Attachment and Loss, vol. 1: Attachment. New York: Basic Books. [1–6, 8, 11, 12]

———. 1973. Attachment and Loss, vol. 2: Separation Anxiety and Anger. New York: Basic Books. [1, 2, 5, 3, 8]

———. 1980. Attachment and Loss, vol. 3: Loss: Sadness and Depression. New York: Basic Books. [1, 5, 6]

———. 1982. Attachment and Loss, vol. 1: Attachment, 2nd edition. New York: Basic Books. [2, 6]

———. 1988a. Developmental Psychiatry Comes of Age. *Am. J. Psychiatry* **14**:1–10. [11]

———. 1988b. A Secure Base: Clinical Applications of Attachment Theory. New York: Routledge. [5]

Boyd, R., and J. B. Silk. 2015. How Humans Evolved. New York: W. W. Norton. [4]

Bratman, M. E. 1987. Intentions, Plans, and Practical Reason. Cambridge, MA: Harvard Univ. Press. [11]

Brazelton, T. B. 1987. What Every Baby Knows. Reading, MA: Addison-Wesley. [14]

Bretherton, I. 1987. New Perspectives on Attachment Relations: Security, Communication and Internal Working Models. In: Handbook of Infant Development, ed. J. Osofsky, pp. 1061–1100. New York: Wiley. [6]

———. 1991. The Roots and Growing Points of Attachment Theory. In: Attachment across the Life Cycle, ed. C. M. Parkes et al., pp. 9–32. London: Routledge. [2, 5]

———. 1992. The Origins of Attachment Theory: John Bowlby and Mary Ainsworth. *Dev. Psychol.* **28**:759–775. [2, 5, 6]

———. 2010. Fathers in Attachment Theory and Research: A Review. *Early Child Dev. Care* **180**:9–23. [5]

———. 2013. Revisiting Mary Ainsworth's Conceptualization and Assessments of Maternal Sensitivity-Insensitivity. *Attach. Hum. Dev.* **15**:460–484. [2]

Bretherton, I., and K. A. Munholland. 2008. Internal Working Models in Attachment Relationships: Elaborating a Central Construct in Attachment Theory. In: Handbook of Attachment: Theory, Research, and Clinical Applications (2nd ed.), ed. J. Cassidy and P. R. Shaver, pp. 102–127. New York: Guilford. [8]

Bretherton, I., S. Seligman, J. Solomon, J. Crowell, and J. McIntosh. 2011. If I Could Tell the Judge Something About Attachment: Perspectives on Attachment Theory in the Family Law Courtroom. *Fam. Court Rev.* **49**:539–548. [14]

Briga, M., I. Pen, and J. Wright. 2012. Care for Kin: Within-Group Relatedness and Allomaternal Care Are Positively Correlated and Conserved Throughout the Mammalian Phylogeny. *Biol. Lett.* **8**:533–536. [6]

Broad, K. D., J. P. Curley, and E. B. Keverne. 2006. Mother-Infant Bonding and the Evolution of Mammalian Social Relationships. *Phil. Trans. R. Soc. B* **361**:2199–2214. [11]

Brockelmann, W. Y., U. Reichard, U. Treesucon, and J. J. Raemaekers. 1998. Dispersal, Pair Formation and Social Structure in Gibbons (*Hylobates lar*). *Behav. Ecol. Sociobiol.* **42**:329–339. [4]

Bronfenbrenner, U. 1977. Toward an Experimental Ecology of Human Development. *Am. Psychol.* **32**:513–531. [2, 7]

Brown, G. R., T. E. Dickins, R. Sear, and K. N. Laland. 2011. Evolutionary Accounts of Human Behavioural Diversity. *Phil. Trans. R. Soc. B* **366**:313–324. [4]

Brown, S. L., and R. M. Brown. 2006. Selective Investment Theory: Recasting the Functional Significance of Close Relationships. *Psychoanal. Inq.* **17**:1–29. [6]

Brüne, M. 2016. Borderline Personality Disorder: Why Fast and Furious? *Evol. Med. Public Health* **1**:52–66. [11]

Buchheim, A., M. Heinrichs, C. George, et al. 2009. Oxytocin Enhances the Experience of Attachment Security. *Psychoneuroendocrinology* **34**:1417–1422. [10]

Buller, D. J. 2005. Adapting Minds: Evolutionary Psychology and the Persistent Quest for Human Nature. Cambridge, MA: MIT Press. [2]

Bullinger, A. F., F. Zimmermann, J. Kaminski, and M. Tomasello. 2011. Different Social Motives in the Gestural Communication of Chimpanzees and Human Children. *Dev. Sci.* **14**:58–68. [4]

Burbank, V. K. 2011. An Ethnography of Stress: The Social Determinants of Health in Aboriginal Australia. New York: Palgrave MacMillan. [11]

Burkart, J. M., O. Allon, F. Amici, et al. 2014. The Evolutionary Origin of Human Hyper-Cooperation. *Nat. Commun.* **5**:4747. [11]

Burkart, J. M., S. B. Hrdy, and C. P. van Schaik. 2009. Cooperative Breeding and Human Cognitive Evolution. *Evol. Anthropol.* **18**:175–186. [11]

Burkart, J. M., and C. P. van Schaik. 2010. Cognitive Consequences of Cooperative Breeding in Primates? *Anim. Cogn.* **13**:1–19. [5]

Burkhardt, R. W., Jr. 2005. Patterns of Behavior: Konrad Lorenz, Niko Tinbergen, and the Founding of Ethology. Chicago: Univ. Chicago Press. [2]

Burnette, D., J. Sun, and F. Sun. 2013. A Comparative Review of Grandparent Care of Children in the U.S. And China. *Ageing Int.* **38**:43–57. [7]

Butcher, R. L., and I. S. Gersch. 2014. Parental Experiences of the "Time Together" Home Visiting Intervention: An Attachment Theory Perspective. *Educ. Psychol. Pract.* **30**:1–18. [14]

Butler, D., and T. Suddendorf. 2014. Reducing the Neural Search Space for Hominid Cognition: What Distinguishes Human and Great Ape Brains from Those of Small Apes? *Psychon. Bull. Rev.* **21**:590–619. [3]

Byng-Hall, J. 1991. An Appreciation of John Bowlby: His Significance for Family Therapy. *J. Fam. Ther.* **13**:5–16. [2]

Byrne, J., T. G. O'Connor, R. S. Marvin, and W. F. Whelan. 2005. Practitioner Review: The Contribution of Attachment Theory to Child Custody Assessments. *J. Child Psychol. Psychiatry* **46**:115–127. [14]

Bystrova, K., V. Ivanova, M. Edhborg, et al. 2009. Early Contact versus Separation: Effects on Mother-Infant Interaction One Year Later. *Birth* **36**:97–109. [10]

Caldji, C., B. Tannenbaum, S. Sharma, et al. 1998. Maternal Care During Infancy Regulates the Development of Neural Systems Mediating the Expression of Fearfulness in the Rat. *PNAS* **95**:5335–5340. [10]

Callaghan, B., and N. Tottenham. 2016a. The Neuro-Environmental Loop of Plasticity: A Cross-Species Analysis of Parental Effects on Emotion Circuitry Development Following Typical and Adverse Caregiving. *Neuropsychopharmacology Rev.* **41**:163–176. [10, 11]

———. 2016b. The Stress Acceleration Hypothesis: Effects of Early-Life Adversity on Emotion Circuits and the Brain. *Curr. Opin. Behav. Sci.* **7**:76–81. [11]

Callaghan, T., H. Moll, H. Rakoczy, et al. 2011. Early Social Cognition in Three Cultural Contexts. In: Monographs of the Society for Research in Child Development, pp. 1–142, vol. 76, W. A. Collins, series ed. Oxford: Wiley-Blackwell. [8]

Campbell, C. J., A. Fuentes, K. MacKinnon, M. Panger, and S. K. Bearder. 2012. Primates in Perspective. Oxford: Oxford Univ. Press. [4]

Campos, J. J., R. Kermoian, and M. R. Zumbahlen. 1992. Socioemotional Transformations in the Family System Following Infant Crawling Onset. *New Dir. Child Adolesc. Dev.* **1992**:25–40. [8]

Capra, F., and P. L. Luisi. 2014. The Systems View of Life: A Unifying Vision. Cambridge: Cambridge Univ. Press. [11]

Carlson, E. A., and B. Egeland. 2004. The Construction of Experience: A Longitudinal Study of Representation and Behavior. *Child Dev.* **75**:66–83. [8]

Carlson, V. J., and R. L. Harwood. 2003. Attachment, Culture, and the Caregiving System: The Cultural Patterning of Everyday Experiences among Anglo and Puerto Rican Mother-Infant Pairs. *Infant Ment. Health J.* **24**:53–73. [6]

———. 2014. The Precursors of Attachment Security: Behavioral Systems and Culture. In: Different Faces of Attachment: Cultural Variations on a Universal Human Need, ed. H. Otto and H. Keller, pp. 278–303. New York: Cambridge Univ. Press. [12]

Caron, E. B., P. Weston-Lee, D. Haggerty, and M. Dozier. 2016. Community Implementation Outcomes of Attachment and Biobehavioral Catch-up. *Child Abuse Neglect* **53**:128–137. [9]

Carpenter, C. R., ed. 1940. A Field Study in Siam of the Behavior and Social Relations of the Gibbon (*Hylobates lar*), vol. 16. Baltimore: Johns Hopkins Press. [4]

Carter, C. S., L. Ahnert, K. E. Grossman, et al., eds. 2005. Attachment and Bonding: A New Synthesis. Dahlem Workshop Report, vol. 92, J. Lupp, series ed. Cambridge, MA: MIT Press. [1]

Casler, L. 1961. Maternal Deprivation: A Critical Review of the Literature. *Soc. Res. Child Dev.* **26**:1–63. [2]

Cassidy, J. 1999. The Nature of the Child's Ties. In: Handbook of Attachment: Theory, Research, and Clinical Applications, ed. J. Cassidy and P. R. Shaver, pp. 3–20. New York: Guilford. [2]

———. 2008. The Nature of the Child's Ties. In: Handbook of Attachment: Theory, Research, and Clinical Applications, 2nd edition, ed. J. Cassidy and P. R. Shaver, pp. 3–22. New York: Guilford. [5, 7]

———. 2016. The Nature of the Child's Ties. In: Handbook of Attachment: Theory, Research, and Clinical Applications (3rd ed.), ed. J. Cassidy and P. R. Shaver, pp. 3–24. New York: Guilford. [2]

Cassidy, J., and P. R. Shaver, eds. 2008. Handbook of Attachment: Theory, Research, and Clinical Applications (2nd ed.). New York: Guilford. [5, 6, 14]

———, eds. 2016. Handbook of Attachment: Theory, Research, and Clinical Applications (3rd ed.). New York: Guilford. [5, 8, 14]

Cassidy, J., S. S. Woodhouse, G. Cooper, et al. 2005. Examination of the Precursors of Infant Attachment Security: Implications for Early Intervention and Intervention Research. In: Enhancing Early Attachments: Theory, Research, Intervention, and Policy, ed. L. J. Berlin et al., pp. 34–60. New York: Guilford. [8]

Castellani, B., R. Rajaram, J. G. Buckwalter, M. Ball, and F. W. Hafferty. 2015. Place and Health as Complex Systems: A Case Study and Empirical Test. Springerbriefs in Public Health. Cham: Springer. [6]

Caulfield, T., and C. Condit. 2012. Science and the Sources of Hype. *Public Health Genomics* **15**:209–217. [2]

Champagne, F., J. Diorio, S. Sharma, and M. J. Meaney. 2001. Naturally Occurring Variations in Maternal Behavior in the Rat Are Associated with Differences in Estrogen-Inducible Central Oxytocin Receptors. *PNAS* **98**:12736–12741. [10]

Chao, R. K., and V. Tseng. 2002. Parenting of Asians. In: Handbook of Parenting (vol. 4): Social Conditions and Applied Parenting, ed. M. H. Bornstein, pp. 59–93. Mahwah, NJ: Lawrence Erlbaum. [6]

Chapin, B. L. 2013a. Attachment in Rural Sri Lanka: The Shape of Caregiver Sensitivity, Communication, and Autonomy. In: Attachment Reconsidered: Cultural Perspectives on a Western Theory, ed. N. Quinn and J. M. Mageo, pp. 143–165. New York: Palgrave Macmillan. [5]

———. 2013b. Caregiver Sensitivity, Communication, and the Shape of Autonomy in a Sri Lankan Village. In: Attachment Reconsidered: Cultural Perspectives on a Western Theory, ed. N. Quinn and J. Mageo, pp. 143–164. Hampshire, UK: SPA Palgrove. [6]

Chapman, C. A., and J. M. Rothman. 2006. Within-Species Differences in Primate Social Structure: Evolution of Plasticity and Phylogenetic Constraints. *Primates* **50**:12–22. [4]

Charnov, E. L. 1991. Evolution of Life History Variation among Female Mammals. *PNAS* **88**:1134–1137. [4]

———. 1993. Life History Invariants: Some Explorations of Symmetry in Evolutionary Ecology. Oxford: Oxford Univ. Press. [4, 11]

Charvet, C. J., and B. L. Finlay. 2012. Embracing Covariation in Brain Evolution: Large Brains, Extended Development, and Flexible Primate Social Systems. *Prog. Brain Res.* **195**:71–87. [4]

Chaudhary, N. 2004. Listening to Culture: Constructing Reality in Everyday Life. New Delhi: Sage. [5, 6, 14]

———. 2011. Peripheral Lives, Central Meanings: Women and Their Place in Indian Society. In: Cultural Dynamics of Women's Lives, pp. 7–30, Mind, Culture and Development: Cultural Psychology within the World, vol. 1, A. C. Bastos et al., series ed. Charlotte, NC: Information Age. [5]

———. 2012. Father's Role in the Indian Family: A Story That Must Be Told. In: The Father's Role: Cross-Cultural Perspectives, ed. D. Shwalb et al., pp. 68–94. New York: Routledge. [5]

———. 2013. Parent Beliefs, Socialisation Practices and Children's Development in Indian Families. University Grants Commission, New Delhi. http://www.nanditachaudhary.com/resource/reportUGCProject.pdf (accessed May 10, 2017). [5]

———. 2015. Self and Identity of the Young Child: Emerging Trends in Contemporary Indian Families. Report of the Indian Council for Social Science Research. http://www.nanditachaudhary.com/resource/ReportICSSR.pdf (accessed July 7, 2017). [5]

———. 2018. Childhood in Indian Families: The Cultural Contexts of Care. New Delhi: Springer, in press. [5]

Chaudhary, N., and J. Valsiner. 2015. Cultural Sensibilities Matter in Child Rearing. http://www.thehindu.com/opinion/op-ed/cultural-sensibilities-matter-in-parenting/article7562231.ece (accessed April 26, 2017). [14]

Chen, X., Y. Bian, T. Xin, L. Wang, and R. K. Silbereisen. 2010. Perceived Social Change and Childrearing Attitudes in China. *Eur. Psychol.* **15**:260–270. [7]

Cheney, D. L., and R. M. Seyfarth. 2007. Baboon Metaphysics. Chicago: Univ. of Chicago Press. [4]

Chevalier, A., and O. Marie. 2013. Economic Uncertainty, Parental Selection and the Criminal Activity of the "Children of the Wall." Research Memorandum 066. Maastricht Univ. Graduate School of Business and Economics (GSBE). http://ftp.iza.org/dp7712.pdf (accessed Nov. 10, 2016). [5]

Child Trends. 2015a. Attitudes About Spanking: Indicators on Children and Youth. Child Trends Databank. http://www.childtrends.org/wp-content/uploads/2012/10/51_Attitudes_Toward_Spanking.pdf (accessed Nov. 9, 2016). [13]

———. 2015b. Family Structure: Indicators on Children and Youth. http://www.childtrends.org/indicators/family-structure/ (accessed Nov. 16, 2016). [6]

Chisholm, J. S. 1996. The Evolutionary Ecology of Attachment Organization. *Hum. Nat.* **7**:1–38. [5, 6, 12]

———. 1999a. Attachment and Time Preference: Relations between Early Stress and Sexual Behavior in a Sample of American University Women. *Hum. Nat.* **10**:51–83. [11]

———. 1999b. Death, Hope, and Sex: Steps to an Evolutionary Ecology of Mind and Morality. Cambridge: Cambridge Univ. Press. [11]

———. 2003. Uncertainty, Contingency and Attachment: A Life History Theory of Theory of Mind. In: From Mating to Mentality: Evaluating Evolutionary Psychology, ed. K. Sterelny and J. Fitness, pp. 125–153. Hove, UK: Psychology Press. [11]

———. 2012. Flourishing, Feelings, and Fitness: An Evolutionary Perspective on Health Capability. In: Pragmatic Evolution: Practical Applications of Evolutionary Theory, ed. A. Poiani, pp. 188–210. Cambridge: Cambridge Univ. Press. [11]

Chisholm, J. S., and D. A. Coall. 2008. Not by Bread Alone: The Role of Psychosocial Stress in Age at First Reproduction and Health Inequalities. In: Evolutionary Medicine and Health, ed. W. Trevathan et al., pp. 134–148. New York: Oxford Univ. Press. [11]

Chisholm, J. S., J. Quinlivan, R. Petersen, and D. A. Coall. 2005. Early Stress Predicts Age at Menarche and First Birth, Adult Attachment and Expected Lifespan. *Hum. Nat.* **16**:233–265. [11]

Chivers, D. J. 1974. The Siamang in Malaya: A Field Study of a Primate in Tropical Rain Forest, vol. 4. Basel: Karger. [4]

———. 1976. Communication Within and between Family Groups of Siamang (*Symphalangus syndactylus*). *Behaviour* **57**:116–135. [4]

Chugani, H. T., M. E. Behen, O. Muzik, et al. 2001. Local Brain Functional Activity Following Early Deprivation: A Study of Postinstitutionalized Romanian Orphans. *NeuroImage* **14**:1290–1301. [9]

Cicchetti, D., and F. A. Rogosch. 2012. Gene × Environment Interaction and Resilience: Effects of Child Maltreatment and Serotonin, Corticotropin Releasing Hormone, Dopamine, and Oxytocin Genes. *Dev. Psychopathol.* **24**:411–427. [8]

Clark, A. 2016. Surfing Uncertainty: Prediction, Action, and the Embodied Mind. Oxford: Oxford Univ. Press. [11]

Clark, A., and D. J. Chalmers. 1998. The Extended Mind. *Analysis* **58**:7–19. [11]

Clarke-Stewart, K. A. 1989. Infant Day Care: Maligned or Malignant. *Am. Psychol.* **44**:266–273. [14]

Clutton-Brock, T. 2002. Breeding Together: Kin Selection and Mutualism in Cooperative Vertebrates. *Science* **296**:69–72. [5]

Clutton-Brock, T., and C. Janson. 2012. Primate Socioecology at the Crossroads: Past, Present, and Future. *Evol. Anthropol.* **21**:136–150. [4]

Coall, D. A., and J. S. Chisholm. 2003. Evolutionary Perspectives on Pregnancy: Maternal Age at Menarche and Infant Birth Weight. *Soc. Sci. Med.* **57**:1771–1781. [11]

Coan, J. A. 2008. Toward a Neuroscience of Attachment. In: Handbook of Attachment, vol. 2, ed. J. Cassidy and P. R. Shaver, pp. 241–265. New York: Guilford Press. [1]

Coccaro, E. F., J. R. Fanning, K. L. Phan, and R. Lee. 2015. Serotonin and Impulsive Aggression. *CNS Spectrums* **20**:295–302. [11]

Coffield, C. N., E. M. Y. Mayhew, J. M. Hayiland-Jones, and A. S. Walker-Andrews. 2014. Adding Odor: Less Distress and Enhanced Attention for 6-Month-Olds. *Infant. Behav. Dev.* **37**:155–161. [10]

Cole, M. 1985. The Zone of Proximal Development: Where Culture and Cognition Create Each Other. In: Culture, Communication, and Cognition: Vygotskian Perspectives, ed. J. V. Wertsch, pp. 146–161. Cambridge: Cambridge Univ. Press. [2]

———. 1996. Cultural Psychology: A Once and Future Discipline. Cambridge: Cambridge Univ. Press. [5]

Cole, S. W., G. Conti, J. M. Arevalo, et al. 2012. Transcriptional Modulation of the Developing Immune System by Early Life Social Adversity. *PNAS* **109**:20578–20583. [10]

Coll, C. T. G., E. L. Bearer, and R. M. Lerner, eds. 2004. Nature and Nurture: The Complex Interplay of Genetic and Environmental Influences on Human Behavior and Development. Mahwah, NJ: Lawrence Erlbaum. [8]

Collaborative Group on Hormonal Factors in Breast Cancer. 2012. Menarche, Menopause, and Breast Cancer Risk: Individual Participant Meta-Analysis, Including 118,964 Women with Breast Cancer from 117 Epidemiological Studies. *Lancet Oncol.* **13**:1141–1151. [11]

Cong, X., S. M. Ludington-Hoe, N. Hussain, et al. 2015. Parental Oxytocin Responses During Skin-to-Skin Contact in Pre-Term Infants. *Early Hum. Dev.* **91**:401–406. [10]

Coqueugniot, H., J. J. Hublin, F. Veillon, F. Houët, and T. Jacob. 2004. Early Brain Growth in Homo Erectus and Implications for Cognitive Ability. *Nature* **431**:299–302. [4]

Cosmides, L., and J. Tooby. 1997. Evolutionary Psychology: A Primer. Center for Evolutionary Psychology. http://www.cep.ucsb.edu/primer.html (accessed May 2, 2017). [2]

Course, M. 2011. Becoming Mapuche: Person and Ritual in Indigenous Chile. Urbana: Univ. of Illinois Press. [6]

Cowan, P. A. 1997. Beyond Meta-Analysis: A Plea for a Family Systems View of Attachment. *Child Dev.* **68**:601–603. [12]

Creswell, J. W., and V. L. Plano-Clark. 2007. Designing and Conducting Mixed Methods Research. Thousand Oaks, CA: Sage. [8]

Crittenden, A. N., and F. W. Marlowe. 2008. Allomaternal Care among the Hadza of Tanzania. *Hum. Nat.* **19**:249–262. [5, 6]

Crittenden, P. 2000. A Dynamic-Maturational Approach to Continuity and Change in Patterns of Attachment. In: The Organization of Attachment Relationships, ed. P. Crittenden and A. Claussen, pp. 343–358. New York: Cambridge Univ. Press. [12]

Croy, I., A. Luong, C. Triscoli, et al. 2016. Interpersonal Stroking Touch Is Targeted to C Tactile Afferent Activation. *Behav. Brain Res.* **297**: [10]

Csibra, G., and G. Gergely. 2011. Natural Pedagogy as Evolutionary Adaptation. *Phil. Trans. R. Soc. B* **366**:1149–1157. [11]

Cyr, C., E. M. Euser, M. J. Bakermans-Kranenburg, and M. H. Van IJzendoorn. 2010. Attachment Security and Disorganization in Maltreating and High-Risk Families: A Series of Meta-Analyses. *Dev. Psychopathol.* **22**:87–108. [13]

Dahl, R. E. 2004. Adolescent Brain Development: A Period of Vulnerabilities and Opportunities. Keynote Address. *Ann. NY Acad. Sci.* **1021**:1–22. [10]

Dally, A. 1982. Inventing Motherhood: The Consequences of an Ideal. New York: Schocken Books. [2]

Daly, A. J. 2010. Social Network Theory and Educational Change. Cambridge, MA: Harvard Univ. Press. [6]

Damasio, A. 1994. Descartes' Error: Emotion, Reason, and the Human Brain. New York: Avon. [11]

———. 1999. The Feeling of What Happens: Body and Emotion in the Making of Consciousness. New York: Harcourt Brace. [11]

———. 2010. Self Comes to Mind: Constructing the Conscious Mind. New York: Pantheon. [11]

D'Andrade, R. 1992. Schemas and Motivation. In: Human Motives and Cultural Models, ed. R. D'Andrade and C. Strauss, pp. 23–44. Cambridge: Cambridge Univ. Press. [11]Darwin, C. 1872/1965. The Expression of the Emotions in Man and Animals, vol. 526. Chicago: Univ. Chicago Press. [4]

Daston, L., and G. Mitman, eds. 2006. Thinking with Animals: New Perspectives on Anthropomorphism. New York: Columbia Univ. Press. [4]

Davila Ross, M., S. Menzler, and E. Zimmermann. 2008. Rapid Facial Mimicry in Orangutan Play. *Biol. Lett.* **4**:27–30. [3]

Dawkins, R. 1976. The Selfish Gene. New York: Oxford Univ. Press. [11]

Deacon, T. W. 2012. Incomplete Nature: How Mind Emerged from Matter. New York: W. W. Norton. [11]

Deater-Deckard, K., K. A. Dodge, J. E. Bates, and G. S. Pettit. 1996. Physical Discipline among African American and European American Mothers: Links to Children's Externalizing Behaviors. *Dev. Psychol.* **32**:1065. [13]

Debbané, M., G. Salaminios, P. Luyten, et al. 2016. Attachment, Neurobiology, and Mentalizing Along the Psychosis Continuum. *Front. Hum. Neurosci.* **10**:406. [11]

De Bellis, M. D., M. S. Keshavan, D. B. Clark, et al. 1999. Developmental Traumatology Part II: Brain Development. *Biol. Psychiatry* **45**:1271–1284. [10]

De Bellis, M. D., M. S. Keshavan, K. Frustaci, et al. 2002. Superior Temporal Gyrus Volumes in Maltreated Children and Adolescents with PTSD. *Biol. Psychiatry* **51**:544–552. [10]

De Bellis, M. D., and M. Kuchibhatla. 2006. Cerebellar Volumes in Pediatric Maltreatment-Related Posttraumatic Stress Disorder. *Biol. Psychiatry* **60**:697–703. [10]

Degler, C. N. 1991. In Search of Human Nature: The Decline and Revival of Darwinism in American Social Thought. Oxford: Oxford Univ. Press. [2]

de León, L. 1998. The Emergent Participant: Interactive Patterns in the Socialization of Tzotzil (Mayan) Infants. *J. Ling. Anthropol.* **8**:131–161. [6, 8]

Del Giudice, M. 2014. An Evolutionary Life History Framework for Psychopathology. *Psychoanal. Inq.* **25**:261–300. [11]

deMause, L. 1975. The History of Childhood. New York: Harper and Row. [2]

Dettmer, A. M., M. A. Novak, S. J. Suomi, and J. S. Meyer. 2012. Physiological and Behavioral Adaptation to Relocation Stress in Differentially Reared Rhesus Monkeys: Hair Cortisol as a Biomarker for Anxiety-Related Responses. *Psychoneuroendocrino* **37**:191–199. [9]

Dettmer, A. M., and S. J. Suomi. 2014. Nonhuman Primate Models of Neuropsychiatric Disorders: Influences of Early Rearing, Genetics, and Epigenetics. *ILAR* **55**:361–370. [10]

Dettwyler, K. A. 2004. When to Wean: Biological Versus Cultural Perspectives. *J. Clin. Gynecol. Obstet.* **47**:712–723. [4]

de Waal, F. B. M. 1998. Chimpanzee Politics: Power and Sex Amongst Apes. Baltimore: Johns Hopkins Univ. Press. [3]

———. 2001. The Ape and the Sushi Master. New York: Basic Books. [1]

———. 2016. Are We Smart Enough to Know How Smart Animals Are? New York: W. W. Norton. [4]

Dewalt, K. M., and B. R. DeWalt. 2010. Participant Observation: A Guide for Fieldworkers. Lanham, MD: Rowman Altamira. [8]

Dewey, K. G. 1997. Energy and Protein Requirements During Lactation. *Annu. Rev. Nutr.* **17**:19–36. [6]

de Wolff, M., and M. H. van IJzendoorn. 1997. Sensitivity and Attachment: A Meta-Analysis on Parental Antecedents of Infant Attachment. *Child Dev.* **68**:571–591. [6, 12]

Diamond, J. 2012. The World until Yesterday: What Can We Learn from Traditional Societies? London: Penguin Books. [3]

Diamond, M. C., M. R. Rosenzweig, E. L. Bennett, B. Lindner, and L. Lyon. 1972. Effects of Environmental Enrichment and Impoverishment on Rat Cerebral Cortex. *J. Neurobiol.* **3**:47–64. [9]

Dielentheis, T. F., E. Zaiss, and T. Geissmann. 1991. Infant Care in a Family of Siamangs (Hylobates syndactylus) with Twin Offspring at Berlin Zoo. *Zoo Biol.* **10**:309–317. [4]

Diener, M. 2000. Gift from the Gods: A Balinese Guide to Early Child Rearing. In: A World of Babies: Imagined Childcare Guides for Seven Societies, ed. J. DeLoache and A. Gottlieb, pp. 199–231. New York: Cambridge Univ. Press. [6]

Dillon, D. G., A. J. Holmes, J. L. Birk, et al. 2009. Childhood Adversity Is Associated with Left Basal Ganglia Dysfunction During Reward Anticipation in Adulthood. *Biol. Psychiatry* **66**:206–213. [9]

Ding, Y. H., X. Xu, Z. Y. Wang, H. R. Li, and W. P. Wang. 2012. Study of Mother–Infant Attachment Patterns and Influence Factors in Shanghai. *Early Hum. Dev.* **88**:295–300. [7]

Ditzen, B., M. Schaer, B. Gabriel, et al. 2009. Intranasal Oxytocin Increases Positive Communication and Reduces Cortisol Levels During Couple Conflict. *Biol. Psychiatry* **65**:728–731. [10]

Divale, W., and M. Harris. 1976. Population, Warfare, and the Male Supremacist Complex. *Am. Anthropol.* **78**:521–538. [11]

Dixson, A. 2012. Primate Sexuality. Oxford: Oxford Univ. Press. [4]

Doi, T. 2014. The Anatomy of Dependence. New York: Kodansha. [5]

Dolen, G., A. Darvishzadeh, K. W. Huang, and R. C. Malenka. 2013. Social Reward Requires Coordinated Activity of Nucleus Accumbens Oxytocin and Serotonin. *Nature* **501**:179–184. [10]

Dozier, M., K. C. Stoval, K. E. Albus, and B. Bates. 2001. Attachment for Infants in Foster Care: The Role of Caregiver State of Mind. *Child Dev.* **72**:1467–1477. [8]

Drago, L., and B. Thierry. 2000. Effects of Six-Day Maternal Separation on Tonkean Macaque Infants. *Primates* **41**:137–145. [4]

Draper, P., and H. Harpending. 1982. Father Absence and Reproductive Strategy: An Evolution-Ary Perspective. *J. Anthropol. Res.* **58**:141–161. [11]

Du Bois, C. 1944. The People of Alor: A Social-Psychological Study of an East Indian Island. Minneapolis: Univ. of Minnesota Press. [5,12]

Dunbar, R. I. M. 1988. Primate Social Systems. London: Chapman and Hall. [4]

———. 1992. Neocortex Size as a Constraint on Group Size in Primates. *J. Hum. Evol.* **22**:469–493. [4]

Dunbar, R. I. M., and S. Schultz. 2007. Evolution in the Social Brain. *Science* **317**:1344–1347. [6]

Dunsworth, H. M., A. G. Warrener, T. Deacon, P. T. Ellison, and H. Pontzer. 2012. Metabolic Hypothesis for Human Altriciality. *PNAS* **109**:15212–15216. [4]

Duranti, A. 2008. Further Reflections on Reading Other Minds. *Anthropol. Q.* **18**:483–494. [5]

Dykas, M. J., and J. Cassidy. 2011. Attachment and the Processing of Social Information across the Life Span: Theory and Evidence. *Psychol. Bull.* **137**:19. [8]

Eberle, M., and P. M. Kappeler. 2008. Mutualism, Reciprocity, or Kin Selection? Cooperative Rescue of a Conspecific from a Boa in a Nocturnal Solitary Forager the Gray Mouse Lemur. *Am. J. Primatol.* **70**:410–414. [3]

Edwards, C. P., and M. Bloch. 2010. The Whitings' Concepts of Culture and How They Have Fared in Contemporary Psychology and Anthropology. *J. Cross-Cult. Psychol.* **41**:485–498. [8]

Edwards, V. J., G. W. Holden, V. J. Felitti, and R. F. Anda. 2003. Relationship between Multiple Forms of Childhood Maltreatment and Adult Mental Health in Community Respondents: Results from the Adverse Childhood Experiences Study. *Am. J. Psychiatry* **160**:1453–1460. [10]

Ehrenreich, B., and D. English. 1978. For Her Own Good: Two Centuries of Experts' Advice to Women. New York: Anchor. [14]

Ein-Dor, T., and G. Hirschberger. 2016. Rethinking Attachment Theory: From a Theory of Relationships to a Theory of Individual and Group Survival. *Curr. Dir. Psychol.* **25**:223–227. [11]

Elder, G. H., J. Modell, and R. D. Parke. 1994. Children in Time and Place: Developmental and Historical Insights. Cambridge: Cambridge Univ. Press. [2]

Ellis, B. J., A. J. Figueredo, B. H. Brumbach, and G. L. Schlomer. 2009. Fundamental Dimensions of Environmental Risk: The Impact of Harsh versus Unpredictable Environments on the Evolution and Development of Life History Strategies. *Hum. Nat.* **20**:204–268. [11]

Emde, R. N., D. P. Wolf, and D. Oppenheim, eds. 2003. Revealing the Inner Worlds of Young Children: The Macarthur Story Stem Battery and Parent-Child Narratives. New York: Oxford Univ. Press. [8]

Emery, N. J. 2000. The Eyes Have It: The Neuroethology, Function and Evolution of Social Gaze. *Neurosci. Biobehav. Rev.* **24**:581–604. [3]

Emery, R. E., R. K. Otto, and W. T. Donaue. 2005. A Critical Assessment of Child Custody Evaluations: Limited Science and a Flawed System. *Psychol. Sci. Publ. Int.* **6**:1–29. [14]

Emmeche, C., and K. Kull, eds. 2011. Towards a Semiotic Biology: Life Is the Action of Signs. London: Imperial College Press. [11]

Engelmann, J., and E. Hermann. 2016. Chimpanzees Trust Their Friends. *Curr. Biol.* **26**:252–256. [3]

Erickson, F. 1995. Ethnographic Microanalysis. In: Sociolinguistics and Language Teaching, ed. S. L. McKay and N. H. Hornberger, pp. 282–306. Cambridge: Cambridge Univ. Press. [8]

Erickson, M. F., and B. Egeland. 2004. Linking Theory and Research to Practice: The Minnesota Longitudinal Study of Parents and Children and the STEEP[tm] Program. *Clin. Psychol.* **8**:5–9. [14]

Esposito, G., P. Setoh, S. Yoshida, and K. O. Kuroda. 2015. The Calming Effect of Maternal Carrying in Different Mammalian Species. *Front. Psychol.* **6**:1–6. [6]

Everett, D. L. 2009. Don't Sleep, There Are Snakes: Life and Language in the Amazonian Jungle. New York: Vintage Books. [5]

———. 2014. Concentric Circles of Attachment among the Pirãha: A Brief Survey. In: Different Faces of Attachment: Cultural Variations on a Universal Human Need, ed. H. Otto and H. Keller, pp. 169–186. Cambridge: Cambridge Univ. Press. [2, 5, 6, 12, 14]

Fadiman, A. 1997. The Spirit Catches You and You Fall Down: A Hmong Child, Her American Doctors, and the Collision of Two Cultures. New York: Farrar, Straus, and Giroux. [13]

Fairbanks, L. A. 1988. Vervet Monkey Grandmothers: Interactions with Infant Grand-offspring. *Int. J. Primatol.* **9**:425–441. [4]

———. 1990. Reciprocal Benefits of Allomothering for Female Vervet Monkeys. *Anim. Behav.* **40**:553–562. [3, 4]

———. 1996. Individual Differences in Maternal Style: Causes and Consequences for Mothers and Offspring. *Adv. Study Behav.* **23**:579–611. [4]

———. 2000. Behavioral Development of Nonhuman Primates and the Evolution of Human Behavioral Ontogeny. In: The Evolution of Behavioral Ontogeny, ed. S. Parker et al., pp. 131–158. Santa Fe: SAR Press. [5]

Fairbanks, L. A., and K. Hinde. 2013. Behavioral Response of Mothers and Infants to Variation in Maternal Condition: Adaptation, Compensation, and Resilience. In: Building Babies: Primate Development in Proximate and Ultimate Perspective, ed. K. B. H. Clancy et al., pp. 281–302. New York: Springer. [4]

Fairbanks, L. A., and M. T. McGuire. 1986. Age, Reproductive Value, and Dominance-Related Behavior in Vervet Monkey Females: Cross-Generational Influences on Social Relationships and Reproduction. *Anim. Behav.* **34**:1710–1721. [4]

———. 1993. Maternal Protectiveness and Response to the Unfamiliar in Vervet Monkeys. *Am. J. Primatol.* **30**:119–129. [4]

Faircloth, C. 2013. Attachment Parenting and Intensive Motherhood in the UK and France. New York: Berghahn Books. [14]

Fareri, D. S., and N. Tottenham. 2016. Effects of Early Life Stress on Amygdala and Striatal Development. *Dev. Cogn. Neurosci.* **19**:233–247. [10]

Farroni, T., G. Csibra, F. Simion, and M. H. Johnson. 2002. Eye Contact Detection in Humans from Birth. *PNAS* **99**:9602–9605. [3]

Farroni, T., M. H. Johnson, E. Menon, et al. 2005. Newborns' Preference for Face-Relevant Stimuli: Effects of Contrast Polarity. *PNAS* **102**:17245–17250. [6]

Fearon, R. M., and J. Belsky. 2016. Precursors of Attachment Security. In: Handbook of Attachment (3rd Ed.), ed. J. Cassidy and P. R. Shaver, pp. 291–313. New York: Guilford. [12]

Feldman, R. 2007a. Parent-Infant Synchrony and the Construction of Shared Timing: Physiological Precursors, Developmental Outcomes, and Risk Conditions. *J. Child Psychol. Psychiatry* **48**:329–354. [6, 10, 11]

———. 2007b. Parent-Infant Synchrony: Biological Foundations and Developmental Outcomes. *Curr. Dir. Psychol.* **16**:340–345. [6]

———. 2012a. Biobehavioral Synchrony: A Model for Integrating Biological and Microsocial Behavioral Processes in the Study of Parenting. *Parent. Sci. Pract.* **12**:154–164. [10]

———. 2012b. Oxytocin and Social Affiliation in Humans. *Horm. Behav.* **61**:380–391. [10]

———. 2012c. Parent-Infant Synchrony: A Biobehavioral Model of Mutual Influences in the Formation of Affiliative Bonds. *Monogr. Soc. Res. Child Dev.* **77**:42–51. [6]

———. 2014. Synchrony and the Neurobiological Basis of Social Affiliation. In: Mechanisms of Social Connection: From Brain to Group, ed. M. Mikulincer and P. R. Shaver, pp. 145–166. Washington, D.C.: American Psychological Association. [6, 11]

———. 2015. The Adaptive Parental Brain: Implications for Children's Social Development. *Trends Neurosci.* **38**:387–399. [10]

———. 2016. The Neurobiology of Mammalian Parenting and the Biosocial Context of Human Caregiving. *Horm. Behav.* **77**:3–17. [10]

———. 2017. The Neurobiology of Human Attachments. *Trends Cogn. Sci.* **21**:80–99. [10]

Feldman, R., I. Gordon, I. Schneiderman, O. Weisman, and O. Zagoory-Sharon. 2010. Natural Variations in Maternal and Paternal Care Are Associated with Systematic Changes in Oxytocin Following Parent-Infant Contact. *Psychoneuroendocrino* **35**:1133–1141. [10]

Feldman, R., S. Masalha, and D. Alony. 2006. Micro-Regulatory Patterns of Family Interactions; Cultural Pathways to Toddlers' Self-Regulation. *J. Fam. Psychol.* **20**:614–623. [10]

Feldman, R., Z. Rosenthal, and A. I. Eidelman. 2014. Maternal-Preterm Skin-to-Skin Contact Enhances Child Physiologic Organization and Cognitive Control across the First 10 Years of Life. *Biol. Psychiatry* **75**:56–64. [8, 10]

Felitti, V. J., R. F. Anda, D. Nordenberg, et al. 1998. Relationship of Childhood Abuse and Household Dysfunction to Many of the Leading Causes of Death in Adults. The Adverse Childhood Experiences (ACE) Study. *Am. J. Prev. Med.* **14**:245–258. [10]

Ferrari, P. F., A. Paukner, C. Ionica, and S. J. Suomi. 2009. Reciprocal Face-to-Face Communication between Rhesus Macaque Mothers and Their Newborn Infants. *Curr. Biol.* **19**:1768–1772. [3]

Ferrari, P. F., E. Visalberghi, A. Paukner, et al. 2006. Neonatal Imitation in Rhesus Macaques. *PLoS Biol.* **4**:1501–1508. [3]

Field, T. 1985. Attachment as Psychobiological Attunement: Being on the Same Wavelength. In: The Psychobiology of Attachment and Separation, ed. M. Reite and T. Field, pp. 415–450. Orlando: Academic Press. [11]

Finch, J. 1987. The Vignette Technique in Survey Research. *Sociology* **21**:105–114. [8]

Finlay, B. L., and A. D. Workman. 2013. Human Exceptionalism. *Trends Cogn. Sci.* **17**:199–201. [4]

Fischer, K. E., and C. F. Landry. 2007. Understanding the Information Behavior of Stay-at-Home Mothers through Affect. In: Information and Emotion: The Emergent Affective Paradigm in Information Research and Theory, ed. D. Nahl and D. Bilal, pp. 211–234. Medford, NJ: Information Today, Inc. [14]

Fisher, R. A. 1930. The Genetical Theory of Natural Selection. Oxford: Oxford Univ. Press. [4]

Fleming, A. S., M. Numan, and R. S. Bridges. 2009. Father of Mothering: Jay S. Rosenblatt. *Horm. Behav.* **55**:484–487. [10]

Fonagy, P., N. Lorenzini, C. Campbell, and P. Luyten. 2014. Why Are We Interested in Attachments? In: The Routledge Handbook of Attachment: Theory, ed. P. Holmes and S. Farnfield, pp. 31–48. New York: Routledge. [6]

Fonagy, P., and M. Target. 2007. The Rooting of the Mind in the Body: New Links between Attachment Theory and Psychoanalytic Thought. *J. Am. Psychoanal. Assoc.* **55**:411–455. [11]

Fonagy, P., M. Target, and G. Gergely. 2000. Attachment and Borderline Personality Disorder: A Theory and Some Evidence. *Psychiatr. Clin. North Am.* **23**:103–122. [11]

Forehand, R., and B. A. Kotchick. 1996. Cultural Diversity: A Wake-up Call for Parent Training. *Behav. Therapy* **27**:187–206. [13]

Fouts, H. N. 2005. Families in Central Africa: A Comparison of Bofi Farmer and Forager Families. In: Families in Global Perspective, ed. J. L. Roopnarine, pp. 347–363. Boston: Allyn and Bacon, Pearson. [5]

Fox, G. J. 1977. Social Dynamics in Siamang. Dissertation, Univ. of Wisconsin, Milwaukee. [4]

Fox, M., R. Sear, J. Beise, et al. 2009. Grandma Plays Favourites: X-Chromosome Relatedness and Sex-Specific Childhood Mortality. *Proc. R. Soc. Lond. B* **277**:567–573. [4]

Fox, S. E., P. Levitt, and C. A. Nelson III. 2010. How the Timing and Quality of Early Experiences Influence the Development of Brain Architecture. *Child Dev. Perspect.* **81**:28–40. [9]

Fracasso, M. P., N. A. Busch-Rossnagel, and C. B. Fisher. 1994. The Relationship of Maternal Behaviour and Acculturation to the Quality of Attachment in Hispanic Infants Living in New York City. *Hisp. J. Behav. Sci.* **16**:143–154. [6]

Fragaszy, D. M., S. Schwarz, and D. Shimosaka. 1982. Longitudinal Observations of Care and Development of Infant Titi Monkeys (*Cullicebus moloch*). *Am. J. Primatol.* **2**:191–200. [3]

Fraley, R. C. 2002. Attachment Stability from Infancy to Adulthood: Meta-Analysis and Dynamic Modeling of Developmental Mechanisms. *Pers. Soc. Psychol. Rev.* **6**:123–151. [12]

Freud, A. 1960. Discussion of Dr. John Bowlby's Paper. *Psychoanal. Study Child.* **15**:53–62. [2]

Friedrich Ebert Stiftung. 2009. Unternehmen Vereinbarkeit: Reif Für Die Neuen Väter? Zusammenfassung Der Konferenz vom 22.4.2009 in Berlin. http://library.fes.de/pdf-files/do/06480-20090625.pdf (accessed Nov. 10, 2016). [5]

Fuentes, A. 2000. Hylobatid Communities: Changing Views on Pair Bonding and Social Organization in Hominoids. *Am. J. Phys. Anthropol.* **113**:33–60. [4]

———. 2009. Evolution of Human Behavior. New York: Oxford Univ. Press. [13]

Galison, P. 2016. Limits of Localism: The Scale of Sight. In: What Reason Promises: Essays on Reason, Nature, and History, ed. W. Doniger et al., pp. 155–170. Berlin: Walter De Gruyter. [2]

Gallese, V. 2007. Before and Below Theory of Mind: Embodied Simulation and the Neural Correlates of Social Cognition. *Phil. Trans. R. Soc. B* **362**:659–669. [11]

Gallese, V., and G. Lakoff. 2005. The Brain's Concepts: The Role of the Sensorimotor System in Conceptual Knowledge. *Cogn. Neuropsychol.* **21**:455–479. [11]

Gangestad, S. W., and J. A. Simpson. 2007. The Evolution of Mind: Fundamental Questions and Controversies. New York: Guilford. [2]

Ganzel, B. L., P. Kim, H. Gilmore, N. Tottenham, and E. Temple. 2013. Stress and the Healthy Adolescent Brain: Evidence for the Neural Embedding of Life Events. *Dev. Psychopathol.* **25**:879–889. [9]

García, M. P., and M. Otero. 2006. Apuntes Sobre la Referencia Expresa Al Ejercicio Compartido de la Guarda y Custodia de Los Hijos en la Ley 15/2005. *Revista Jurídica de Castilla y León* **8**:69–105. [14]

Garwicz, M., M. Christensson, and E. Psouni. 2009. A Unifying Model for Timing of Walking Onset in Humans and Other Mammals. *PNAS* **106**:21889–21893. [4]

Gaskins, S. 1999. Children's Daily Lives in a Mayan Village: A Case Study of Culturally Constructed Roles and Activities. In: Children's Engagement in the World, ed. A. Göncu, pp. 25–61. Cambridge: Cambridge Univ. Press. [6]

———. 2006. Cultural Perspectives on Infant–Caregiver Interaction. In: The Roots of Human Sociality: Culture, Cognition, and Human Interaction, ed. N. J. Enfield and S. C. Levinson, pp. 279–298. Oxford: Berg. [8]

———. 2013. The Puzzle of Attachment: Unscrambling Maturational and Cultural Contributions to the Development of Early Emotional Bonds. In: Attachment Reconsidered: Cultural Perspectives on a Western Theory, ed. N. Quinn and J. Mageo, pp. 33–64. New York: Palgrave. [5, 8, 9, 12]

———. 2014. Childhood Practices across Cultures: Play and Household Work. In: The Oxford Handbook of Human Development and Culture, ed. L. Jensen, pp. 187–197. Oxford: Oxford Univ. Press. [9, 13]

————. 2015. Childhood Practices across Cultures: Play and Household Work. In: The Oxford Handbook of Human Development and Culture, ed. L. A. Jensen, pp. 185–198. Oxford: Oxford Univ. Press. [5]

————. 2017. The Cultural Organization of Children's Everyday Learning. In: Culture and Developmental Systems, ed. M. Sera et al., pp. 223–274, Minnesota Symposium on Child Psychology, vol. 38. Hoboken, NJ: Wiley, in press. [8]

Gee, D. G., L. J. Gabard-Durnam, J. Flannery, et al. 2013. Early Developmental Emergence of Amygdala-Prefrontal Connectivity after Maternal Deprivation. *PNAS* **110**:15638–15643. [9–11]

Geertz, C. 1973/1966. The Interpretation of Cultures. New York: Basic Books. [2, 11]

Georgas, J. 2006. Families and Family Change. In: Families across Cultures: A 30 Nation Psychological Study, ed. J. Georgas et al., pp. 3–50. New York: Cambridge Univ. Press. [5]

George, C., and J. Solomon. 1989. Internal Working Models of Caregiving and Security of Attachment at Age Six. *Infant Ment. Health J.* **10**:222–237. [5]

Gergely, G., O. Koós, and J. S. Watson. 2010. Contingent Parental Reactivity in Early Socio-Emotional Development. In: The Embodied Self: Dimensions, Coherence, and Disorders, ed. T. Fuchs et al., pp. 141–169. Stuttgart: Schattauer GmbH. [11]

Gergely, G., and J. S. Watson. 1996. The Social Biofeedback Theory of Parental Affect-Mirroring: The Development of Emotional Self-Awareness and Self-Control in Infancy. *Int. J. Psychoanal.* **77**:1181–1212. [11]

Gergen, K., A. Gulerce, A. Lock, and G. Misra. 1996. Psychological Science in Cultural Context. *Am. Psychol.* **51**:496–503. [5]

Gerhart, J., and M. Kirschner. 1997. Cells, Embryos and Evolution. Malden, MA: Blackwell Science. [4]

Gernhardt, A., H. Keller, and H. Rübeling. 2016. Children's Family Drawings as Expressions of Attachment Representations across Cultures: Possibilities and Limitations. *Child Dev.* **25**:1069–1078. [5]

Gershoff, E. T. 2013. Spanking and Child Development: We Know Enough Now to Stop Hitting Our Children. *Child Dev. Perspect.* **7**:133–137. [13]

Gibson, E. J., C. J. Owsley, A. Walker, and J. Megaw-Nyce. 1979. Development of the Perception of Invariants: Substance and Shape. *Perception* **8**:609–619. [4]

Gilbert, R., C. S. Widom, K. Browne, et al. 2009. Burden and Consequences of Child Maltreatment in High-Income Countries. *Lancet Infect. Dis.* **373**:68–81. [10]

Ginsberg, S. D., P. R. Hof, W. T. McKinney, and J. H. Morrison. 1993. The Noradrenergic Innervation Density of the Monkey Paraventricular Nucleus Is Not Altered by Early Social Deprivation. *Neurosci. Lett.* **158**:130–134. [10]

Gittins, S. P., and J. J. Raemaekers. 1980. Siamang, lar, and Agile Gibbons. In: Malayan Forest Primates: Ten Years' Study in Tropical Rain Forest, ed. D. J. Chivers, pp. 63–105. New York: Plenum. [4]

Gleason, M. M., N. A. Fox, S. S. Drury, et al. 2014. Indiscriminate Behaviors in Previously Institutionalized Young Children. *Pediatrics* **133**:e657–665. [9]

Goff, B., D. G. Gee, E. H. Telzer, et al. 2013. Reduced Nucleus Accumbens Reactivity and Adolescent Depression Following Early-Life Stress. *Neurosci. Biobehav. Rev.* **249**:129–138. [9]

Goff, B., and N. Tottenham. 2015. Early-Life Adversity and Adolescent Depression: Mechanisms Involving the Ventral Striatum. *CNS Spectrums* **20**:337–345. [9]

Goldberg, T. L., and R. W. Wrangham. 1997. Genetic Correlates of Social Behaviour in Wild Chimpanzees: Evidence from Mitochondrial DNA. *Anim. Behav.* **54**:559–570. [4]

Golden, D. 2017. Childrearing in the New Country: Advice for Immigrant Mothers in Israel. In: A World of Babies: Imagined Childcare Guides for Eight Societies, ed. A. Gottlieb and J. DeLoache, pp. 123–151. New York: Cambridge Univ. Press. [14]

Goldstein, J., A. Freud, and A. J. Solnit. 1973. Beyond the Best Interests of the Child. New York: Free Press. [14]

Goodall, J. 1977. Infant Killing and Cannibalism in Free-Living Chimpanzees. *Folia Primatol.* **28**:259–282. [4]

———. 1986. The Chimpanzees of Gombe: Patterns of Behavior. Cambridge, MA: Belknap Press. [3, 4]

Goodson, J. L., A. K. Evans, L. Lindberg, and C. D. Allen. 2005. Neuro-Evolutionary Patterning of Sociality. *Proc. R. Soc. Lond. B* **272**:227–235. [4]

Gopnik, A. 2000. Explanation as Orgasm and the Drive for Causal Understanding: The Evolution, Function, and Phenomenology of the Theory-Formation System. In: Explanation and Cognition, ed. F. C. Keil and R. C. Wilson, pp. 299–323. Cambridge, MA: MIT Press. [11]

———. 2009. The Philosophical Baby: What Children's Minds Tell Us About Truth, Love, and the Meaning of Life. New York: Farrar, Straus and Giroux. [11]

Gordon, I., and R. Feldman. 2008. Synchrony in the Triad: A Microlevel Process Model of Coparenting and Parent-Child Interactions. *Fam. Process* **47**:465–479. [6]

Gordon, I., O. Zagoory, J. F. Leckman, and R. Feldman. 2010. Oxytocin and the Development of Parenting in Humans. *Biol. Psychiatry* **68**:377–382. [10]

Gottlieb, A. 2004. The Afterlife Is Where We Come From: The Culture of Infancy in West Africa. Chicago: Univ. Chicago Press. [1, 2, 6, 12]

———. 2014. Is It Time to Detach from Attachment Theory? Perspectives from the West African Rain Forest. In: Different Faces of Attachment: Cultural Variations of a Universal Human Need, ed. H. Otto and H. Keller, pp. 187–214. Cambridge: Cambridge Univ. Press. [2, 6, 10]

Gottlieb, A., and J. DeLoache. 2017. Raising a World of Babies: Parenting in the Twenty-First Century. In: A World of Babies: Imagined Childcare Guides for Eight Societies, ed. A. Gottlieb and J. DeLoache, pp. 1–32. New York: Cambridge Univ. Press. [1, 6, 14]

Gottlieb, G. 2007. Probabilistic Epigenesis. *Dev. Sci.* **10**:1–11. [9, 10]

Grafen, A. 1988. On the Use of Data on Lifetime Reproductive Success. In: Reproductive Success, ed. T. H. Clutton-Brock, pp. 454–471. Chicago: Chicago Univ. Press. [4]

Grant, P. R. 1986. Ecology and Evolution of Darwin's Finches. Princeton: Princeton Univ. Press. [2]

Grant, P. R., and B. R. Grant. 2008. How and Why Species Multiply: The Radiation of Darwin's Finches. Princeton: Princeton Univ. Press. [2]

Gratier, M. 2003. Expressive Timing and Interactional Synchrony between Mothers and Infants: Cultural Similarities, Cultural Differences, and the Immigration Experience. *Cogn. Dev.* **18**:533–554. [6, 10]

Greenough, W. T., J. E. Black, and C. S. Wallace. 1987. Experience and Brain Development. *Child Dev.* **58**:539–559. [9]

Gregory, R., H. Cheng, H. A. Rupp, D. R. Sengelaub, and J. R. Heiman. 2015. Oxytocin Increases Vta Activation to Infant and Sexual Stimuli in Nulliparous and Postpartum Women. *Horm. Behav.* **69C**:82–88. [10]

Groh, A. M., R. P. Fearon, M. J. Bakermans-Kranenburg, et al. 2014. The Significance of Attachment Security for Children's Social Competence with Peers: A Meta-Analytic Study. *Attach. Hum. Dev.* **16**:103–136. [12]

Grossmann, K., K. E. Grossmann, G. Spangler, G. Suess, and L. Unzner. 1985. Maternal Sensitivity and Newborns' Orientation Responses as Related to Quality of Attachment in Northern Germany. *Monogr. Soc. Res. Child Dev.* **50**:233–256. [7]

Grossmann, K. E. 1999. Old and New Internal Working Models of Attachment: The Organization of Feelings and Language. *Attach. Hum. Dev.* **1**:253–269. [12]

Grossmann, K. E., and K. Grossmann. 1990. The Wider Concept of Attachment in Cross-Cultural Research. *Human Dev.* **33**:31–47. [2]

———. 2005. Universality of Human Social Attachment as an Adaptive Process. In: Attachment and Bonding: A New Synthesis, ed. C. S. Carter et al., pp. 199–228, Dahlem Workshop Report, vol. 92, J. Lupp, series ed. Cambridge, MA: MIT Press. [2]

Gunnar, M. R., L. Broderson, M. Nachmias, K. Buss, and J. Rigatuso. 1996. Stress Reactivity and Attachment Security. *Dev. Psychobiol.* **29**:191–204. [11]

Gunnar, M. R., J. R. Doom, and E. A. Esposito. 2015a. Psychoneuroendocrinology of Stress. In: Handbook of Child Psychology and Developmental Science, vol. 3: Socioemotional Processes, ed. M. Lamb and C. Garcia Coll, pp. 1–46. New York: Wiley. [10]

Gunnar, M. R., C. E. Hostinar, M. M. Sanchez, N. Tottenham, and R. M. Sullivan. 2015b. Parental Buffering of Fear and Stress Neurobiology: Reviewing Parallels across Rodent, Monkey, and Human Models. *Soc. Neurosci.* **10**:474–478. [10]

Gutknecht, D., M. Holodynski, and H. Schöler. 2012. Bildung in Der Kinderkrippe: Wege Zur Professionellen Responsivität. Stuttgart: Kohlhammer. [13]

Guyer, A. E., J. Kaufman, H. B. Hodgdon, et al. 2006. Behavioral Alterations in Reward System Function: The Role of Childhood Maltreatment and Psychopathology. *J. Am. Acad. Child Adolesc. Psychiatry* **45**:1059–1067. [9]

Haber, S. N., and B. Knutson. 2010. The Reward Circuit: Linking Primate Anatomy and Human Imaging. *Neuropsychopharmacology* **35**:4–26. [9]

Hackman, D. A., and M. J. Farah. 2009. Socioeconomic Status and the Developing Brain. *Trends Cogn. Sci.* **13**:65–73. [10]

Haderthauer, C., and H. Zehetmair, eds. 2013. Bildung Braucht Bindung. Munich: Hanns Seidel Stiftung. [14]

Hagan, J. F., J. S. Shaw, and P. M. Duncan, eds. 2008. Bright Futures: Guidelines for Health Supervision of Infants, Children, and Adolescents (3rd ed.). Elk Grove Village, IL: American Academy of Pediatrics. [13]

Hair, N. L., J. L. Hanson, B. L. Wolfe, and S. D. Pollak. 2015. Association of Child Poverty, Brain Development, and Academic Achievement. *JAMA Pediatr.* **169**:822–829. [10]

Halgunseth, L. C., J. M. Ispa, and D. Rudy. 2006. Parental Control in Latino Families: An Integrated Review of the Literature. *Child Dev.* **77**:1282–1297. [6]

Hallowell, A. I. 1955. Culture and Experience. Philadelphia: Univ. of Pennsylvania Press. [6]

Hamilton, W. 1964. The Genetical Evolution of Social Behaviour, I. *J. Theor. Biol.* **7**:1–16. [6]

Hamilton, W. 1975. Innate Social Aptitudes of Man: An Approach from Evolutionary Genetics. In: Asa Studies 4: Biosocial Anthropology, ed. R. Fox, pp. 133–153. London: Malaby Press. [6]

Hanson, J. L., D. Albert, A.-M. R. Iselin, et al. 2015a. Cumulative Stress in Childhood Is Associated with Blunted Reward-Related Brain Activity in Adulthood. *Soc. Cogn. Affect. Neurosci.* **11**:405–412. [9]

Hanson, J. L., A. Chandra, B. L. Wolfe, and S. D. Pollak. 2011. Association between Income and the Hippocampus. *PLoS One* **6**:e18712. [10]

Hanson, J. L., A. R. Hariri, and D. E. Williamson. 2015b. Blunted Ventral Striatum Development in Adolescence Reflects Emotional Neglect and Predicts Depressive Symptoms. *Biol. Psychiatry* **78**:598–605. [9]

Hanson, M., and P. Gluckman. 2016. Developing the Future: Life Course Epidemiology, Dohad, and Evolutionary Medicine. *Int. J. Epidemiol.* **45**:993–996. [11]

Harkness, S. 2015. The Strange Situation of Attachment Research: A Review of Three Books. *Rev. Anthropol.* **44**:178–197. [2]

Harkness, S., and C. M. Super. 1996. Parent's Cultural Belief Systems. Their Origins, Experssions and Consequences. New York: Guilford Press. [1]

Harkness, S., P. O. Zylicz, C. M. Super, et al. 2011. Children's Activities and Their Meanings for Parents: A Mixed-Methods Study in Six Western Cultures. *J. Fam. Psychol.* **25**:799. [8]

Harlow, H. F. 1958. The Nature of Love. *Am. Psychol.* **13**:573–685. [10]

Harlow, H. F., and M. K. Harlow. 1965. Effects of Various Mother-Infant Relationships on Rhesus Monkey Behaviors. In: Determinants of Infant Behaviour, ed. B. M. Foss, pp. 15–36. London: Methuen [10]

Harlow, H. F., and S. J. Suomi. 1974. Induced Depression in Monkeys. *Behav. Biol.* **12**:273–296. [4]

Harms, W. 2004. Information and Meaning in Evolutionary Processes. New York: Cambridge Univ. Press. [11]

Harris, G. G. 1989. Concepts of Individual, Self, and Person in Description and Analysis. *Am. Anthropol.* **91**:599–612. [6]

Harris, J. R. 1995. Where Is the Child's Environment? A Group Socialization Theory of Development. *Psychol. Rev.* **102**:458–489. [4]

Harris, P. L. 2012. Trusting What You Are Told: How Children Learn from Others. Cambridge: Belknap Press. [2]

Harrison, M. E., and A. J. Marshall. 2011. Strategies for the Use of Fallback Foods in Apes. *Int. J. Primatol.* **32**:531–563. [4]

Harwood, R. L., J. G. Miller, and N. L. Irizarry. 1995. Culture and Attachment: Perceptions of the Child in Context. New York: Guilford. [1, 2, 5, 6, 8]

Hasanović, M., O. Sinanović, Z. Selimbasić, I. Pajević, and E. Avdibegović. 2006. Psychological Disturbances of War-Traumatized Children from Different Foster and Family Settings in Bosnia and Herzegovina. *Croat. Med. J.* **47**:85–94. [13]

Haun, D. B. M., Y. Rekers, and M. Tomasello. 2014. Children Conform to the Behavior of Peers: Other Great Apes Stick with What They Know. *Psychol. Sci.* **25**:2160–2167. [4]

Hawkes, K. 2003. Grandmothers and the Evolution of Human Longevity. *Am. J. Hum. Biol.* **15**:380–400. [4]

———. 2004. The Grandmother Effect. *Nature* **428**:128–129. [11]

———. 2006. Life History Theory and Human Evolution: A Chronicle of Ideas and Findings. In: The Evolution of Human Life History, ed. K. Hawkes and R. R. Paine, pp. 45–93. Santa Fe: SAR Press. [4]

———. 2010. How Grandmother Effects Plus Individual Variation in Frailty Shape Fertility and Mortality: Guidance from Human-Chimpanzee Comparisons. *PNAS* **107**:8977–8984. [4]

———. 2014. Primate Sociality to Human Cooperation: Why Us and Not Them? *Hum. Nat.* **25**:28–48. [2, 4, 11]

Hawkes, K., and J. E. Coxworth. 2013. Grandmothers and the Evolution of Human Longevity: A Review of Findings and Future Directions. *Evol. Anthropol.* **22**:294–302. [4]

Hawkes, K., J. F. O'Connell, and N. G. Blurton Jones. 1997. Hadza Women's Time Allocation, Offspring Provisioning, and the Evolution of Long Postmenopausal Life Spans. *Curr. Anthropol.* **38**:551–557. [5]

Hawkes, K., J. F. O'Connell, N. G. Blurton Jones, H. Alvarez, and E. L. Charnov. 1998. Grandmothering, Menopause, and the Evolution of Human Life Histories. *PNAS* **95**:1336–1339. [4]

Hawks, J., E. T. Wang, G. Cochran, H. C. Harpending, and R. K. Moyzis. 2007. Recent Acceleration of Human Adaptive Evolution. *PNAS* **104**:20753–20758. [2]

Hay, M. C. 2015. Methods That Matter: Integrating Mixed Methods for More Effective Social Science Research. Chicago: Univ. Chicago Press. [5]

Hays, S. 1996. The Cultural Contradictions of Motherhood. New Haven: Yale Univ. Press. [14]

Hebb, D. O. 1949. The Organization of Behavior. New York: Wiley. [11]

Heim, C., L. J. Young, D. J. Newport, et al. 2009. Lower Csf Oxytocin Concentrations in Women with a History of Childhood Abuse. *Mol. Psychiatry* **14**:954–958. [10]

Hein, G., G. Silani, K. Preuschoff, C. D. Batson, and T. Singer. 2010. Neural Responses to Ingroup and Outgroup Members' Suffering Predict Individual Differences in Costly Helping. *Neuron* **68**:149–160. [10]

Heinicke, C. M. 1995. Expanding the Study of the Formation of the Child's Relationships. *Monogr. Soc. Res. Child Dev.* **60**:300–309. [5]

Henrich, J., H. J. Heine, and A. Norenzayan. 2010. The Weirdest People in the World. *Behav. Brain Sci.* **33**:61–83. [1, 2, 5, 8, 9]

Henrich, J., and R. McElreath. 2003. The Evolution of Cultural Evolution. *Evol. Anthropol.* **12**:123–135. [11]

Hensch, T. K. 2005. Critical Period Mechanisms in Developing Visual Cortex. *Curr. Top. Dev. Biol.* **69**:215–237. [9]

Herman, E. 1995. The Romance of American Psychology: Political Culture in the Age of Experts. Berkeley: Univ. of California Press. [2]

Herman, S. P. 1997. Practice Parameters for Child Custody Evaluation. *J. Am. Acad. Child Adolesc. Psychiatry* **36**:57S–68S. [14]

Hewlett, B. S. 1991. Intimate Fathers: The Nature and Context of Aka Pygmy Paternal Infant Care. Ann Arbor: Univ. of Michigan Press. [5]

———. 2004. Fathers in Forager, Farmer, and Pastoral Cultures. In: The Role of the Father in Child Development (4th ed.), ed. M. E. Lamb, pp. 182–195. New York: Wiley. [5]

Hewlett, B. S., and M. E. Lamb, eds. 2005. Hunter-Gatherer Childhoods: Evolutionary, Developmental, and Cultural Perspectives. New Brunswick, NJ: Aldine Transaction. [1, 6, 8, 10, 12]

Hewlett, B. S., M. E. Lamb, B. Leyendecker, and A. Schölmerich. 2000. Internal Working Models, Trust, and Sharing among Foragers. *Curr. Anthropol.* **41**:287–297. [12]

Hill, K. R., and A. M. Hurtado. 1996. Ache Life History: The Ecology and Demography of a Foraging People. New York: Walter de Gruyter. [5]

Hill, K. R., R. S. Walker, M. Bozicevic, et al. 2011. Co-Residence Patterns in Hunter-Gatherer Societies Show Unique Human Social Structure. *Science* **331**:1286–1289. [4, 6]

Hinde, R. A. 1956. Ethological Models and the Concept of Drive. *Br. J. Philos. Sci.* **6**:321–331. [2]

Hinde, R. A. 1976. Interactions, Relationships and Social Structure. *Man* **11**:1. [4]

————. 1982. Attachment: Some Conceptual and Biological Issues. In: The Place of Attachment in Human Behavior, ed. C. Murray Parkes and J. Stevenson-Hinde, pp. 60–76. New York: Basic Books. [2, 12]

————. 1991. Relationships, Attachment, and Culture: A Tribute to John Bowlby. *Infant Ment. Health J.* **12**:154–163. [2, 3]

————. 1999. Commentary: Aspects of Relationships in Child Development. In: Relationships as Developmental Contexts, ed. W. A. Collins and B. Laursen, pp. 323–329, Minnesota Symposium on Child Psychology, vol. 30. New York: Erlbaum. [6]

Hinde, R. A., and Y. Spencer-Booth. 1971a. Effects of Brief Separation from Mother on Rhesus Monkeys. *Science* **173**:111–118. [3]

————. 1971b. Towards Understanding Individual Differences in Rhesus Mother-Infant Interaction. *Anim. Behav.* **19**:165–173. [4]

Hinde, R. A., and J. Stevenson-Hinde. 1990. Attachment: Biological, Cultural and Individual Desiderata. *Human Dev.* **33**:62–72. [2]

Hiraiwa, M. 1981. Maternal and Alloparental Care in a Troop of Free-Ranging Japanese Monkeys. *Primates* **22**:309–329. [4]

Hobson, P. 2002. The Cradle of Thought. Oxford: Oxford Univ. Press. [11]

Hoffman, K. A., S. P. Mendoza, M. B. Hennessy, and W. A. Mason. 1995. Responses of Infant Titi Monkeys, Callicebus moloch, to Removal of One or Both Parents: Evidence for Paternal Attachment. *Dev. Psychobiol.* **28**:399–407. [3]

Hofmann, H. A., A. K. Beery, D. T. Blumstein, et al. 2014. An Evolutionary Framework for Studying Mechanisms of Social Behavior. *Trends Ecol. Evol.* **29**:581–589. [4]

Holland, D. 1992a. Cross-Cultural Differences in the Self. *J. Anthropol. Res.* **48**:283–300. [6]

Holland, J. H. 1992b. Complex Adaptive Systems Theory. *Daedalus* **121**:17–30. [11]

————. 1995. Hidden Order: How Adaptation Builds Complexity. Reading, MA: Addison-Wesley. [11]

————. 2012. Signals and Boundaries: Building Blocks for Complex Adaptive Systems. Cambridge, MA: MIT Press. [11]

Hopkins, W. D., A. C. Keebaugh, L. A. Reamer, et al. 2014a. Genetic Influences on Receptive Joint Attention in Chimpanzees (*Pan troglodytes*). *Sci. Rep.* **4**:3774. [1]

Hopkins, W. D., M. Misiura, L. Reamer, et al. 2014b. Poor Receptive Joint Attention Skills Are Associated with Atypical Gray Matter Asymmetry in the Posterior Superior Temporal Gyrus of Chimpanzees (*Pan troglodyes*). *Front. Psychol.* **5**:7. [1]

Houston, A., T. Szekely, and J. McNamara. 2005. Conflict between Parents over Care. *Trends Ecol. Evol.* **20**:33–38. [4]

Howard, A. 1985. Ethnopsychology and the Prospects for a Cultural Psychology. In: Person, Self and Experience: Exploring Pacific Ethnopsychologies, ed. G. M. White and J. Kirkpatrick, pp. 401–420. Berkley: Univ. of California Press. [6]

Howes, C. 2009. Culture and Child Development in Early Childhood Programs: Practices for Quality Education and Care. New York: Teachers College Press. [13]

Howes, C., C. C. Matheson, and C. E. Hamilton. 1994. Maternal, Teacher, and Child Care History Correlates of Children's Relationships with Peers. *Child Dev.* **65**:264–273. [12]

Howes, C., and S. Spieker. 2016. Attachment Relationships in the Context of Multiple Caregivers. In: Handbook of Attachment: Theory, Research, and Clinical Applications , 2nd edition, ed. J. Cassidy and P. R. Shaver, pp. 317–334. New York: Guilford. [12, 14]

Howes, C., and A. G. Wishard Guerra. 2009. Networks of Attachment Relationships in Low-Income Children of Mexican Heritage: Infancy through Preschool. *Soc. Dev.* **18**:896–914. [6]

Hrdy, S. B. 1976. Care and Exploitation of Nonhuman Primate Infants by Conspecifics Other Than the Mother. In: Advances in the Study of Behavior, ed. J. Rosenblatt et al., pp. 101–158, vol. 6. New York: Academic Press. [4]

———. 1977. The Langurs of Abu: Female and Male Strategies of Reproduction. Cambridge, MA: Harvard Univ. Press. [4]

———. 1999. Mother Nature: Maternal Instincts and How They Shape the Human Species. New York: Ballantine Books. [5, 4, 11]

———. 2005a. Comes the Child before Man: How Cooperative Breeding and Prolonged Postweaning Dependence Shaped Human Potentials. In: Hunter-Gatherer Childhoods: Evolutionary, Developmental and Cultural Perspectives, ed. B. S. Hewlett and M. E. Lamb, pp. 65–91. New Brunswick, NJ: Transaction Publishers. [6]

———. 2005b. Cooperative Breeders with an Ace in the Hole. In: Grandmotherhood: The Evolutionary Significance of the Second Half of Female Life, ed. E. Voland et al., pp. 295–317. New Brunswick: Rutgers Univ. Press. [5]

———. 2005c. Evolutionary Context of Human Development: The Cooperative Breeding Model. In: Attachment and Bonding: A New Synthesis, ed. C. S. Carter et al., pp. 9–32. Dahlem Workshop Report, vol. 92, J. Lupp, series ed. Cambridge, MA: MIT Press. [2]

———. 2008. Cooperative Breeding and the Paradox of Facultative Fathering. In: The Neurobiology of the Parental Mind, ed. R. Bridges, pp. 405–414. New York: Academic Press. [4]

———. 2009. Mothers and Others: The Evolutionary Origins of Mutual Understanding. Cambridge, MA: Harvard Univ. Press. [2–5, 8, 11,]

———. 2016a. Development Plus Social Selection in the Emergence of "Emotionally Modern" Humans. In: Childhood: Origins, Evolution, and Implications, ed. C. L. Meehan and A. N. Crittenden, pp. 11–44. Santa Fe: School for Advanced Research Press. [11]

———. 2016b. Variable Postpartum Responsiveness among Humans and Other Primates with "Cooperative Breeding": A Comparative and Evolutionary Perspective. *Horm. Behav.* **77**:272–283. [11]

Hulbert, A. 2003. Raising America: Experts, Parents, and a Century of Advice About Children. New York: Vintage Books. [2]

Humphreys, K. L., C. H. Zeanah, C. A. Nelson III, N. A. Fox, and S. S. Drury. 2015. Serotonin Transporter Genotype (5HTTLPR) Moderates the Longitudinal Impact of Atypical Attachment on Externalizing Behavior. *J. Dev. Behav. Ped.* **36**:409–416. [9]

Hurlemann, R., and D. Scheele. 2015. Dissecting the Role of Oxytocin in the Formation and Loss of Social Relationships. *Biol. Psychiatry* [10]

Hurley, S., and M. Nudds, eds. 2006. Rational Animals? Oxford: Oxford Univ. Press. [4]

Hurst, J. E., and P. S. Kavanagh. 2017. Life History Strategies and Psychopathology: The Faster the Life Strategies, the More Symptoms of Psychopathology. *Evol. Hum. Behav.* **38**:1–8. [11]

Huseynov, A., C. P. E. Zollikofer, W. Coudyzer, et al. 2016. Developmental Evidence for Obstetric Adaptation of the Human Female Pelvis. *PNAS* **113**:5227–5232. [4]

Iacoboni, M., R. P. Woods, M. Brass, et al. 1999. Cortical Mechanisms of Human Imitation. *Science* **286**:2526–2528. [3]

Ichise, M., D. C. Vines, T. Gura, et al. 2006. Effects of Early Life Stress on [11C]Dasb Positron Emission Tomography Imaging of Serotonin Transporters in Adolescent Peer- and Mother-Reared Rhesus Monkeys. *J. Neurosci.* **26**:4638–4643. [10]

Isler, K., and C. P. van Schaik. 2014. How Humans Evolved Large Brains. *Evol. Anthropol.* **23**:65–75. [4]

Ispa, J. M., M. A. Fine, L. C. Halgunseth, et al. 2004. Maternal Intrusiveness, Maternal Warmth, and Mother-Toddler Relationship Outcomes: Variations across Low-Income Ethnic and Acculturation Groups. *Child Dev.* **75**:1613–1631. [6]

Jabbi, M., M. Swart, and C. Keysers. 2007. Empathy for Positive and Negative Emotions in the Gustatory Cortex. *NeuroImage* **34**:1744–1753. [10]

Jablonka, E., and M. J. Lamb. 2005. Evolution in Four Dimensions: Genetic, Epigenetic, Behavioral, and Symbolic Variation in the History of Life (Rev. ed.). Cambridge, MA: Bradford. [1, 11]

————. 2007. Bridging the Gap: The Developmental Aspects of Evolution. *Behav. Brain Sci.* **30**:353–365. [1]

————. 2014. Evolution in Four Dimensions: Genetic, Epigenetic, Behavioral, and Symbolic Variation in the History of Life (Rev. ed.). Cambridge, MA: MIT Press. [2]

Jablonka, E., and G. Raz. 2009. Transgenerational Epigenetic Inheritance: Prevalence, Mechanisms, and Implications for the Study of Heredity and Evolution. *Q. Rev. Biol.* **84**:131–176. [11]

Jensen, L. A. 2011. Introduction: Changing Our Scholarship for a Changing World. In: Bridging Cultural and Developmental Approaches to Psychology: New Syntheses in Theory, ed. L. A. Jensen, pp. 3–25. Oxford: Oxford Univ. Press. [13]

Johnson, M. 2007. The Meaning of the Body. Chicago: Univ. Chicago Press. [11]

Johnson, M. H. 2005. Subcortical Face Processing. *Nat. Rev. Neurosci.* **6**:766–774. [4]

Johnson, S. M. 2001. Emergence: The Connected Lives of Ants, Brains, Cities, and Software. New York: Scribner. [11]

————. 2008. Couple and Family Therapy: An Attachment Perspective. In: Handbook of Attachment: Theory, Research, and Clinical Applications, ed. J. Cassidy and P. Shaver, pp. 811–829. New York: Guilford. [14]

Johow, J., E. Voland, and K. P. Willführ. 2013. Reproductive Strategies in Female Post-generative Life. In: Darwinian Perspectives on the Nature of Women, ed. M. L. Fisher et al., pp. 243–259. Oxford: Oxford Univ. Press. [4]

Jones, J. H. 2009. The Force of Selection on the Human Life Cycle. *Evol. Hum. Behav.* **30**:305–314. [11]

Juffer, F., M. J. Bakermans-Kranenburg, and M. H. van IJzendoorn. 2008. Promoting Positive Parenting: An Introduction. In: Promoting Positive Parenting: An Attachment-Based Intervention, ed. F. Juffer et al., pp. 1–10. New York: Lawrence Erlbaum. [14]

Julius, H. 2009. Bindungsgeleitete Intervention in Der Schulischen Erziehungshilfe. In: Bindungen Im Kindesalter: Diagnostik und Interventionen, ed. H. Julius et al., pp. 293–315. Göttingen: Hogrefe. [14]

Jusczyk, P. W., D. B. Pisoni, M. A. Reed, A. Fernald, and M. Myers. 1983. Infants' Discrimination of the Duration of a Rapid Spectrum Change in Nonspeech Signals. *Science* **222**:175–177. [4]

Kagitcibasi, C. 1970. Social Norms and Authoritarianism: A Turkish-American Comparison. *J. Pers. Soc. Psychol.* **16**:444–451. [6]

————. 2005. Autonomy and Relatedness in Cultural Context: Implications for Self and Family. *J. Cross-Cult. Psychol.* **36**:403–422. [6]

Kahn, R. L., and T. C. Antonucci. 1980. Convoys over the Life Course: Attachment, Roles, and Social Support. In: Life-Span Development and Behavior, vol. 2, ed. P. B. Baltes and O. Brim, pp. 254–283. New York: Academic Press. [5]

Kahneman, D. 2011. Thinking, Fast and Slow. New York: Farrar, Straus and Giroux. [11]

Kajikawa, S., S. Amano, and T. Kondo. 2004. Speech Overlap in Japanese Mother-Child Conversations. *J. Child Lang.* **31**:215–230. [6]

Kaplan, H. S., K. R. Hill, J. B. Lancaster, and A. M. Hurtado. 2000. A Theory of Human Life History Evolution: Diet, Intelligence, and Longevity. *Evol. Anthropol.* **9**:156–185. [4]

Kaplan, H. S., T. Mueller, S. Gangestad, and J. B. Lancaster. 2003. Neural Capital and Life Span Evolution among Primates and Humans. In: Brain and Longevity, ed. C. E. Finch et al., pp. 69–97. Berlin: Springer. [4]

Kappeler, P. M. 1997. Determinants of Primate Social Organization: Comparative Evidence and New Insights from Malagasy Lemurs. *Biol. Rev. Camb. Philos. Soc.* **72**:111–151. [3]

Kappeler, P. M., and C. P. van Schaik. 2002. Evolution of Primate Social Systems. *Int. J. Primatol.* **23**:707–740. [4]

Karen, R. 1994. Becoming Attached: Unfolding the Mystery of the Infant–Mother Bond and Its Impact on Later Life. New York: Warner Books. [7]

———. 1998. Becoming Attached: First Relationships and How They Shape Our Capacity to Love. New York: Oxford Univ. Press. [2]

Kärtner, J., H. Keller, and N. Chaudhary. 2010a. Cognitive and Social Influences on Early Prosocial Behavior in Two Sociocultural Contexts. *Dev. Psychol.* **46**:905–914. [6]

Kärtner, J., H. Keller, B. Lamm, et al. 2008. Similarities and Differences in Contingency Experiences of 3-Month-Olds across Sociocultural Contexts. *Infant. Behav. Dev.* **31**:488–500. [6]

Kärtner, J., H. Keller, and R. Yovsi. 2010b. Mother–Infant Interaction During the First 3 months: The Emergence of Culture-Specific Contingency Patterns. *Child Dev.* **81**:540–554. [6]

Kaufman, I. C., and L. A. Rosenblum. 1969. Effects of Separation from Mother on the Emotional Behavior of Infant Monkeys. *Ann. N.Y. Acad. Sci.* **159**:681–695. [4]

Keller, E. F. 2014a. From Gene Action to Reactive Genomes. *J. Physiol.* **592**:2423–2429. [2]

———. 2016a. Thinking About Biology and Culture: Can the Natural and Human Sciences Be Integrated? *Sociol. Rev.* **64**:26–41. [2]

Keller, H. 2003. Socialization for Competence: Cultural Models of Infancy. *Human Dev.* **46**:288–311. [6]

———. 2007. Cultures of Infancy. Mahwah, NJ: Erlbaum. [1, 2, 5, 6, 8, 10, 12]

———. 2008. Attachment Past and Present: But What About the Future? *Integr. Psychol. Behav. Sci.* **42**:406–415. [2, 7]

———. 2012. Autonomy and Relatedness Revisited: Cultural Manifestations of Universal Human Needs. *Child Dev. Perspect.* **6**:12–18. [6]

———. 2013a. Attachment and Culture. *J. Cross-Cult. Psychol.* **44**:175–194. [2, 8, 9, 14]

———. 2013b. Culture and Development: Developmental Pathways to Psychological Autonomy and Hierarchical Relatedness (2). International Association for Cross-Cultural Psychology. http://scholarworks.gvsu.edu/orpc/vol6/iss1/1/ (accessed Oct. 28, 2016). [6]

———. 2013c. Infancy and Well-Being. In: Handbook of Child Well-Being, ed. A. Ben-Arieh et al., pp. 1605–1627. Dordrecht: Springer. [5]

Keller, H. 2014b. Understanding Relationships: What We Would Need to Know to Conceptualize Attachment as the Cultural Solution of a Universal Developmental Task. In: Different Faces of Attachment: Cultural Variations on a Universal Human Need, ed. H. Otto and H. Keller, pp. 1–24. Cambridge: Cambridge Univ. Press. [2]

———. 2015. Attachment: A Pancultural Need but a Cultural Construct. *Curr. Opin. Psychol.* **8**:59–63. [5, 6]

———. 2016b. Psychological Autonomy and Hierarchical Relatedness as Organizers of Developmental Pathways. *Phil. Trans. R. Soc. B.* **371**:20150070. [6]

Keller, H., C. Demuth, and R. Yovis. 2012. The Multi-Voicedness of Independence and Interdependence: The Case of Cameroonian Nso. *Cult. Psychol.* **14**:115–144. [6]

Keller, H., E. Hentschel, R. Yovsi, et al. 2004a. The Psycho-Linguistic Embodiment of Parental Ethnotheories: A New Avenue to Understanding Cultural Processes in Parental Reasoning. *Cult. Psychol.* **10**:293–330. [8]

Keller, H., and J. Kärtner. 2013. Development: The Culture–Specific Solution of Universal Developmental Tasks. In: Advances in Culture and Psychology, ed. M. L. Gelfand et al., pp. 63–116, vol. 3. Oxford: Oxford Univ. Press. [5, 6, 8]

Keller, H., A. Lohaus, P. Kuensemueller, et al. 2004b. The Bio-Culture of Parenting: Evidence from Five Cultural Communities. *Parent. Sci. Pract.* **4**:25–50. [3]

Keller, H., A. Lohaus, S. Volker, M. Cappenberg, and A. Chasiotis. 1999. Temporal Contingency as an Independent Component of Parenting Behavior. *Child Dev.* **70**:474–485. [6]

Keller, H., and H. Otto. 2009. The Cultural Socialization of Emotion Regulation During Infancy. *J. Cross-Cult. Psychol.* **40**:996–1011. [6, 8]

Keller, H., H. Otto, B. Lamm, R. Yovsi, and J. Kärtner. 2008. The Timing of Verbal/Vocal Communications between Mothers and Their Infants: A Longitudinal Cross-Cultural Comparison. *Infant Behav. Dev.* **31**:217–226. [6]

Keller, H., A. Schölmerich, and I. Eibl-Eibesfeldt. 1988. Communication Patterns in Adult-Infant Interactions in Western and Non-Western Cultures. *J. Cross-Cult. Psychol.* **19**:427–445. [1]

Kelly, J. B., and M. E. Lamb. 2000. Using Child Development Research to Make Appropriate Custody and Access Decisions for Young Children. *Fam. Concil. Courts Rev.* **38**:297–311. [14]

Kelso, J. A. S. 1995. Dynamic Patterns: The Self-Organization of Brain and Behavior. Cambridge, MA: MIT Press. [11]

Kendal, J. R. 2011. Cultural Niche Construction and Human Learning Environments: Investigating Sociocultural Perspectives. *Biol. Theory* **6**:241–250. [11]

Kendrick, K. M., E. B. Keverne, and B. A. Baldwin. 1987. Intracerebroventricular Oxytocin Stimulates Maternal Behaviour in the Sheep. *Neuroendocrinology* **46**:56–61. [10]

Kennedy, G. 2005. From the Ape's Dilemma to the Weanling's Dilemma: Early Weaning and Its Evolutionary Context. *J. Hum. Evol.* **48**:123–145. [4]

Kenyon, M., C. Roos, V. T. Binh, and D. J. Chivers. 2011. Extrapair Paternity in Golden-Cheeked Gibbons (*Nomascus gabriellae*) in the Secondary Lowland Forest of Cat Tien National Park, Vietnam. *Folia Primatol.* **82**:154–164. [4]

Kermoian, R., and P. H. Leiderman. 1986. Infant Attachment to Mother and Child Caretaker in an East African Community. Special Issue: Cross-Cultural Human Development. *Int. J. Behav. Dev.* **9**:455–469. [5, 12]

Kessel, F. S. 2009. Research on Child Development: Historical Perspectives. In: The Child: An Encyclopedic Companion, ed. R. A. Shweder et al., pp. 828–833. Chicago: Univ. Chicago Press. [2]

Kessen, W. 1979. The American Child and Other Cultural Inventions. *Am. Psychol.* **34**:815–820. [2]

Kessler, S. E., and L. T. Nash. 2010. Grandmothering in Galago *Senegalensis braccatus*. *Afr. Primates* **7**:42–49. [3]

Keysers, C., and V. Gazzola. 2014. Hebbian Learning and Predictive Mirror Neurons for Actions, Sensations and Emotions. *Phil. Trans. R. Soc. B.* **369**:20130175. [10]

Kilner, R. M., and C. A. Hinde. 2012. Parent-Offspring Conflict. In: The Evolution of Parental Care, ed. N. J. Royle et al., pp. 119–132. Oxford: Oxford Univ. Press. [11]

Kim, P. S., J. E. Coxworth, and K. Hawkes. 2012. Increased Longevity Evolves from Grandmothering. *Proc. R. Soc. Lond. B* **279**:4880–4884. [4]

Kim, P. S., J. S. McQueen, J. E. Coxworth, and K. Hawkes. 2014. Grandmothering Drives the Evolution of Longevity in a Probabilistic Model. *J. Theor. Biol.* **353**:84–94. [4]

Kitayama, S., S. Duffy, and Y. Uchida. 2007. Self as a Cultural Mode of Being. In: Handbook of Cultural Psychology, ed. S. Kitayama and D. Cohen, pp. 136–174. New York: Guilford. [6]

Kitayama, S., M. Karasawa, and B. Mesquita. 2004. Collective and Personal Processes in Regulating Emotions: Emotion and Self in Japan and the U.S. In: The Regulation of Emotion, ed. P. Philippot and R. S. Feldman, pp. 251–273. Mahwah, NJ: Lawrence Erlbaum. [6]

Knipe, E. 2015. Families and Households. London: Office for National Statistics. [6]

Knutson, B., C. M. Adams, G. W. Fong, and D. Hommer. 2001. Anticipation of Increasing Monetary Reward Selectively Recruits Nucleus Accumbens. *J. Neurosci.* **21**:RC159. [9]

Knutson, B., G. W. Fong, S. M. Bennett, C. M. Adams, and D. Hommer. 2003. A Region of Mesial Prefrontal Cortex Tracks Monetarily Rewarding Outcomes: Characterization with Rapid Event-Related fMRI. *NeuroImage* **18**:263–272. [9]

Kobak, R., and S. Madsen. 2008. Disruptions in Attachment Bonds: Implications for Theory, Research, and Clinical Intervention. In: Handbook of Attachment: Theory, Research, and Clinical Applications (2nd edition), ed. J. Cassidy and P. Shaver, pp. 21–43. New York: Guilford. [7]

Kochanska, G. 2002. Mutually Responsive Orientation between Mothers and Their Young Children: A Context for the Early Development of Conscience. *Curr. Dir. Psychol.* **11**:191–195. [8]

Kohda, M. 1985. Allomothering Behaviour of New and Old World Monkeys. *Primates* **26**:28–44. [3]

Kokal, I., V. Gazzola, and C. Keysers. 2009. Acting Together in and Beyond the Mirror Neuron System. *NeuroImage* **47**:2046–2206. [10]

Kokko, H., and M. Jennions. 2003. It Takes Two to Tango. *Trends Ecol. Evol.* **18**:103–104. [4]

Koller, S. H., M. Raffaelli, and G. Carlo. 2012. Conducting Research About Sensitive Subjects: The Case of Homeless Youth. *Universitas Psychologica* **11**:55–65. [13]

Kondo-Ikemura, K., and E. Waters. 1995. Maternal Behavior and Infant Security in Old Monkeys: Conceptual Issues and a Methodological Bridge between Human and Nonhuman Primate Research. *Monogr. Soc. Res. Child Dev.* **60**:97–110. [8]

Konner, M. J. 1973. Newborn Walking: Additional Data. *Science* **179**:307. [6]

———. 1976. Maternal Care, Infant Behavior and Development among The !Kung. In: Kalahari Hunter-Gatherers: Studies of The !Kung San and Their Neighbors, ed. R. B. Lee and I. DeVore, pp. 218–245. Cambridge, MA: Harvard Univ. Press. [6, 8]

Konner, M. J. 2005. Hunter-Gatherer Infancy and Childhood. In: Hunter-Gatherer Childhoods: Evolutionary, Developmental, and Cultural Perspectives, ed. B. S. Hewlett and M. E. Lamb, pp. 19–64. New Brunswick, NJ: Transaction Publishers. [1]

———. 2010. The Evolution of Childhood: Relationships, Emotion, Mind. Cambridge, MA: Harvard Univ. Press. [4, 6, 8, 11]

Koops, W., and M. Zuckerman, eds. 2003. Beyond the Century of the Child: Cultural History and Developmental Psychology. Philadelphia: Univ. of Pennsylvania Press. [2]

Kosfeld, M., M. Heinrichs, P. J. Zak, U. Fischbacher, and E. Fehr. 2005. Oxytocin Increases Trust in Humans. *Nature* **435**:673–676. [8, 10]

Kostan, K. M., and C. T. Snowdon. 2002. Attachment and Social Preferences in Cooperatively-Reared Cotton-Top Tamarins. *Am. J. Primatol.* **57**:131–139. [3]

Kraemer, G. W., and J. Bachevalier. 1998. Cognitive Changes Associated with Persisting Behavioral Effects of Early Psychosocial Stress in Rhesus Monkeys: A View from Psychobiology. In: Adversity, Stress, and Psychopathology, ed. B. P. Dohrenwend, pp. 438–462. New York: Oxford Univ. Press. [10]

Kramer, K. L. 2005. Children's Help and the Pace of Reproduction: Cooperative Breeding in Humans. *Evol. Anthropol.* **14**:224–237. [4]

Kramer, K. L., and P. T. Ellison. 2010. Pooled Energy Budgets: Resituating Human Energy Allocation Trade-Offs. *Evol. Anthropol.* **19**:136–147. [6]

Kramer, K. L., and R. Greaves. 2011. Postmarital Residence and Bilateral Kin Associations among Hunter-Gatherers: Pumé Foragers Living in the Best of Both Worlds. *Hum. Nat.* **22**:41–63. [6]

Krebs, V. E. 2002. Mapping Networks of Terrorist Cells. *Connections* **24**:43–52. [6]

Kuczynski, L., C. M. Parkin, and R. Pitman. 2015. Socialization as Dynamic Process: A Dialectical, Transactional Perspective. In: Handbook of Socialization: Theory and Research, ed. J. E. Grusec and P. D. Hastings, pp. 135–157. New York: Guilford. [6]

Kulks, D. 1999. Komponenten Des Elternverhaltens Im Kulturellen Kontext: Costa Rica–Deutschland. Unpublished thesis, Universität Osnabrück, Germany. [6]

Kurtz, S. M. 1992. All the Mothers Are One: Hindu India and the Cultural Reshaping of Psychoanalysis. New York: Columbia Univ. Press. [14]

Kushnick, G. 2012. Helper Effects on Breeder Allocations to Direct Care. *Am. J. Hum. Biol.* **24**:545–550. [4]

Kusserow, A. 2004. American Individualisms: Child Rearing and Social Class in Three Neighborhoods. New York: Palgrave MacMillan. [14]

Kuwahata, H., I. Adachi, K. Fujita, M. Tomonaga, and T. Matsuzawa. 2004. Development of Schematic Face Preference in Macaque Monkeys. *Behav. Processes* **66**:17–21. [3]

Kwong, K., H. Chung, L. Sun, J. C. Chou, and A. Taylor-Shih. 2009. Factors Associated with Reverse-Migration Separation among a Cohort of Low-Income Chinese Immigrant Families in New York City. *Soc. Work Health Care* **48**:348–359. [7]

Labile, D., R. A. Thompson, and J. Froimson. 2015. Early Socialization: The Influences of Close Relationships. In: Handbook of Socialization, ed. J. E. Grusec and P. Hastings, pp. 35–59. New York: Guilford. [6]

Ladd-Taylor, M., and L. Umansky, eds. 1998. Bad Mothers: The Politics of Blame in Twentieth-Century America. New York: NYU Press. [2]

Laewen, H. J., B. Andres, and É. Hédervári-Heller. 2006. Ohne Eltern Geht Es Nicht: Die Eingewöhnung Von Kindern in Krippen und Tagespglegestellen. Berlin: Cornelsen. [14]

————. 2011. Die Ersten Tage: Ein Modell Zur Eingewöhnung in Krippe und Tagespflege. Berlin: Cornelsen. [13]

Lakoff, G., and M. Johnson. 1999. Philosophy in the Flesh: The Embodied Mind and Its Challenge to Western Thought. New York: Basic Books. [11]

Laland, K. N., and G. R. Brown. 2002. Sense and Nonsense: Evolutionary Perspectives on Human Behavior. Oxford: Oxford Univ. Press. [2]

————. 2011. Sense and Nonsense: Evolutionary Perspectives on Human Behavior. Oxford: Oxford Univ. Press. [4]

Laland, K. N., B. Matthews, and M. Feldman. 2016. An Introduction to Niche Construction Theory. *Evol. Ecol.* **30**:191–202. [11]

Laland, K. N., T. Uller, M. W. Feldman, et al. 2015. The Extended Evolutionary Synthesis: Its Structure, Assumptions and Predictions. *Proc. R. Soc. Lond. B* **282**:20151019. [11]

Lam, C. M. 1997. A Cultural Perspective on the Study of Chinese Adolescent Development. *Child Adolesc. Social Work J.* **14**:95–113. [7]

Lamb, M. E. 2005. Attachments, Social Networks, and Developmental Contexts. *Human Dev.* **48**:108–112. [2]

Lamb, M. E., W. P. Gardner, E. R. Charnov, R. A. Thompson, and D. Estes. 1984a. Studying the Security of Infant-Adult Attachment: A Reprise. *Behav. Brain Sci.* **7**:163–167. [1, 2]

Lamb, M. E., R. A. Thompson, W. Gardner, E. L. Charnov, and J. P. Connell. 1985. Infant-Mother Attachment: The Origins and Developmental Significance of Individual Differences in Strange Situation Behavior. Hillsdale, NJ: Erlbaum. [2, 12]

Lamb, M. E., R. A. Thompson, W. Gardner, E. L. Charnov, and D. Estes. 1984b. Security of Infantile Attachment as Assessed in the Strange Situation: Its Study and Biological Interpretation. *Behav. Brain Sci.* **7**:127–147. [2, 5, 12]

Lamm, B., and H. Keller. 2012. Väter in Verschiedenen Kulturen. In: Das Väter-Handbuch, ed. H. Walter and A. Eickhorst, pp. 77–88. Giessen: Psychosozial-Verlag. [5]

Lamm, B., and J. Teiser. 2013. Intergenerationeller Wandel. In: Die Entwicklung Der Psyche in Der Geschichte Der Menschheit, ed. G. Jüttemann, pp. 152–163. Lengerich: Pabst Science Publisher. [5]

Lancaster, J. B. 1971. Play-Mothering: The Relations between Juvenile Females and Young Infants among Free-Ranging Vervet Monkeys (*Cercopithecus aethiops*). *Folia Primatol.* **15**:161–182. [3, 4]

Lancaster, J. B., H. S. Kaplan, K. Hill, and A. M. Hurtado. 2000. The Evolution of Life History, Intelligence and Diet among Chimpanzees and Human Foragers. In: Evolution, Culture, and Behavior, ed. F. Tonneau and N. S. Thompson, pp. 47–72. New York: Plenum. [5]

Lancy, D. F. 2008. The Anthropology of Childhood: Cherubs, Chattel, Changelings. Cambridge: Cambridge Univ. Press. [1]

————. 2014. Babies Aren't Persons: A Survey of Delayed Personhood. In: Different Faces of Attachment: Cultural Variations of a Universal Human Need, ed. H. Keller and H. Otto, pp. 66–112. Cambridge: Cambridge Univ. Press. [2, 8]

————. 2015. The Anthropology of Childhood: Cherubs, Chattel, Changelings (2nd edition). New York: Cambridge Univ. Press. [1, 2, 5, 6, 8]

————. 2016. Playing with Knives: The Socialization of Self-Initiated Learners. *Child Dev.* **87**:654–665. [5]

Lancy, D. F., J. Bock, and S. Gaskins, eds. 2010. The Anthropology of Learning in Childhood. Walnut Creek, CA: AltaMira Press. [8]

Lane, D., D. Pumain, S. E. van der Leeuw, and G. West, eds. 2009. Complexity Perspectives in Innovation and Social Change. Methodos Series, vol. 7. New York: Springer. [6]

Langford, D. J., S. E. Crager, Z. Shehzad, et al. 2006. Social Modulation of Pain as Evidence for Empathy in Mice. *Science* **312**:1967–1970. [3]

Lansford, J. E. 2010. The Special Problem of Cultural Differences in Effects of Corporal Punishment. *Law Contemp. Probl.* **73**:89–106. [13]

Lappan, S. 2008. Male Care of Infants in a Siamang (*Symphalangus syndactylus*) Population Including Socially Monogamous and Polyandrous Groups. *Behav. Ecol. Sociobiol.* **62**:1307–1317. [4]

Latzman, R. D., H. D. Freeman, S. J. Schapiro, and W. D. Hopkins. 2015. The Contribution of Genetics and Early Rearing Experiences to Hierarchical Personality Dimensions in Chimpanzees (*Pan troglodytes*). *J. Pers. Soc. Psychol.* **109**:889–900. [10]

Lawson, D. W., and R. Mace. 2011. Parental Investment and the Optimization of Human Family Size. *Phil. Trans. R. Soc. B* **366**:333–343. [4]

Leach, P. 1986. Your Baby and Child: From Birth to Age Five. New York: A. A. Knopf. [14]

Leclère, C., S. Viaux, M. Avril, et al. 2014. Why Synchrony Matters During Mother-Child Interactions: A Systematic Review. *PLoS One* **9**:e113571. [6, 11]

Lee, N., L. Mikesell, A. D. L. Joaquin, A. W. Mates, and J. H. Schumann. 2009. The Interactional Instinct: The Evolution and Acquisition of Language. New York: Oxford Univ. Press. [6]

Lee, P. C. 1996. The Meanings of Weaning: Growth, Lactation, and Life History. *Evol. Anthropol.* **5**:87–98. [4]

Lehmann, S., O. E. Havik, T. Havik, and E. R. Heiervang. 2013. Mental Disorders in Foster Children: A Study of Prevalence, Comorbidity and Risk Factors. *Child Adolesc. Psychiatry Ment. Health* **7**:39. [10]

Lehner, P. N. 1998. Handbook of Ethological Methods. Cambridge: Cambridge Univ. Press. [8]

Lehrman, D. S. 1953. A Critique of Konrad Lorenz's Theory of Instinctive Behavior. *Q. Rev. Biol.* **28**:337–363. [2]

Lehtonen, J., G. Parker, and L. Schärer. 2016. Why Anisogamy Drives Ancestral Sex Roles. *Evol. Hum. Behav.* **70**:1129–1135. [4]

Leiderman, P. H., and G. F. Leiderman. 1977. Economic Changes and Infant Care in an East African Agricultural Community. In: Culture and Infancy: Variations in the Human Experience, ed. P. H. Leiderman and S. Tulkin, pp. 405–438. New York: Academic Press. [6]

Leighton, D. R. 1987. Gibbons: Territoriality and Monogamy. In: Primate Societies, ed. B. B. Smuts et al., pp. 135–145. Chicago: Univ. Chicago Press. [4]

Leinaweaver, J. 2010. Outsourcing Care: How Peruvian Migrants Meet Transnational Family Obligations. *Lat. Am. Perspect.* **37**:67–87. [7]

———. 2014. Informal Kinship-Based Fostering around the World: Anthropological Findings. *Child Dev. Perspect.* **8**:131–136. [7]

Leonetti, D. L., D. C. Nath, and N. S. Hemam. 2007. In-Law Conflict. *Curr. Anthropol.* **48**:861–890. [4]

Levin, S. 2003. Complex Adaptive Systems: Exploring the Known and the Unknowable. *Bull. Am. Math. Soc.* **40**:3–19. [11]

LeVine, R. A. 1974. Parental Goals: A Cross-Cultural Review. *Teachers College Rec.* **76**:226–239. [6]

———. 1980. A Cross-Cultural Perspective on Parenting. In: Parenting in a Multicultural Society, ed. M. D. Fantini and R. Cardenas, pp. 17–26. New York: Longman. [8]

———. 2001. Culture and Personality Studies 1918–1960: Myth and History. *J. Pers.* **69**:803–818. [2]

———. 2004. Challenging the Expert: Findings from an African Study of Infant Care and Development. In: Childhood and Adolescence in Cross-Cultural Perspectives, ed. U. P. Gielen and J. L. Roopnarine, pp. 149–165. Westport, CT: Praeger. [6]

———. 2007. Ethnographic Studies of Childhood: A Historical Overview. *Am. Anthropol.* **109**:247–260. [2, 8]

———. 2010. The Six Cultures Study: Prologue to a History of a Landmark Project. *J. Cross-Cult. Psychol.* **41**:513–521. [2]

———. 2014. Attachment Theory as Cultural Ideology. In: Different Faces of Attachment: Cultural Variations of a Universal Human Need, ed. H. Otto and H. Keller, pp. 50–65. Cambridge: Cambridge Univ. Press. [2, 6, 12]

LeVine, R. A., S. Dixon, S. LeVine, et al. 1994. Child Care and Culture: Lessons from Africa. New York: Cambridge Univ. Press. [1, 2, 5, 6]

LeVine, R. A., and S. LeVine. 2016. Do Parents Matter? Why Japanese Babies Sleep Soundly, Mexican Siblings Don't Fight, and American Families Should Just Relax. New York: PublicAffairs. [1, 5]

LeVine, R. A., and P. M. Miller. 1990. Commentary. *Human Dev.* **33**:73–80. [2]

LeVine, R. A., and K. Norman. 2001. The Infant's Acquisition of Culture: Early Attachment Reexamined in Anthropological Perspective. In: The Psychology of Cultural Experience, ed. C. C. Moore and H. F. Mathews, pp. 83–104. Cambridge: Cambridge Univ. Press. [2, 5, 6, 8]

Levinson, S. C., and R. D. Gray. 2012. Tools from Evolutionary Biology Shed New Light on the Diversification of Languages. *Trends Cogn. Sci.* **16**:167–173. [5]

Levitt, M. J. 2005. Social Relations in Childhood and Adolescence: The Convoy Model Perspective. *Human Dev.* **48**:28–47. [6]

Levy, F., M. Keller, and P. Poindron. 2004. Olfactory Regulation of Maternal Behavior in Mammals. *Horm. Behav.* **46**:284–302. [10]

Lewis, M. 1994. Does Attachment Imply a Relationship or Multiple Relationships? *Psychoanal. Inq.* **5**:47–51. [5]

———. 2005. The Child and Its Family: The Social Network Model. *Human Dev.* **48**:8–27. [2, 12]

Lewis, M., and K. Takahashi. 2005. Beyond the Dyad: Conceptualization of Social Networks. *Human Dev.* **48**:5–7. [6]

Lewontin, R. C. 1995/1982. Human Diversity. New York: Scientific American Library. [2]

Liedloff, J. 1997. The Continuum Concept: In Search of Happiness Lost. New York: Addison-Wesley. [14]

Lillas, C., and M. A. Marchel. 2015. Moving Away from Weird: Systems-Based Shifts in Research, Diagnosis, and Clinical Practice. *Perspect. Infant Men. Health* **23**:10–15. [13]

Locke, J. L., and B. Bogin. 2006. Language and Life History: A New Perspective on the Development and Evolution of Human Language. *Behav. Brain Sci.* **29**:259–325. [4]

Lorenz, K. 1966. On Aggression. New York: Harcourt, Brace and World. [2]

Lovejoy, C. O. 1981. The Origin of Man. *Science* **211**:341–350. [11]

Lyons, D. E., A. G. Young, and F. C. Keil. 2007. The Hidden Structure of Overimitation. *PNAS* **104**:19751–19756. [11]

Lyons, D. M., K. J. Parker, M. Katz, and A. F. Schatzberg. 2009. Developmental Cascades Linking Stress Inoculation, Arousal Regulation, and Resilience. *Front. Behav. Neurosci.* **3**:32. [10]

Maas, H. S. 1970. Review of Attachment and Loss, vol. 1: Attachment by John Bowlby, Basic Books. *Psychiatry* **33**:412–414. [2]

Maccoby, E. E., and R. H. Mnookin. 1993. Dividing the Child. Cambridge, MA: Harvard Univ. Press. [14]

MacDonald, K. 1992. Warmth as a Developmental Construct: An Evolutionary Analysis. *Child Dev.* **63**:753–773. [6]

Mace, R., and R. Sear. 2005. Are Humans Cooperative Breeders? In: Grandmotherhood: The Evolutionary Significance of the Second Half of Female Life, ed. E. Voland et al., pp. 143–159. New Brunswick: Rutgers Univ. Press. [5]

Machado, C. J., and J. Bachevalier. 2003. Non-Human Primate Models of Childhood Psychopathology: The Promise and the Limitations. *J. Child Psychol. Psychiatry* **44**:64–87. [10]

MacKenzie, M. J., E. Nicklas, J. Brooks-Gunn, and J. Waldfogel. 2011. Who Spanks Infants and Toddlers? Evidence from the Fragile Families and Child Well-Being Study. *Child Youth Serv. Rev.* **33**:1364–1373. [13]

MacKinnon, J. R. 1977. A Comparative Ecology of Asian Apes. *Primates* **18**:747–772. [4]

MacKinnon, J. R., and K. S. MacKinnon. 1980. Niche Differentiation in a Primate Community. In: Malayan Forest Primates: Ten Years' Study in Tropical Rain Forest, ed. D. J. Chivers, pp. 167–190. New York: Plenum. [4]

Macksoud, M. S., and J. L. Aber. 1996. The War Experiences and Psychosocial Development of Children in Lebanon. *Child Dev.* **67**:70–88. [13]

MacLean, E. L., B. Hare, C. L. Nunn, et al. 2014. The Evolution of Self-Control. *PNAS* **111**:E2140–E2148. [4]

Madianou, M., and D. Miller. 2012. Migration and the New Media: Transnational Families and Polymedia. New York: Routledge. [7]

Maestripieri, D. 1994. Social Structure, Infant Handling, and Mothering Styles in Group-Living Old World Monkeys. *Int. J. Primatol.* **15**:531–553. [4]

———. 1998. Parenting Styles of Abusive Mothers in Group-Living Rhesus Macaques. *Anim. Behav.* **55**:1–11. [4]

———. 2003. Attachment. In: Primate Psychology, ed. D. Maestripieri, pp. 108–143. Cambridge, MA: Harvard Univ. Press. [3]

———. 2005. Early Experience Affects the Intergenerational Transmission of Infant Abuse in Rhesus Monkeys. *PNAS* **102**:9726–9729. [4]

Maestripieri, D., and K. A. Carroll. 1998. Risk Factors for Infant Abuse and Neglect in Group-Living Rhesus Monkeys. *Psychol. Sci.* **9**:143–145. [4]

Maestripieri, D., K. Wallen, and K. A. Carroll. 1997a. Genealogical and Demographic Influences on Infant Abuse and Neglect in Group-Living Sooty Mangabeys (*Cercocebus atys*). *Dev. Psychobiol.* **31**:175–180. [4]

———. 1997b. Infant Abuse Runs in Families of Group-Living Pigtail Macaques. *Child Abuse Neglect* **21**:465–471. [4]

Mageo, J. M. 2013. Toward a Cultural Psychodynamics of Attachment: Samoa and U.S. Comparisons. In: Attachment Reconsidered: Cultural Perspectives on a Western Theory, ed. N. Quinn and J. M. Mageo, pp. 191–214. New York: Palgrave Macmillan. [2, 6]

Main, M. 1981. Avoidance in the Service of Attachment. In: Behavioral Development, ed. K. Immelman et al., pp. 651–669. Cambridge: Cambridge Univ. Press. [11]

———. 1999. Mary D. Salter Ainsworth: Tribute and Portrait. *Psychoanal. Inq.* **19**:682–776. [5]

Main, M., E. Hesse, and S. Hesse. 2011. Attachment Theory and Research: Overview with Suggested Applications to Child Custody. *Fam. Court Rev.* **49**:426–463. [14]

Main, M., and J. Solomon. 1990. Procedures for Identifying Infants as Disorganized/Disoriented During the Ainsworth Strange Situation. In: Attachment in the Preschool Years: Theory, Research, and Intervention, ed. M. Greenberg et al., pp. 121–160. Chicago: Univ. Chicago Press. [2]

Malone, N., A. Fuentes, and F. J. White. 2012. Variation in the Social Systems of Extant Hominoids: Comparative Insight into the Social Behavior of Early Hominins. *Int. J. Primatol.* **33**:1251–1277. [4]

Manson, J. H., R. W. Wrangham, J. L. Boone, et al. 1991. Intergroup Aggression in Chimpanzees and Humans [and Comments and Replies]. *Curr. Anthropol.* **32**:369–390. [4]

Markus, H., and S. Kitayama. 1991. Culture and the Self: Implications for Cognition, Emotion, and Motivation. *Psychol. Rev.* **98**:224–253. [6]

———. 2010. Cultures and Selves: A Cycle of Mutual Constitution. *Perspect. Psychol. Sci.* **5**:420–430. [6]

Marlin, B. J., M. Mitre, A. D'Amour J, M. V. Chao, and R. C. Froemke. 2015. Oxytocin Enables Maternal Behaviour by Balancing Cortical Inhibition. *Nature* [10]

Marlowe, F. W. 2004. Marital Residence among Foragers. *Curr. Anthropol.* **45**:277–284. [4]

Martin, L. J., D. M. Spicer, M. H. Lewis, J. P. Gluck, and L. C. Cork. 1991. Social Deprivation of Infant Rhesus Monkeys Alters the Chemoarchitecture of the Brain: I. Subcortical Regions. *J. Neurosci.* **11**:3344–3358. [10]

Martin, R. D. 1990. Primate Origins and Evolution: A Phylogenetic Reconstruction. Princeton: Princeton Univ. Press. [4]

Martini, M., and J. Kirkpatrick. 1981. Early Interactions in the Marquesas Islands. In: Culture and Early Interactions, ed. T. M. Field et al., pp. 189–213. Hillsdale, NJ: Erlbaum. [8]

Marvin, R. S., C. Glen, H. Kent, and P. Bert. 2002. The Circle of Security Project: Attachment-Based Intervention with Caregiver-Preschool Child Dyads. *Attach. Hum. Dev.* **4**:107–124. [14]

Marvin, R. S., T. L. Van Devender, M. Iwanaga, S. LeVine, and R. A. LeVine. 1977. Infant-Caregiver Attachment among the Hausa in Nigeria. In: Ecological Factors in Human Development, ed. H. M. McGurk, pp. 247–260. Amsterdam: North Holland. [1, 5]

Mascaro, J. S., P. D. Hackett, and J. K. Rilling. 2014. Differential Neural Responses to Child and Sexual Stimuli in Human Fathers and Non-Fathers and Their Hormonal Correlates. *Psychoneuroendocrino* **46**:153–163. [10]

Maslow, A. H. 1968. Toward a Psychology of Being. Princeton: Van Nostrand Reinhold. [6]

Massart, R., M. Suderman, N. Provencal, et al. 2014. Hydroxymethylation and DNA Methylation Profiles in the Prefrontal Cortex of the Nonhuman Primate Rhesus Macaque and the Impact of Maternal Deprivation on Hydroxymethylation. *Neuroscience* **268**:139–148. [10]

Matsumura, S. 1999. The Evolution of "Egalitarian" and "Despotic" Social Systems among Macaques. *Primates* **40**:23–31. [4]

Matsuzawa, T. 2006. Evolutionary Origins of the Human Mother-Infant Relationship. In: Cognitive Development in Chimpanzees, ed. T. Matsuzawa et al., pp. 127–141. Tokyo: Springer. [3, 4]

Matsuzawa, T., M. Tomonaga, and M. Tanaka, eds. 2006. Cognitive Development in Chimpanzees. Tokyo: Springer. [4]

Mauss, M. 1985. A Category of the Human Mind: The Notion of Person, the Notion of Self. In: The Category of the Person: Anthropology, Philosophy, History, ed. M. Carrithers et al., pp. 1–25. Cambridge: Cambridge Univ. Press [6]

Mayes, L. C., and S. Lassonde. 2014. A Girl's Childhood: Psychological Development, Social Change, and the Yale Child Study Center. New Haven: Yale Univ. Press. [2]

Maynard Smith, J. 1982. Evolution and the Theory of Games. Cambridge: Cambridge Univ. Press. [11]

McCall, R. B., C. J. Groark, L. Fish, et al. 2016. Characteristics of Children Transitioned to Intercountry Adoption, Domestic Adoption, Foster Care, and Biological Families from Institutions in St Petersburg, Russian Federation. *International Social Work* **59**:778–790. [9]

McConkey, K. R., F. Aldy, A. Ario, and D. J. Chivers. 2002. Selection of Fruit by Gibbons (Hylobates Muelleri X agilis) in the Rain Forests of Central Borneo. *Int. J. Primatol.* **23**:123–145. [4]

McElwain, N. L., and C. Booth-LaForce. 2006. Maternal Sensitivity to Infant Distress and Nondistress as Predictors of Infant-Mother Attachment Security. *J. Fam. Psychol.* **20**:247–255. [6]

McEwen, C. A., and B. S. McEwen. 2017. Social Structure, Adversity, Toxic Stress, and Inter-Generational Poverty: An Early Childhood Model. *Annu. Rev. Sociol.* **43**:29.21–29.28. [11]

McGowan, P. O., A. Sasaki, A. C. D'Alessio, et al. 2009. Epigenetic Regulation of the Glucocorticoid Receptor in Human Brain Associates with Childhood Abuse. *Nat. Neurosci.* **12**:342–348. [11]

McGrew, W. C. 2004. The Cultured Chimpanzee: Reflections on Cultural Primatology. Cambridge: Cambridge Univ. Press. [4]

McIntosh, J. 2011. Guest Editor's Introduction to Special Issue on Attachment Theory, Separation, and Divorce: Forging Coherent Understandings for Family Law. *Fam. Court Rev.* **49**:418–425. [12]

McKenna, J. J. 1979. The Evolution of Allomothering Behavior among Colobine Monkeys: Function and Opportunism in Evolution. *Am. Anthropol.* **81**:818–840. [4]

McLaughlin, K. A., N. A. Fox, C. H. Zeanah, et al. 2010. Delayed Maturation in Brain Electrical Activity Partially Explains the Association between Early Environmental Deprivation and Symptoms of Attention-Deficit/Hyperactivity Disorder. *Biol. Psychiatry* **68**:329–336. [9]

McLaughlin, K. A., M. A. Sheridan, and H. Lambert. 2014. Childhood Adversity and Neural Development: Deprivation and Threat as Distinct Dimensions of Early Experience. *Neurosci. Biobehav. Rev.* **47**:578–591. [9]

McLaughlin, K. A., M. A. Sheridan, F. Tibu, et al. 2015. Causal Effects of the Early Caregiving Environment on Development of Stress Response Systems in Children. *PNAS* **112**:5637–5642. [9]

McLaughlin, K. A., M. A. Sheridan, W. Winter, et al. 2013. Widespread Reductions in Cortical Thickness Following Severe Early-Life Deprivation: A Neurodevelopmental Pathway to Attention-Deficit/Hyperactivity Disorder. *Biol. Psychiatry* **76**:629–663. [9]

McLaughlin, K. A., C. H. Zeanah, N. A. Fox, and C. A. Nelson III. 2011. Attachment Security as a Mechanism Linking Foster Care Placement to Improved Mental Health Outcomes in Previously Institutionalized Children. *J. Child Psychol. Psychiatry* **53**:46–55. [9]

Mead, M. 1934. Mind, Self, and Society. Chicago: Univ. Chicago Press. [5]

———. 1954. Some Theoretical Considerations on the Problem of Mother-Child Separation. *Am. J. Orthopsychiatry* **24**:471–483. [1, 2]

———. 1955. Theoretical Settings: 1954. In: Childhood in Contemporary Cultures, ed. M. Mead and M. Wolfenstein, pp. 3–20. Chicago: Univ. Chicago Press. [2]

———. 1962. A Cultural Anthropologist's Approach to Maternal Deprivation. In: Deprivation of Maternal Care: A Reassessment of Its Effects, pp. 45–62, Public Health Papers 14. Geneva: World Health Organization. [2, 6]

Meehan, C. L., and S. Hawks. 2013. Cooperative Breeding and Attachment among the Aka Foragers. In: Attachment Reconsidered: Cultural Perspectives on a Western Theory, ed. N. Quinn and J. M. Mageo, pp. 85–113. New York: Palgrave. [1, 2, 5, 6, 8, 9, 12]

Mehler, J., P. W. Jusczyk, G. Lambertz, et al. 1988. A Precursor of Language Acquisition in Young Infants. *Cognition* **29**:143–178. [4]

Mehta, M. A., N. I. Golembo, C. Nosarti, et al. 2009a. Amygdala, Hippocampal and Corpus Callosum Size Following Severe Early Institutional Deprivation: The English and Romanian Adoptees Study Pilot. *J. Child Psychol. Psychiatry* **50**:943–951. [9]

Mehta, M. A., E. Gore-Langton, N. Golembo, et al. 2009b. Hyporesponsive Reward Anticipation in the Basal Ganglia Following Severe Institutional Deprivation Early in Life. *J. Cogn. Neurosci.* **22**:2316–2325. [9]

Meins, E., and C. Fernyhough. 2006. Mind-Mindedness Coding Manual, Version 2.2. A Working Paper, University of York. https://www.york.ac.uk/media/psychology/mind-mindedness/MM%20manual%20version%202.2-2.pdf (accessed Dec. 15, 2016). [5]

Meltzoff, A. N. 2005. Imitation and Other Minds: The Like Me Hypothesis. In: Perspectives on Imitation: From Neuroscience to Social Science, vol. 2, ed. S. Hurley and N. Chater, pp. 55–77. Cambridge, MA: MIT Press. [11]

Meltzoff, A. N., and M. K. Moore. 1977. Imitation of Facial and Manual Expressions by Human Neonates. *Science* **198**:75–78. [3, 11]

———. 1983. Newborn Infants Imitate Adult Facial Gestures. *Child Dev.* **54**:702–709. [3]

Ménard, N. 2004. Do Ecological Factors Explain Variation in Social Organization? In: Macaque Societies: A Model for the Study of Social Organization, ed. B. Thierry et al., pp. 237–262. Cambridge University Press: Cambridge. [4]

Menary, R., ed. 2010. The Extended Mind. Cambridge, MA: MIT Press. [11]

Mendle, J., and J. Ferrero. 2012. Detrimental Psychological Outcomes Associated with Pubertal Timing in Adolescent Boys. *Dev. Rev.* **32**:49–66. [11]

Mendoza, S. P., and W. A. Mason. 1986. Parental Division of Labour and Differentiation of Attachments in a Monogamous Primate (*Callicebus moloch*). *Anim. Behav.* **34**:1336–1347. [3]

Mercer, J. 2009. Child Custody Evaluations, Attachment Theory, and an Attachment Measure: The Science Remains Limited. *Sci. Rev. Mental Health Prac.* **7**:37–54. [14]

Mesman, J., T. Minter, and A. Angnged. 2016a. Received Sensitivity: Adapting Ainsworth's Scale to Capture Sensitivity in a Multiple-Caregiver Context. *Attach. Hum. Dev.* **18**:101–114. [5]

Mesman, J., M. H. van IJzendoorn, K. Behrens, et al. 2015. Is the Ideal Mother a Sensitive Mother? Beliefs About Early Childhood Parenting in Mothers across the Globe. *Int. J. Beh. Dev.* **40**:385–397. [5, 6, 8]

Mesman, J., M. H. van IJzendoorn, and A. Sagi-Schwartz. 2016b. Cross-Cultural Patterns of Attachment: Universals and Contextual Dimensions. In: Handbook of Attachment, 3rd edition, ed. J. Cassidy and P. R. Shaver, pp. 790–815. New York: Guilford. [2, 5, 8, 12]

Mikulincer, M., and P. R. Shaver. 2007. Attachment in Adulthood: Structure, Dynamics, and Change. New York: Guilford. [12]

Mikulincer, M., and P. R. Shaver. 2014. Mechanisms of Social Connection from Brain to Group. Washington, D.C.: American Psychological Association. [6]

Miller, F. G. 2009. The Randomized Controlled Trial as a Demonstration Project: An Ethical Perspective. *Am. J. Psychiatry* **166**:743–745. [9]

Miller, L., ed. 2002. Eat or Be Eaten: Predator Sensitive Foraging among Primates. Cambridge: Cambridge Univ. Press. [4]

Miller, P. M., and M. L. Commons. 2010. The Benefits of Attachment Parenting for Infants and Children: A Behavioral Developmental View. *Behav. Dev. Bull.* **16**:1. [14]

Mills, D. A., C. P. Windle, H. F. Baker, and R. M. Ridley. 2004. Analysis of Infant Carrying in Large, Well-Established Family Groups of Captive Marmosets (*Callithrix jacchus*). *Primates* **45**:259–265. [3]

Minnis, H., S. Macmillan, R. Pritchett, et al. 2013. Prevalence of Reactive Attachment Disorder in a Deprived Population. *Br. J. Psychiatry* **202**:342–346. [10]

Mintz, S. 2009. Huck's Raft: A History of American Childhood. Ann Arbor: Univ. of Michigan Press. [2]

Mitani, J., J. Call, P. Kappeleer, R. Palombit, and J. B. Silk, eds. 2012. The Evolution of Primate Societies. Chicago: Univ. Chicago Press. [4]

Mizuno, K., S. Takiguchi, M. Yamazaki, et al. 2015. Impaired Neural Reward Processing in Children and Adolescents with Reactive Attachment Disorder: A Pilot Study. *Asian J. Psychiatr.* **17**:89–93. [10]

Mizushima, S. G., T. X. Fujisawa, S. Takiguchi, et al. 2015. Effect of the Nature of Subsequent Environment on Oxytocin and Cortisol Secretion in Maltreated Children. *Front. Psychiatry* **6**: [10]

Mjelde-Mossey, L. A. 2007. Cultural and Demographic Changes and Their Effects Upon the Traditional Grandparent Role for Chinese Elders. *J. Hum. Behav. Soc. Environ.* **16**:107–120. [5]

Moffett, M. 2013. Human Identity and the Evolution of Societies. *Hum. Nat.* **24**:219–267. [11]

Moll, J., and R. de Oliveira-Souza. 2009. Extended Attachment and the Human Brain: Internalized Cultural Values and Evolutionary Implications. In: The Moral Brain: Essays on the Evolutionary and Neuroscientific Aspects of Morality, ed. J. Verplaetse et al., pp. 69–85. Houten, NL: Springer. [11]

Møller, A. P., and R. Thornhill. 1998. Male Parental Care, Differential Parental Investment by Females and Sexual Selection. *Anim. Behav.* **55**:1507–1515. [5]

Montagu, A. 1968. Man and Aggression. New York: Oxford Univ. Press. [2]

Montesi, M. 2015. Información y Crianza Con Apego en España. In: XII Congreso ISKO España y II Congreso ISKO España-Portugal, Murcia: Universidad de Murcia. http://www.iskoiberico.org/wp-content/uploads/2015/11/204_Montesi.pdf (accessed Nov. 16, 2016). [14]

Moore, E. R., G. C. Anderson, N. Bergman, and T. Dowswell. 2012. Early Skin-to-Skin Contact for Mothers and Their Healthy Newborn Infants. *Cochrane Db Syst Rev* **5**:CD003519. [10]

Morelli, G. A. 2015. The Evolution of Attachment Theory and Cultures of Human Attachment in Infancy and Early Childhood. In: The Oxford Handbook of Human Development and Culture, ed. L. A. Jensen, pp. 149–164. New York: Oxford Univ. Press. [5, 6, 7]

Morelli, G. A., and P. I. Henry. 2013. Afterword: Cross-Cultural Challenges to Attachment Theory. In: Attachment Reconsidered: Cultural Perspectives on a Western Theory, ed. N. Quinn and J. M. Mageo, pp. 241–249. New York: Palgrave. [2, 6, 12]

Morelli, G. A., P. I. Henry, and S. Foerster. 2014. Relationships and Resource Uncertainty: Cooperative Development of Efe Hunter-Gatherer Infants and Toddlers. In: Ancestral Landscapes in Human Evolution: Culture, Childrearing and Social Well-being, ed. D. F. Narvaez et al., pp. 69–103. New York: Oxford Univ. Press. [5, 6]

Morelli, G. A., P. I. Henry, E. Tronick, and H. Baldwin. 2002a. Quality of Care, Consistency of Care, and Relationships in Efe Infants. *Anthropol. News* **434**:20. [6]

Morelli, G. A., B. Rogoff, and C. Angelillo. 2002b. Cultural Variation in Children's Access to Work or Involvement in Specialized Child-Focused Activities. *Int. J. Behav. Dev.* **27**:264–274. [6]

Morelli, G. A., and F. Rothbaum. 2007. Situating the Child in Context: Attachment Relationships and Self-Regulation in Different Cultures. In: Handbook of Cultural Psychology, ed. S. Kitayama and D. Cohen, pp. 500–527. New York: Guilford. [6]

Moriceau, S., and R. M. Sullivan. 2005. Neurobiology of Infant Attachment. *Dev. Psychobiol.* **47**:230–242. [9]

Morishita, H., and T. K. Hensch. 2008. Critical Period Revisited: Impact on Vision. *Curr. Opin. Neurobiol.* **18**:101–107. [9]

Morris, D. 1967. The Naked Ape: A Zoologist's Study of the Human Animal. New York: McGraw Hill. [2]

Morss, J. R. 1990. The Biologising of Childhood: Developmental Psychology and the Darwinian Myth. Hove, UK: Erlbaum. [2]

Morton, N., and K. D. Browne. 1998. Theory and Observation of Attachment and Its Relation to Child Maltreatment: A Review. *Child Abuse Neglect* **22**:1093–1104. [9]

Mota, M. T. D., C. R. Franci, and M. B. de Sousa. 2006. Hormonal Changes Related to Paternal and Alloparental Care in Common Marmosets (*Callithrix jacchus*). *Horm. Behav.* **49**:293–302. [3]

Mottolese, R., J. Redoute, N. Costes, D. Le Bars, and A. Sirigu. 2014. Switching Brain Serotonin with Oxytocin. *PNAS* **111**:8637–8642. [10]

Munroe, R. H., and R. L. Munroe. 1971. Household Density and Infant Care in an East African Society. *J. Soc. Psychol.* **83**:3–13. [5]

Muroyama, Y. 1994. Exchange of Grooming for Allomothering in Female Patas Monkeys. *Behaviour* **128**:103–119. [3]

Murray, M. 2013. Staying with the Baby: Intensive Mothering and Social Mobility in Santiago de Chile. In: Parenting in Global Perspective: Negotiating Ideologies of Kinship, Self and Politics, ed. C. Faircloth et al., pp. 256–284. London: Routledge. [6]

———. 2014. Back to Work? Childcare Negotiations and Intensive Mothering in Santiago de Chile. *J. Fam. Issues* **36**:1–21. [14]

Myowa, M. 1996. Imitation of Facial Gestures by an Infant Chimpanzee. *Primates* **37**:207–213. [3]

———. 2012. Mane Ga Hagukumu Hito No Kokoro (Imitation and Social Mind). Tokyo: Iwanami Syoten. [3]

Myowa-Yamakoshi, M., and M. Tomonaga. 2001a. Development of Face Recognition in an Infant Gibbon (*Hylobates agilis*). *Infant Behav. Dev.* **24**:215–227. [3]

———. 2001b. Perceiving Eye Gaze in an Infant Gibbon (*Hylobates agilis*). *Psychologia* **44**:24–30. [3]

Myowa-Yamakoshi, M., M. Tomonaga, M. Tanaka, and T. Matsuzawa. 2004. Imitation in Neonatal Chimpanzees (*Pan troglodytes*). *Dev. Sci.* **7**:437–442. [3]

Myowa-Yamakoshi, M., M. Tomonaga, M. K. Yamauchi, M. Tanaka, and T. Matsuzawa. 2005. Development of Face Recognition in Infant Chimpanzees (*Pan troglodytes*). *Cogn. Dev.* **20**:49–63. [3]

Nakagawa, N. 1995. A Case of Infant Kidnapping and Allomothering by Members of Neighboring Group in Patas Monkeys. *Folia Primatol.* **64**:62–68. [3]

Nakamichi, M., A. Silldorff, C. Bingham, and P. Sexton. 2007. Spontaneously Occurring Mother-Infant Swapping and the Relationships of the Infants with Their Biological and Foster Mothers in a Captive Group of Lowland Gorillas (*Gorilla gorilla gorilla*). *Infant Behav. Dev.* **30**:399–408. [3]

National Scientific Council on the Developing Child. 2004. Young Children Develop in an Environment of Relationships: Working Paper 1. Centre on the Developing Child. Harvard Univ. http://developingchild.harvard.edu/resources/wp1/ (accessed Nov. 11, 2016). [5]

Neely, R. 1984. The Primary Caretaker Parent Rule: Child Custody and the Dynamics of Greed. *Yale Law Policy Rev.* **3**:168–186. [12]

Negayama, K., and S. Honjo. 1986. An Experimental Study on Developmental Changes of Maternal Discrimination of Infants in Crab-Eating Monkeys (*Macaca fascicularis*). *Dev. Psychobiol.* **19**:49–56. [3]

Nelson, C. A., III, K. Bos, M. R. Gunnar, and E. J. Songuga-Barke. 2011. The Neurobiological Toll of Early Human Deprivation. *Monogr. Soc. Res. Child Dev.* **76**:127–146. [10]

Nelson, C. A., III and M. A. Sheridan. 2011. Lessons from Neuroscience Research for Understanding Causal Links between Family and Neighborhood Characteristics and Educational Outcomes. In: Whither Opportunity: Rethinking the Role of Neighborhoods and Families on Schools and School Outcomes for American Children, ed. G. J. Duncan and R. J. Murnane, pp. 27–46. New York: Russel Sage. [9]

Nelson, C. A., III, C. H. Zeanah, N. A. Fox, et al. 2007. Cognitive Recovery in Socially Deprived Young Children: The Bucharest Early Intervention Project. *Science* **318**:1937–1940. [9]

Nelson, E. E., K. N. Herman, C. E. Barrett, et al. 2009. Adverse Rearing Experiences Enhances Responding to Both Aversive and Rewarding Stimuli in Juvenile Rhesus Monkeys. *Biol. Psychiatry* **66**:702–704. [10]

Nelson, E. E., and J. T. Winslow. 2009. Non-Human Primates: Model Animals for Developmental Psychopathology. *Neuropsychopharmacology* **34**:90–105. [10]

NICHD Early Child Care Research Network, ed. 2005. Child Care and Child Development: Results from the Nichd Study of Early Child Care and Youth Development. New York: Guilford. [13]

Nichter, M., and M. Nichter. 2010. A Tale of Simeon: Reflections on Raising a Child While Conducting Field Work in Rural South India. In: Children in the Field, ed. J. Cassell, pp. 65–90. Philadelphia: Temple Univ. Press. [5]

Nicolas, G., A. Bejarano, and D. L. Lee, eds. 2015. Contemporary Parenting: A Global Perspective. New York: Routledge. [5]

Niedenthal, P., A. Wood, and M. Rychlowska. 2014. Embodied Emotion Concepts. In: The Routledge Handbook of Embodied Cognition, ed. L. Shapiro, pp. 240–249. London: Routledge. [11]

Niela-Vilén, H., A. Axelin, S. Salantera, and H. L. Melender. 2014. Internet-Based Peer Support for Parents: A Systematic Integrative Review. *Int. J. Nurs. Stud.* **51**:1524–1537. [14]

Nisbett, R. E., and A. Norenzayan. 2002. Culture and Cognition. In: Stevens' Handbook of Experimental Psychology, Third edition, ed. D. L. Medin. New York: Wiley. [1]

Nishida, T. 1983. Alloparental Behavior in Wild Chimpanzees of the Mahale Mountains, Tanzania. *Folia Primatol.* **41**:1–33. [3]

Nishida, T., N. Corp, M. Hamai, et al. 2003. Demography, Female Life History, and Reproductive Profiles among the Chimpanzees of Mahale. *Am. J. Primatol.* **59**:99–121. [3]

Noble, K. G., S. M. Houston, E. Kan, and E. R. Sowell. 2012. Neural Correlates of Socioeconomic Status in the Developing Human Brain. *Dev. Sci.* **15**:516–527. [10]

Nolan, J. L., Jr. 1998. The Therapeutic State: Justifying Government at Century's End. New York: NYU Press. [14]

Nowak, M. A. 2006. Five Rules for the Evolution of Cooperation. *Science* **314**:1560–1563. [11]

Nowak, M. A., and R. Highfield. 2011. Super Cooperators: Altruism, Evolution, and Why We Need Each Other to Succeed. New York: Free Press. [6]

Nowak, M. A., C. E. Tarnita, and E. O. Wilson. 2010. The Evolution of Eusociality. *Nature* **466**:1057–1062. [11]

Nsamenang, A. B. 2006. Human Ontogenesis: An Indigenous African View on Development and Intelligence. *Int. J. Psychol.* **41**:293–297. [6]

Numan, M., and L. J. Young. 2015. Neural Mechanisms of Mother-Infant Bonding and Pair Bonding: Similarities, Differences, and Broader Implications. *Horm. Behav.* [10]

Nunn, C. L., and S. M. Altizer. 2006. Infectious Diseases in Primates: Behavior, Ecology and Evolution. Oxford: Oxford Univ. Press. [4]

Nussbaum, M. 2001. Upheavals of Thought: The Intelligence of Emotions. Cambridge: Cambridge Univ. Press. [11]

Nutter-El-Ouardani, C. 2014. Childhood and Development in Rural Morocco: Cultivating Reason and Strength. In: Everyday Life in the Muslim Middle East (3rd ed.), ed. D. L. Bowen et al., pp. 24–38. Bloomington: Indiana Univ. Press. [13]

Ochs, E. 1988. Culture and Language Development: Language Acquisition and Socialization in a Samoan Village. Cambridge: Cambridge Univ. Press. [5]

Ochs, E., and B. Schieffelin. 1984. Language Acquisition and Socialization: Three Developmental Stories and Their Implications. In: Culture Theory: Essays on Mind, Self and Emotion, ed. R. Shweder and R. LeVine, pp. 276–320. Chicago: Univ. of Chicago Press. [6]

O'Connor, T. G., D. Croft, and H. Steele. 2000. The Contributions of Behavioural Genetic Studies to Attachment Theory. *Attach. Hum. Dev.* **2**:107–122. [5]

Odling-Smee, J. F., K. N. Laland, and M. W. Feldman. 2003. Niche Construction: The Neglected Process in Evolution. Princeton: Princeton Univ. Press. [11]

Olff, M., J. L. Frijling, L. D. Kubzansky, et al. 2013. The Role of Oxytocin in Social Bonding, Stress Regulation and Mental Health: An Update on the Moderating Effects of Context and Interindividual Differences. *Psychoneuroendocrino* **38**:1883–1894. [10]

Oostenbroek, J., T. Suddendorf, M. Nielsen, et al. 2016. Comprehensive Longitudinal Study Challenges the Existence of Neonatal Imitation in Humans. *Curr. Biol.* **26**:1334–1338. [3]

Oppenheim, D., A. Sagi, and M. E. Lamb. 1988. Infant-Adult Attachments on the Kibbutz and Their Relation to Socioemotional Development 4 Years Later. *Dev. Psychol.* **24**:427–433. [12]

Otto, H. 2014. Don't Show Your Emotions! Emotion Regulation and Attachment in the Cameroonian Nso. In: Different Faces of Attachment: Cultural Variations on a Universal Human Need, ed. H. Otto and H. Keller, pp. 215–229. Cambridge: Cambridge Univ. Press. [2, 8, 10]

Otto, H., and H. Keller, eds. 2014. Different Faces of Attachment: Cultural Variations on a Universal Human Need. Cambridge: Cambridge Univ. Press. [1, 2, 4–6, 8, 12]

Otto, H., and H. Keller 2015. Is There Something Like German Parenting? In: Contemporary Parenting: A Glogal Perspective, ed. G. Nicolas et al., pp. 81–94. London: Routledge. [5]

Palm, G. 2014. Attachment Theory and Fathers: Moving from Being There to Being With. *J. Fam. Theory Rev.* **6**:282–297. [5]

Palombit, R. A. 1994. Extra-Pair Copulations in a Monogamous Ape. *Anim. Behav.* **47**:721–723. [4]

Panksepp, J. 1998. Affective Neuroscience: The Foundations of Human and Animal Emotions. New York: Oxford Univ. Press. [1, 4, 11]

Panksepp, J., and L. Bevin. 2012. The Archaeology of Mind: Neuroevolutionary Origins of Human Emotions. New York: W. W. Norton. [11]

Parke, R. D., and R. Buriel. 2006. Socialization in the Family: Ethnic and Ecological Perspectives. In: Handbook of Child Psychology (6th edition), vol. 3. Social, Emotional, and Personality Development (N. Eisenberg, vol. ed.), ed. W. Damon and R. M. Lerner, pp. 429–504. New York: Wiley. [12]

Parker, K. J., C. L. Buckmaster, K. Sundlass, A. F. Schatzberg, and D. M. Lyons. 2006. Maternal Mediation, Stress Inoculation, and the Development of Neuroendocrine Stress Resistance in Primates. *PNAS* **103**:3000–3005. [4]

Parma, V., M. Bulgheroni, R. Tirindelli, and U. Castiello. 2013. Body Odors Promote Automatic Imitation in Autism. *Biol. Psychiatry* **74**:220–226. [10]

Parreñas, R. 2005. Long Distance Intimacy: Class, Gender and Intergenerational Relations between Mothers and Children in Filipino Transnational Families. *Global networks* **5**:317–336. [7]

Passingham, R. E. 1985. Rates of Brain Development in Mammals Including Man. *Brain Behav. Evol.* **26**:167–175. [4]

Paul, D. 1992. Fitness: Historical Perspectives. In: Keywords in Evolutionary Biology, ed. F. Keller and E. Lloyd, pp. 112–114. Cambridge, MA: Harvard Univ. Press. [6]

Paus, T., D. L. Collins, A. C. Evans, L. B. Pike, and A. Zijdenbos. 2001. Maturation of White Matter in the Human Brain: A Review of Magnetic Resonance Studies. *Brain Res. Bull.* **54**:255–266. [11]

Pechtel, P., and D. A. Pizzagalli. 2011. Effects of Early Life Stress on Cognitive and Affective Function: An Integrated Review of Human Literature. *Psychopharmacology* **214**:55–70. [9]

———. 2013. Disrupted Reinforcement Learning and Maladaptive Behavior in Women with a History of Childhood Sexual Abuse: A High-Density Event-Related Potential Study. *JAMA Psychiatry* **70**:499–507. [9]

Pedersen, C. A., J. A. Ascher, Y. L. Monroe, and A. J. Prange, Jr. 1982. Oxytocin Induces Maternal Behavior in Virgin Female Rats. *Science* **216**:648–650. [10]

Peirce, C. S. 1958. Collected Papers of Charles Sanders Peirce, vols. I and II: Principles of Philosophy and Elements of Logic, ed. C. Hartshorne and P. Weiss. Cambridge, MA: Harvard Univ. Press. [11]

Pence, A. 2013. Voices Less Heard: The Importance of Critical and Indigenous Perspectives. In: Handbook of Early Childhood Development Research and Its Impact on Global Policy, ed. P. R. Britto et al., pp. 161–181. New York: Oxford Universtiy Press. [8, 13]

Pereira, M. E., and M. K. Izard. 1989. Lactation and Care for Unrelated Infants in Forest-Living Ringtailed Lemurs. *Am. J. Primatol.* **18**:101–108. [3]

Phelps, E. A. 2010. From Endophenotypes to Evolution: Social Attachment, Sexual Fidelity and the avpr1a Locus. *Curr. Opin. Neurobiol.* **20**:795–802. [1]

Pianka, E. R. 1970. On r- and K-Selection. *Am. Nat.* **104**:592–597. [11]

Pigliucci, M., and L. Finkelman. 2014. The Extended (Evolutionary) Synthesis Debate: Where Science Meets Philosophy. *Bioscience* **64**:511–516. [11]

Pigliucci, M., and G. Müller. 2010. Evolution: The Extended Synthesis. Cambridge, MA: MIT Press. [11]

Pinneau, S. R. 1955. The Infantile Disorders of Hospitalism and Anaclitic Depression. *Psychol. Bull.* **52**:429–452. [2]

Pitman, C. A., and R. W. Shumaker. 2009. Does Early Care Affect Joint Attention in Great Apes (Pan troglodytes, Pan paniscus, Pongo Abelii, Pongo Pygmaeus, Gorilla Gorilla)? *J. Comp. Psychol.* **123**:334. [8]

Plant, R. J. 2010. Mom: The Transformation of Motherhood in Modern America. Chicago: Univ. Chicago Press. [2, 6]

Plooij, F. X. 1984. The Behavioral Development of Free-Living Chimpanzee Babies and Infants. Norwood, NJ: Ablex. [4]

Plotkin, H. 2004. Evolutionary Thought in Psychology: A Brief History. Oxford: Blackwell. [2]

Polan, H. J., and M. A. Hofer. 2008. Psychobiological Origins of Infant Attachment and Its Role in Development. In: Handbook of Attachment: Theory, Research, and Clinical Applications, ed. J. Cassidy and P. R. Shaver, pp. 158–172. New York: Guilford. [11]

Porter, R. H., J. M. Cernoch, and F. J. McLaughlin. 1983. Maternal Recognition of Neonates through Olfactory Cues. *Physiol. Behav.* **30**:151–154. [10]

Posada, G., Y. Goa, F. Wu, et al. 1995. The Secure-Base Phenomenon across Cultures: Children's Behavior, Mothers' Preferences, and Experts' Concepts. *Monogr. Soc. Res. Child Dev.* **60**:27–47. [5]

Posada, G., A. Jacobs, M. K. Richmond, et al. 2002. Maternal Caregiving and Infant Security in Two Cultures. *Dev. Psychol.* **38**:67–78. [6]

Pritchett, R., J. Pritchett, E. Marshall, C. Davidson, and H. Minnis. 2014. Reactive Attachment Disorder in the General Population: A Hidden Essence Disorder. *Scientific World Journal* **2013**:818157. [10]

Promislow, D., and P. Harvey. 1990. Living Fast and Dying Young: A Comparative Analysis of Life-History Variation in Mammals. *J. Zool.* **220**:417–437. [11]

———. 1991. Mortality Rates and the Evolution of Life Histories. *Acta Oecol.* **12**:94–101. [11]

Provencal, N., M. J. Suderman, C. Cuillemin, et al. 2012. Signature of Maternal Rearing in the Methylome in Rhesus Macaque Prefrontal Cortex and T Cells. *J. Neurosci.* **32**:15626–15642. [10]

Queller, D. C. 1997. Why Do Females Care More Than Males? *Proc. R. Soc. Lond. B* **264**:1555–1557. [4]

Quiatt, D. 1979. Aunts and Mothers: Adaptive Implications of Allomaternal Behavior of Nonhuman Primates. *Am. Anthropol.* **81**:310–319. [4]

Quidel, J., and J. Pichinao. 2002. Haciendo Crecer Personas Pequeñas en el Pueblo Mapuche. Temuco, Chile: Ministerio del Educación. [6]

Quinlan, R. J., and M. B. Quinlan. 2007. Parenting and Cultures of Risk: A Comparative Analysis of Infidelity, Aggression, and Witchcraft. *Am. Anthropol.* **109**:164–179. [11]

Quinn, N. 2003. Cultural Selves. *Ann. NY Acad. Sci.* **1001**:145–176. [6]

———. 2005. Universals of Child Rearing. *Anthropol. Theory* **5**:477–516. [6]

Quinn, N., and D. Holland, eds. 1987. Cultural Models in Language and Thought. Cambridge: Cambridge Univ. Press. [11]

Quinn, N., and J. M. Mageo, eds. 2013. Attachment Reconsidered: Cultural Perspectives on a Western Theory. New York: Palgrave Macmillan. [1, 2, 4–6, 8, 12]

Rabain-Jamin, J. 1994. Language and Socialization of the Child in African Families Living in France. In: Cross-Cultural Roots of Minority Child Development, ed. P. Greenfield and R. Cocking, pp. 147–166. Hillsdale, NJ: Lawrence Erlbaum. [6]

Radcliffe-Brown, A. 1940. On Joking Relationships. *J. Int. Afr. Inst.* **13**:195–210. [5]

Raeff, C. 2006. Multiple and Inseparable: Conceptualizing the Development of Independence and Interdependence. *Human Dev.* **49**:96–121. [6]

Raffety, E. 2017. From Cultural Revolution to Childcare Revolution: Conflicting Advice on Childrearing in Contemporary China. In: A World of Babies: Imagined Childcare Guides for Eight Societies, ed. A. Gottlieb and J. DeLoache, pp. 71–92. New York: Cambridge Univ. Press. [6]

Raikes, H. A., and R. A. Thompson. 2005. Links between Risk and Attachment Security: Models of Influence. *J. Appl. Dev. Psychol.* **26**:440–455. [12]

Rakic, P., J.-P. Bourgeois, M. F. Eckenhoff, N. Zecevic, and P. S. Goldman-Rakic. 1986. Concurrent Overproduction of Synapses in Diverse Regions of the Primate Cerebral Cortex. *Science* **232**:232–235. [4]

Rao, M., E. M. Blass, M. M. Brignol, L. Marino, and L. Glass. 1993. Effects of Crying on Energy Metabolism in Human Neonates. *Pediatr. Res.* **33**:309. [6]

Ravignani, A. 2015. Evolving Perceptual Biases for Antisynchrony: A Form of Temporal Coordination Beyond Synchrony. *Front. Neurosci.* **9**:339. [6]

Reddy, V. 2008. How Infants Know Minds. Cambridge, MA: Harvard Univ. Press. [11]

Reichard, U. H. 1995. Extra-Pair Copulations in a Monogamous Gibbon (*Hylobates lar*). *Ethology* **100**:99–112. [4]

———. 2009. The Social Organization and Mating System of Khao Yai White-Handed Gibbons: 1992–2006. In: The Gibbons: New Perspectives on Small Ape Socioecology and Population Biology, ed. S. Lappan and D. J. Whittaker, pp. 347–384, Developments in Primatology: Progress and Prospects, S. T. Rosen, series ed. New York: Springer Science + Business Media. [4]

Reichard, U. H., M. Ganpanakngan, and C. Barelli. 2012. White-Handed Gibbons of Khao Yai: Social Flexibility, Complex Reproductive Strategies, and a Slow Life History. In: Long-Term Field Studies of Primates, ed. P. Kappeler and D. P. Watts, pp. 237–258. Berlin: Springer Science + Business Media. [4]

Reichard, U. H., and V. Sommer. 1997. Group Encounters in Wild Gibbons (Hylobates Lar): Agonism, Affiliation, and the Concept of Infanticide. *Behaviour* **134**:1135–1174. [4]

Reiches, M. W., P. T. Ellison, S. F. Lipson, et al. 2009. Pooled Energy Budget and Human Life History. *Am. J. Hum. Biol.* **21**:421–429. [4]

Reis, H. T. 2000. Caregiving, Attachment, and Relationships. *Psychoanal. Inq.* **11**:120–123. [6]

———. 2014. Responsiveness: Affective Interdependence in Close Relationships. In: Mechanisms of Social Connection: From Brain to Group, ed. M. Mikulincer and P. R. Shaver, pp. 255–271. Washington, D.C.: American Psychological Association. [6]

Reis, H. T., W. A. Collins, and E. Berscheid. 2000. The Relationship Context of Human Behavior and Development. *Psychol. Bull.* **126**:844–872. [6]

Repetti, R. L., B. M. Reynolds, and M. S. Sears. 2015. Families under the Microscope: Repeated Sampling of Perceptions, Experiences, Biology, and Behavior. *J. Marriage Fam.* **77**:126–146. [8]

Reynolds, J. D., N. B. Goodwin, and R. P. Freckleton. 2002. Evolutionary Transitions in Parental Care and Live Bearing in Vertebrates. *Phil. Trans. R. Soc. B* **357**:269–281. [4]

Reznick, D., M. J. Bryant, and F. Bashey. 2002. r- and K-Selection Revisited: The Role of Population Regulation in Life-History Evolution. *Ecology* **83**:1509–1520. [11]

Richerson, P. J., R. Baldini, A. V. Bell, et al. 2016. Cultural Group Selection Plays an Essential Role in Explaining Human Cooperation: A Sketch of the Evidence. *Behav. Brain Sci.* **39**:e58. [11]

Richerson, P. J., and R. Boyd. 2005. Not by Genes Alone: How Culture Transformed Human Evolution. Chicago: Univ. of Chicago Press. [2, 11]

Riem, M. M., M. J. Bakermans-Kranenburg, S. Pieper, et al. 2011. Oxytocin Modulates Amygdala, Insula, and Inferior Frontal Gyrus Responses to Infant Crying: A Randomized Controlled Trial. *Biol. Psychiatry* **70**:291–297. [10]

Riem, M. M., M. J. Bakermans-Kranenburg, and I. M. H. van IJzendoorn. 2016. Intranasal Administration of Oxytocin Modulates Behavioral and Amygdala Responses to Infant Crying in Females with Insecure Attachment Representations. *Attach. Hum. Dev.* **18**:213–234. [10]

Riem, M. M., M. H. van IJzendoorn, M. Tops, et al. 2012. No Laughing Matter: Intranasal Oxytocin Administration Changes Functional Brain Connectivity During Exposure to Infant Laughter. *Neuropsychopharmacology* **37**:1257–1266. [10]

Riley, P. 2013. Attachment Theory, Teacher Motivation and Pastoral Care: A Challenge for Teachers and Academics. *Pastor. Care Educ.* **31**:112–129. [14]

Rippeyoung, P. 2013. Governing Motherhood: Who Pays and Who Profits? Canadian Centre for Policy Alternatives Nova Scotia Office. http://www.policyalternatives. ca/sites/default/files/uploads/publications/Nova%20Scotia%20Office/2013/01/Governing%20Motherhood.pdf (accessed Nov. 1, 2016). [14]

Rippeyoung, P., and M. C. Noonan. 2012. Is Breastfeeding Truly Cost Free? Income Consequences of Breastfeeding for Women. *Am. Sociol. Rev.* **77**:244–267. [14]

Robbins, A. M., M. M. Robbins, N. Gerald-Steklis, and H. D. Steklis. 2006. Age-Related Patterns of Reproductive Success among Female Mountain Gorillas. *Am. J. Phys. Anthropol.* **131**:511–521. [4]

Robertson, J., and J. Robertson. 1971. Young Children in Brief Separation: A Fresh Look. *Psychoanal. Study Child.* **26**:264–315. [2]

Robins, K. F. 2017. Quechua or Spanish? Farm or School? New Paths for Andean Children in Post-Civil-War Peru. In: A World of Babies: Imagined Childcare Guides for Eight Societies, 2nd edition, ed. A. Gottlieb and J. DeLoache, pp. 225–260. Cambridge: Cambridge Univ. Press. [6]

Robson, S. L., C. P. Van Schaik, and K. Hawkes. 2006. The Derived Features of Human Life History. In: The Evolution of Human Life History, ed. K. Hawkes and R. R. Paine, pp. 17–44. Santa Fe: SAR Press. [4]

Roepstorff, A. 2008. Things to Think With: Words and Objects as Material Symbols. *Phil. Trans. R. Soc. B* **363**:2049–2054. [11]

Rogoff, B. 2003. The Cultural Nature of Human Development. Oxford: Oxford Univ. Press. [2, 6]

Rogoff, B., R. Mejía-Arauz, and M. Correa-Chávez. 2015. A Cultural Paradigm: Learning by Observing and Pitching. *Adv. Child. Dev. Behav.* **49**:1–22. [6]

Rogoff, B., J. Mistry, A. Goncu, et al. 1993. Guided Participation in Cultural Activity by Toddlers and Caregivers. *Monogr. Soc. Res. Child Dev.* **58**:1–179. [1]

Rogoff, B., L. Moore, M. Correa-Chávez, and A. Dexter. 2014. Children Develop Cultural Repertoires through Engaging in Everyday Routines and Practices. In: Handbook of Socialization, ed. J. E. Grusec and P. D. Hastings, pp. 472–498. Guilford. [6]

Roisman, G. I., C. Booth-Laforce, J. Belsky, K. B. Burt, and A. M. Groh. 2013. Molecular-Genetic Correlates of Infant Attachment: A Cautionary Tale. *Attach. Hum. Dev.* **15**:384–406. [6]

Roisman, G. I., and R. C. Fraley. 2013. Developmental Mechanisms Underlying the Legacy of Childhood Experiences. *Child Dev. Perspect.* **7**:149–154. [12]

Roland, A. 2005. Multiple Mothering and the Familial Self. In: Freud along the Ganges: Psychoanalytic Reflections on the People and Cultures of India, ed. S. Akhtar, pp. 79–90. New York: Other Press. [14]

Romero-Fernandez, W., D. O. Borroto-Escuela, L. F. Agnati, and K. Fuxe. 2012. Evidence for the Existence of Dopamine D2-Oxytocin Receptor Heteromers in the Ventral and Dorsal Striatum with Facilitatory Receptor-Receptor Interactions. *Mol. Psychiatry* [10]

Roopnarine, J. L. 2011. Cultural Variation in Beliefs About Play, Parent-Child Play, and Children's Play: Meaning for Childhood Development. In: The Oxford Handbook of the Development of Play, ed. A. D. Pellegrini, pp. 19–37. New York: Oxford Univ. Press. [6]

———, ed. 2015. Fathers across Cultures: The Importance, Roles, and Diverse Practices of Dads. Santa Barbara, CA: Praeger. [5]

Roopnarine, J. L., E. Talukder, D. Jain, P. Joshi, and P. Srivastav. 1992. Characteristics of Holding, Patterns of Play, and Social Behaviours between Parents and Infants in New Delhi, India. *Dev. Psychol.* **26**:867–873. [5]

Rosabal-Coto, M. 2012. Creencias y Prácticas de Crianza: el Estudio del Parentaje en el Contexto Costarricense. *Revista Costarricense Psicolog.* **31**:1–2. [6]

Rosenberg, K., and W. Trevathan. 1995. Bipedalism and Human Birth: The Obstetrical Dilemma Revisited. *Evol. Anthropol.* **4**:161–168. [4]

Rothbaum, F., M. Kakinuma, R. Nagaoka, and H. Azuma. 2007. Attachment and Amae Parent: Child Closeness in the United States and Japan. *J. Cross-Cult. Psychol.* **38**:465–486. [7]

Rothbaum, F., G. Morelli, and N. Rusk. 2011. Attachment, Learning, and Coping: The Interplay of Cultural Similarities and Differences. In: Advances in Culture and Psychology, ed. M. J. Gelfrand et al., pp. 153–215. Oxford: Oxford Univ. Press. [6]

Rothbaum, F., R. Nagaoka, and I. C. Ponte. 2006. Caregiver Sensitivity in Cultural Context: Japanese and U.S. Teachers' Beliefs About Anticipating and Responding to Children's Needs. *J. Res. Child. Educ.* **21**:23–40. [6]

Rothbaum, F., M. Pott, H. Azuma, K. Miyake, and J. Weisz. 2000a. The Development of Close Relationships in Japan and the United States: Paths of Symbiotic Harmony and Generative Tension. *Child Dev.* **71**:1121–1142. [6]

Rothbaum, F., K. Rosen, T. Ujiie, and U. N. 2002. Family Systems Theory, Attachment Theory, and Culture. *Fam. Process* **41**:328–350. [1]

Rothbaum, F., J. Weisz, M. Pott, K. Miyke, and G. A. Morelli. 2000b. Attachment and Culture: Security in the United States and Japan. *Am. Psychol.* **55**:1093–1104. [2, 12]

Röttger-Rössler, B. 2014. Bonding and Belonging Beyond WEIRD Worlds: Rethinking Attachment Theory on the Basis of Cross-Cultural Anthropological Data. In: Different Faces of Attachment: Cultural Variations on a Universal Human Need, ed. H. Otto and H. Keller, pp. 141–168. Cambridge: Cambridge Univ. Press. [2, 6]

Roy, J., M. Gray, T. Stoinski, M. M. Robbins, and L. Vigilant. 2014. Fine-Scale Genetic Structure Analyses Suggest Further Male Than Female Dispersal in Mountain Gorillas. *BMC Ecol.* **14**:21. [4]

Royle, N. J., A. F. Russell, and A. J. Wilson. 2014. The Evolution of Flexible Parenting. *Science* **345**:776–781. [4]

Royle, N. J., P. T. Smiseth, and M. Kölliker, eds. 2012. The Evolution of Parental Care. Oxford: Oxford Univ. Press. [4, 11]

Rubin, K. H., W. M. Bukowski, and B. Laursen, eds. 2009. Handbook of Peer Interactions, Relationships, and Groups. New York: Guilford. [6]

Rudy, D., and J. E. Grusec. 2006. Authoritarian Parenting in Individualistic and Collectivist Groups: Associations with Maternal Emotion and Cognition and Children's Self Esteem. *J. Fam. Psychol.* **20**:68–78. [6]

Russell, C. L., K. A. Bard, and A. B. Lauren. 1997. Social Referencing by Young Chimpanzees (*Pan troglodytes*). *J. Comp. Psychol.* **111**:185–191. [3]

Russon, A. E., K. A. Bard, and S. T. Parker. 1996. Reaching into Thought: The Minds of the Great Apes. Cambridge: Cambridge Univ. Press. [4]

Ryan, R. M., and E. L. Deci. 2000. Self-Determination Theory and the Facilitation of Intrinsic Motivation, Social Development, and Well-Being. *Am. Psychol.* **55**:68–78. [6]

Sacher, G. A. 1959. Relation of Lifespan to Brain Weight and Body Weight in Mammals. In: CIBA Foundation Colloquia on Ageing, vol. 5, The Lifespan of Animals, ed. G. Wolstenholme and M. O'Connor, pp. 115–133. London: Churchill. [4]

———. 1975. Maturation and Longevity in Relation to Cranial Capacity in Hominid Evolution. In: Primate Functional Morphology and Evolution, ed. R. H. Tuttle, pp. 417–441, World Anthropology. Mouton: The Hague. [4]

Sackett, G. P. 1965. Effects of Rearing Conditions Upon the Behavior of Rhesus Monkeys (*Macaca mulatta*). *Child Dev. Perspect.* **36**:855–868. [10]

Sadler, M., and A. Obach. 2006. Pautas de Crianza Mapuche: Estudio Significaciones, Actitudes y Prácticas de Familias Mapuches en Relación a la Crianza y Cuidado Infantil de Los Niños y Niñas Desde la Gestación Hasta Los Cinco Años. Chile: CIEG, Universidad de Chile and CIGES, Universidad de La Frontera. [6]

Saffran, J. R., R. N. Aslin, and E. L. Newport. 1996. Statistical Learning by 8-Month-Old Infants. *Science* **274**:1926–1928. [4]

Sagi-Schwartz, A. 1990. Attachment Theory and Research from a Cross-Cultural Perspective. *Human Dev.* **33**:10–23. [2]

Saito, A., A. Izumi, and K. Nakamura. 2008. Food Transfer in Common Marmosets: Parents Change Their Tolerance Depending on the Age of Offspring. *Am. J. Primatol.* **70**:999–1002. [3]

Sakai, T., M. Matsui, A. Mikami, et al. 2013. Developmental Patterns of Chimpanzee Cerebral Tissues Provide Important Clues for Understanding the Remarkable Enlargement of the Human Brain. *Proc. R. Soc. Lond. B* **280**:20122398. [4]

Salomo, D., and U. Liszkowski. 2013. Sociocultural Settings Influence the Emergence of Prelinguistic Deictic Gestures. *Child Dev.* **84**:1296–1307. [8]

Salter, M. D. 1940. An Evaluation of Adjustment Based Upon the Concept of Security. University of Toronto Studies, Child Development Series, No. 18. Toronto: Univ. of Toronto Press. [5]

Saltman, B. 2016. Can Attachment Theory Explain All Our Relationships? http://ny-mag.com/thecut/2016/06/attachment-theory-motherhood-c-v-r.html (accessed Oct. 28, 2016). [14]

Sanchez, M. M., E. F. Hearn, D. Do, J. K. Rilling, and J. G. Herndon. 1998. Differential Rearing Affects Corpus Callosum Size and Cognitive Function of Rhesus Monkeys. *Brain Res.* **812**:38–49. [10]

Saraswathi, T. S., and S. Pai. 1997. Socialization in the Indian Context. In: Asian Perspectives on Psychology, ed. H. S. R. Kao and D. Sinha, pp. 74–92. New Delhi: Sage. [5]

Sawhill, I. V. 2013. Family Structure: The Growing Importance of Class. Brookings Institution. https://www.brookings.edu/articles/family-structure-the-growing-importance-of-class/ (accessed Nov. 1, 2016). [6]

Scheele, D., K. M. Kendrick, C. Khouri, et al. 2014. An Oxytocin-Induced Facilitation of Neural and Emotional Responses to Social Touch Correlates Inversely with Autism Traits. *Neuropsychopharmacology* **39**:2078–2085. [10]

Scheele, D., J. Plota, B. Stoffel-Wagner, W. Maier, and R. Hurlemann. 2016. Hormonal Contraceptives Suppress Oxytocin-Induced Brain Reward Responses to the Partner's Face. *Soc. Cogn. Affect. Neurosci.* **11**:767–774. [10]

Scheele, D., A. Wille, K. M. Kendrick, et al. 2013. Oxytocin Enhances Brain Reward System Responses in Men Viewing the Face of Their Female Partner. *PNAS* **110**:20308–20313. [10]

Scheidecker, G. 2017. Kindheit, Kultur und Moralische Emotionen: Zur Sozialisation Von Furcht und Wut Im Ländlichen Madagaskar. Bielefeld: Transcript Verlag. [5, 6]

Scheper-Hughes, N. 1985. Culture, Scarcity, and Maternal Thinking: Maternal Detachment and Infant Survival in a Brazilian Shantytown. *Ethos* **13**:291–317. [8]

———. 1992. Death without Weeping: The Violence of Everyday Life in Brazil. Berkeley: Univ. of California Press. [2, 6, 11]

———. 2014. Family Life as Bricolage: Reflections on Intimacy and Attachment in Death without Weeping. In: Different Faces of Attachment: Cultural Variations on a Universal Human Need, ed. H. Otto and H. Keller, pp. 230–260. Cambridge: Cambridge Univ. Press. [2, 5, 6]

Schneiderman, I., O. Zagoory-Sharon, J. F. Leckman, and R. Feldman. 2012. Oxytocin During the Initial Stages of Romantic Attachment: Relations to Couples' Interactive Reciprocity. *Psychoneuroendocrino* **37**:1277–1285. [10]

Schön, R., and M. Silvén. 2007. Natural Parenting: Back to Basics in Infant Care. *Evol. Psychol.* **5**:102–183. [14]

Schore, A. 1994. Affect Regulation and the Origin of the Self: The Neurobiology of Emotional Development. Hillsdale, NJ: Lawrence Erlbaum. [11]

———. 2013. Bowlby's Environment of Evolutionary Adaptiveness: Recent Studies on the Interpersonal Neurobiology of Attachment and Emotional Development. In: Evolution, Early Experience, and Human Development: From Research to Practice and Policy, ed. D. Narvaez et al., pp. 31–73. New York: Oxford Univ. Press. [11]

Schore, A., and J. McIntosh. 2011. Family Law and the Neuroscience of Attachment, Part I. *Fam. Court Rev.* **49**:501–512. [14]

Schradin, C., and G. Anzenberger. 2003. Mothers, Not Fathers, Determine the Delayed Onset of Male Carrying in Goeldi's Monkey (*Callimico Goeldii*). *J. Hum. Evol.* **45**:389–399. [3]

Schug, M. 2017. Equal Children Play Best: Raising Independent Children in a Nordic Welfare State. In: A World of Babies: Imagined Childcare Guides for Eight Societies, ed. A. Gottlieb and J. DeLoache, pp. 261–292. New York: Cambridge Univ. Press. [6, 14]

Schultz, A. H. 1949. Sex Differences in the Pelves of Primates. *Am. J. Phys. Anthropol.* **7**:401–424. [4]

Schultz, W., P. Dayan, and P. R. Montague. 1997. A Neural Substrate of Prediction and Reward. *Science* **275**:1593–1599. [9]

Schwarz, B., E. Schafermeier, and G. Trommsdorff. 2005. Value Orientation, Child-Rearing Goals, and Parenting. In: Culture and Human Development, ed. W. Friedlmeier et al., pp. 203–230. New York: Psychology Press. [6]

Scott, M. E., W. B. Wilcox, R. Ryberg, and L. DeRose. 2015. World Family Map: Mapping Family Change and Child Well-Being Outcomes. New York: Social Trends Institute and Institute for Family Studies. [6]

Sear, R. 2016. Beyond the Nuclear Family: An Evolutionary Perspective on Parenting. *Curr. Opin. Psychol.* **7**:98–103. [5, 6, 11]

Sear, R., and R. Mace. 2008. Who Keeps Children Alive? A Review of the Effects of Kin on Child Survival. *Evol. Hum. Behav.* **29**:1–18. [5, 4]

Sears, W. 1983. Creative Parenting: How to Use the New Continuum Concept to Raise Children Successfully from Birth through Adolescence. New York: Dodd Mead. [14]

———. 2011. The Payoff: Our 6 Observations on How Ap Kids Turn Out. https://www.askdrsears.com/topics/parenting/attachment-parenting/payoff-our-6-observations-how-ap-kids-turn-out (accessed April 26, 2017). [14]

———. 2016. Observations on Attachment Parenting Outcomes, AskDrSears. http://www.askdrsears.com/topics/parenting/attachment-parenting/payoff-our-6-observations-how-ap-kids-turn-out (accessed July 7, 2017). [14]

Sears, W., and M. Sears. 1993. The Baby Book: Everything You Need to Know About Your Baby from Birth to Age Two. New York: Little, Brown & Co. [13]

———. 2001. The Attachment Parenting Book: A Commonsense Guide to Understand and Nurturing Your Baby. New York: Little, Brown & Co. [14]

Seay, B., and H. F. Harlow. 1965. Maternal Separation in the Rhesus Monkey. *J. Nerv. Ment. Dis.* **140**:434–441. [4]

Seidl-de-Moura, M. L., D. Fernandes Mendes, M. Leal, L. Fontes Pessôa, and R. Vera Cruz de Carvalho. 2012. Social Interaction and Development Lab: Parenting and Trajectories of Self Development in Brazil. *ISSBD Bull.* **2**:33–36. [6]

Sellen, D. W. 2001. Comparison of Infant Feeding Patterns Reported for Nonindustrial Populations with Current Recommendations. *J. Nutri.* **131**:2707–2715. [4]

Sengupta, S. 1999. Women Keep Garment Jobs by Sending Babies to China. *NY Times* **Sept. 15**:1, 4. [7]

Serpell, R., and A. B. Nsamenang. 2014. Locally Relevant and Quality Ecce Programmes: Implications of Research in Indigenous African Child Development and Socialization. In: Early Childhood Care and Education (ECCE) Working Papers Series, vol. 3, Paris: UNESCO. http://unesdoc.unesco.org/images/0022/002265/226564e.pdf (accessed Nov. 9, 2016). [13, 14]

Serra, M., N. De Pisapia, P. Rigo, et al. 2015. Secure Attachment Status Is Associated with White Matter Integrity in Healthy Young Adults. *Neuroreport* **26**:1106–1111. [9]

Setchell, J. M., and D. J. Curtis. 2003. Field and Laboratory Methods in Primatology. Cambridge: Cambridge Univ. Press. [4]

Seymour, S. C. 1999. Women, Family and Child Care in India: A World in Transition. Cambridge: Cambridge Univ. Press. [5]

Seymour, S. C. 2004. Multiple Caretaking of Infants and Young Children: An Area in Critical Need of a Feminist Psychological Anthropology. *Ethos* **32(4)**:538–556. [14]

———. 2013. "It Takes a Village to Raise a Child": Attachment Theory and Multiple Child Care in Alor, Indonesia, and in North India. In: Attachment Reconsidered: Cultural Perspectives on a Western Theory, ed. N. Quinn and J. M. Mageo, pp. 115–139. New York: Palgrave Macmillan. [6, 12]

Shah, C. 2014. Collaborative Information Seeking. *J. Assoc. Inf. Sci. Technol.* **65**:215–236. [14]

Shai, D., and P. Fonagy. 2014. Beyond Words: Parental Embodied Mentalizing and the Parent-Infant Dance. In: Mechanisms of Social Connection: From Brain to Group, ed. M. Mikulincer and P. R. Shaver, pp. 185–203. Washington, D.C.: American Psychological Association. [6]

Shamay-Tsoory, S. G. 2011. The Neural Bases for Empathy. *Neuroscientist* **17**:18–24. [3]

Sharifzadeh, V. S. 1998. Families with Middle Eastern Roots. In: Developing Cross-Cultural Competence: A Guide for Working with Children and Their Families, ed. E. W. Lynch and M. J. Hanson, pp. 441–482. Baltimore: Paul H. Brookes. [6]

Sharma, D. 2003. Infancy in Childhood in India: A Review. In: Childhood, Family and Sociocultural Change in India: Reinterpreting the Inner World, ed. D. Sharma, p. 13–47. New Dehli: Oxford. [14]

Sharp, C., and P. Fonagy. 2008. The Parent's Capacity to Treat the Child as a Pychological Agent: Constructs, Measures and Implications for Developmental Psychopathology. *Soc. Dev.* **17**:737–754. [6]

Shea, N. 2012. New Thinking, Innateness, and Inherited Representation. *Phil. Trans. R. Soc. B* **367**:2234–2244. [11]

Sheridan, M. A., N. A. Fox, C. H. Zeanah, K. A. McLaughlin, and C. A. Nelson III. 2012a. Variation in Neural Development as a Result of Exposure to Institutionalization Early in Childhood. *PNAS* **109**:12927–12932. [9]

Sheridan, M. A., and K. A. McLaughlin. 2014. Dimensions of Early Experience and Neural Development: Deprivation and Threat. *Trends Cogn. Sci.* **18**:580–585. [9, 10]

Sheridan, M. A., K. Sarsour, D. Jutte, M. D'Esposito, and W. T. Boyce. 2012b. The Impact of Social Disparity on Prefrontal Function in Childhood. *PLoS One* **7**:e35744. [10]

Shimada, K., S. Takiguchi, S. Mizushima, et al. 2015. Reduced Visual Cortex Grey Matter Volume in Children and Adolescents with Reactive Attachment Disorder. *NeuroImage Clin.* **9**:13–19. [10]

Shostak, M. 1981. Nisa: The Life and Words of A !Kung Woman. Cambridge, MA: Harvard Univ. Press. [6]

Shwalb, D. W., B. J. Shwalb, and M. E. Lamb, eds. 2013. Fathers in Cultural Context. New York: Routledge. [5]

Shweder, R. A., J. J. Goodnow, G. Hatano, et al. 2000. The Cultural Psychology of Development: One Mind, Many Mentalities. In: Handbook of Child Psychology, vol. 1: Theoretical Models of Human Development, ed. W. Damon and R. M. Lerner, pp. 865–937. New York: Wiley. [6]

Shweder, R. A., N. C. Much, M. Mahapatra, and L. Park. 1997. The "Big Three" of Morality (Autonomy, Community, Divinity) and the "Big Three" Explanations of Suffering. In: Morality and Health, ed. A. M. Brandt and P. Rozin, pp. 119 –169. New York: Routledge. [6]

Shweder, R. A., and M. A. Sullivan. 1993. Cultural Psychology: Who Needs It? *Annu. Rev. Psychol.* **44**:497–523. [2]

Siegel, D., and J. McIntosh. 2011. Family Law and the Neuroscience of Attachment, Part II. *Fam. Court Rev.* **49**:513–520. [14]

Siegel, S. J., S. D. Ginsberg, P. R. Hof, et al. 1993. Effects of Social Deprivation in Prepubescent Rhesus Monkeys: Immunohistochemical Analysis of the Neurofilament Protein Triplet in the Hippocampal Formation. *Brain Res.* **619**:299–305. [10]

Silk, J. B. 1980. Kidnapping and Female Competition among Captive Bonnet Macaques. *Primates* **21**:100–110. [3]

Simpson, E. A., V. Sclafani, A. Paukner, et al. 2014. Inhaled Oxytocin Increases Positive Social Behaviors in Newborn Macaques. *PNAS* **111**:6922–6927. [10]

Simpson, J. A. 1999. Attachment Theory in Modern Evolutionary Perspective. In: Handbook of Attachment: Theory, Research, and Clinical Applications, ed. J. Cassidy and P. R. Shaver, pp. 115–140. New York: Guilford. [2]

Simpson, J. A., and J. Belsky. 2010. Attachment Theory within a Modern Evolutionary Framework. In: Handbook of Attachment: Theory, Research, and Clinical Applications (2nd ed.), ed. J. Cassidy and P. R. Shaver, pp. 131–157. New York: Guilford. [2]

———. 2016. Attachment Theory within a Modern Evolutionary Framework. In: Handbook of Attachment: Theory, Research, and Clinical Applications (3rd ed.), ed. J. Cassidy and P. R. shaver, pp. 91–116. New York: Guilford. [2, 12]

Simpson, M. J. A., and S. B. Datta. 1991. Predicting Infant Enterprise from Early Relationships in Rhesus Macaques. *Behaviour* **116**:42–62. [4]

Singer, T., B. Seymour, J. O'Doherty, et al. 2004. Empathy for Pain Involves the Affective but Not Sensory Components of Pain. *Science* **303**:1157–1162. [10]

Skyrms, B. 2010. Signals: Evolution, Learning, and Information. New York: Oxford Univ. Press. [11]

Slade, A. 2008. The Implications of Attachment Theory and Research for Adult Psychotherapy Research and Clinical Perspectives. In: Handbook of Attachment: Theory, Research, and Clinical Applications, ed. J. Cassidy and P. Shaver, pp. 762–782. New York: Guilford. [14]

———. 2016. The Implications of Attachment and Adult Psychotherapy Theory, Research and Practice. In: Handbook of Attachment: Theory, Research, and Clinical Applications, ed. J. Cassidy and P. Shaver. New York: Guilford. [14]

Slater, A., and P. C. Quinn. 2001. Face Recognition in the Newborn Infant. *Infant Child Dev.* **10**:21–24. [6]

Smith, B. L. 2012. The Case against Spanking. *APA Monitor* **43**:60. [13]

Smith, K. P., and N. A. Christakis. 2008. Social Networks and Health. *Annu. Rev. Sociol.* **34**:405–429. [6]

Smith, K. R., and G. P. Mineau. 2003. Genealogies in Demographic Research. In: Encyclopedia of Population, ed. P. Demeny and G. McNicoll, pp. 448–451. New York: MacMillan. [4]

Smuts, B. B., D. L. Cheney, R. M. Seyfarth, R. W. Wrangham, and T. T. Struhsaker, eds. 1987. Primate Societies. Chicago: Univ. of Chicago Press. [4]

Smyke, A. T., S. F. Koga, D. E. Johnson, et al. 2007. The Caregiving Context in Institution-Reared and Family-Reared Infants and Toddlers in Romania. *J. Child Psychol. Psychiatry* **48**:210–218. [9]

Smyke, A. T., C. H. Zeanah, N. A. Fox, and C. A. Nelson III. 2009. A New Model of Foster Care for Young Children: The Bucharest Early Intervention Project. *Child Adolesc. Psychiatr. Clin. N. Am.* **18**:721–734. [9]

Smyke, A. T., C. H. Zeanah, N. A. Fox, C. A. Nelson III, and D. Guthrie. 2010. Placement in Foster Care Enhances Quality of Attachment among Young Institutionalized Children. *Child Dev.* **81**:212–223. [9]

Sober, E., and D. S. Wilson. 1998. Unto Others: The Evolution and Psychology of Unselfish Behavior. Cambridge, MA: Harvard Univ. Press. [2, 11]

Solomon, J., and C. George. 2008. The Measurement of Attachment Security and Related Constructs in Infancy and Early Childhood. In: Handbook of Attachment: Theory, Research, and Clinical Applications (2nd ed.), ed. J. Cassidy and P. R. Shaver, pp. 383–416. New York: Guilford. [8]

Solovey, M. 2013. Shaky Foundations: The Politics-Patronage-Social Science Nexus in Cold War America. New Brunswick, NJ: Rutgers Univ. Press. [2]

Sommer, V. 1989. Infant Mistreatment in Langur Monkeys: Sociobiology Tackled from the Wrong End? In: The Sociobiology of Sexual and Reproductive Strategies in Animals and Humans, ed. A. E. O. Rasa et al., pp. 110–127. London: Chapman and Hall. [4]

———. 1996. Heilige Egoisten: Die Soziobiologie Indischer Tempelaffen. Munich: C. H. Beck. [4]

Sommer, V. 2000. The Holy Wars About Infanticide. Which Side Are You on? And Why?. In: Infanticide by Males and Its Implications, ed. C. v. Schaik and C. Janson, pp. 9–26. Cambridge: Cambridge Univ. Press. [1]

Sommer, V., and U. Reichard. 2000. Rethinking Monogamy: The Gibbon Case. In: Primate Males: Causes and Consequences of Variation in Group Composition, ed. P. Kappeler, pp. 159–168. Cambridge:: Cambridge Univ. Press. [4]

Sorenson, E. R. 1979. Early Tactile Communication and the Patterning of Human Organization: A New Guinea Case Study. In: Before Speech: The Beginnings of Interpersonal Communication, ed. M. Bullowa, pp. 289–305. New York: Cambridge Univ. Press. [1]

Spencer-Booth, Y., and R. A. Hinde. 1967. The Effects of Separating Rhesus Monkey Infants from Their Mothers for Six Days. *J. Child Psychol. Psychiat.* **7**:179–197. [4]

Spinelli, S., S. Chefer, S. J. Suomi, et al. 2009. Early-Life Stress Induces Long-Term Morphologic Changes in Primate Brain. *Arch. Gen. Psychiatry* **66**:658–665. [9, 10]

Spock, B. 1968. Baby and Child Care. New York: Pocket Books. [14]

Spoor, J. R., and J. R. Kelly. 2004. The Evolutionary Significance of Affect in Groups: Communication and Group Bonding. *Group Process. Intergroup. Relat.* **7**:398–412. [6]

Spradley, J. P. 1979. The Ethnographic Interview. New York: Holt, Rinehart and Winston. [8]

Sroufe, L. A. 1979. The Coherence of Individual Development. *Am. Psychol.* **34**:834–841. [6]

———. 2005. Attachment and Development: A Prospective, Longitudinal Study from Birth to Adulthood. *Attach. Hum. Dev.* **7**:349–367. [12]

Sroufe, L. A., B. Coffino, and E. A. Carlson. 2010. Conceptualizing the Role of Early Experience: Lessons from the Minnesota Longitudinal Study. *Dev. Rev.* **30**:36–51. [8]

Sroufe, L. A., and J. Fleeson. 1986. Attachment and the Construction of Relationships. In: Relationships and Development, ed. W. Hartup and Z. Rubin, pp. 51–71. Hillsdale, NJ: Erlbaum. [8]

Sroufe, L. A., and E. Waters. 1977. Attachment as an Organizational Construct. *Child Dev.* **48**:1184–1199. [8, 11, 12]

Stamp-Dawkins, M. 2012. Why Animals Matter: Animal Consciousness, Animal Welfare, and Human Well-Being. Oxford: Oxford Univ. Press. [4]

Stanford, C. B. 1992. Costs and Benefits of Allomothering in Wild Capped Langurs (*Presbytis pileata*). *Behav. Ecol. Sociobiol.* **30**:29–34. [4]

Stearns, P. N. 2003. Anxious Parents: A History of Modern Childrearing in America. New York: NYU Press. [2, 5]

Stearns, S. C. 1982. The Role of Development in the Evolution of Life Histories. In: Evolution and Development, ed. J. T. Bonner, pp. 237–258. New York: Springer. [11]

———. 1992. The Evolution of Life Histories. Oxford: Oxford Univ. Press. [4, 11]

Sterck, E. H. M., D. P. Watts, and C. P. van Schaik. 1997. The Evolution of Female Social Relationships in Nonhuman Primates. *Behav. Ecol. Sociobiol.* **41**:291–309. [4]

Sterelny, K. 2013. Cooperation in a Complex World: The Role of Proximate Factors in Ultimate Explanations. *Biol. Theory* 7:358–367. [11]

Stevens, J. R. 2014. Evolutionary Pressures on Primate Intertemporal Choice. *Proc. R. Soc. Lond. B* **281**:20140499. [4]

Stotz, K. 2010. Human Nature and Cognitive-Developmental Niche Construction. *Phenom. Cogn. Sci.* **9**:483–501. [11]

———. 2014. Extended Evolutionary Psychology: The Importance of Transgenerational Developmental Plasticity. *Front. Psychol.* **5**:1–14. [11]

Strathearn, L., P. Fonagy, J. Amico, and P. R. Montague. 2009. Adult Attachment Predicts Maternal Brain and Oxytocin Response to Infant Cues. *Neuropsychopharmacology* **34**:2655–2666. [10]

Strauss, C., and N. Quinn. 1997. A Cognitive Theory of Cultural Meaning. Cambridge: Cambridge Univ. Press. [11]

Striepens, N., A. Matusch, K. M. Kendrick, et al. 2014. Oxytocin Enhances Attractiveness of Unfamiliar Female Faces Independent of the Dopamine Reward System. *Psychoneuroendocrino* **39**:74–87. [10]

Strier, K. B. 1994. Myth of the Typical Primate. *Am. J. Phys. Anthropol.* **37**:233–271. [4]

———. 2016. Primate Behavioral Ecology, 5th edition. London: Routledge. [4]

Sturge-Apple, M. L., J. H. Suor, P. T. Davies, et al. 2016. Vagal Tone and Children's Delay of Gratification: Differential Sensitivity in Resource-Poor and Resource-Rich Environments. *Psychol. Sci.* **27**:885–893. [11]

Su, S., X. Li, D. Lin, X. Xu, and M. Zhu. 2012. Psychological Adjustment among Left-Behind Children in Rural China: The Role of Parental Migration and Parent–Child Communication. *Child Care Health Dev.* **39**:162–170. [7]

Suddendorf, T. 2013. The Gap: The Science of What Separates Us from Other Animals. New York: Basic Books. [3]

Suddendorf, T., and A. Whiten. 2001. Mental Evolution and Development: Evidence for Secondary Representation in Children, Great Apes and Other Animals. *Psychol. Bull.* **127**:629–650. [3]

Sugiyama, Y. 2004. Demographic Parameters and Life History of Chimpanzees at Bossou, Guinea. *Am. J. Phys. Anthropol.* **124**:154–165. [4]

Suh, E. M. 2000. Self, the Hyphen between Culture and Subjective Well-Being. In: Culture and Subjective Well-Being, ed. E. Diener and E. M. Suh, pp. 63–86. Cambridge, MA: MIT Press. [6]

Suomi, S. J. 1987. Genetic and Maternal Contributions to Individual Differences in Rhesus Monkey Biobehavioral Development. In: Perinatal Development: A Psychobiological Perspective, ed. N. A. Krasnegor et al., pp. 397–419. Orlando: Academic Press. [10]

———. 2008. Attachment in Rhesus Monkeys. In: Handbook of Attachment: Theory, Research, and Clinical Applications (2nd ed.), ed. J. Cassidy and P. R. Shaver, pp. 173–191. New York: Guilford. [1, 5]

Suomi, S. J., and C. Ripp. 1983. A History of Motherless Mother Monkey Mothering at the University of Wisconsin Primate Laboratory. In: Child Abuse: The Nonhuman Primate Data, ed. M. Reite and N. G. Caine, pp. 49–78. New York: Alan R. Liss. [4]

Suomi, S. J., F. C. P. van der Horst, and R. van der Veer. 2008. Rigorous Experiments on Monkey Love: An Account of Harry F. Harlow's Role in the History of Attachment Theory. *Integr. Psychol. Behav. Sci.* **42**:354. [1, 9]

Super, C. M., and S. Harkness. 1986. The Developmental Niche: A Conceptualization at the Interface of Child and Culture. *Int. J. Behav. Dev.* **9**:545–569. [2, 6, 8, 11]

Sutcliffe, A., R. I. M. Dunbar, J. Binder, and H. Arrow. 2012. Relationships and the Social Brain: Integrating Psychological and Evolutionary Perspectives. *Br. J. Psychol.* **103**:149–168. [6]

Symons, D. K. 2010. A Review of the Practice and Science of Child Custody and Access Assessment in the United States and Canada. *Prof. Psychol. Res. Pr.* **41**:267–273. [14]

Takada, A. 2005. Mother-Infant Interactions among the !Xun: Analysis of Gymnastic and Breastfeeding Behaviors. In: Hunter-Gatherer Childhoods: Evolutionary, Developmental, and Cultural Perspectives, ed. B. S. Hewlett and M. E. Lamb, pp. 289–308. New Brunswick, NJ: Transaction Publishers. [6]

Takada, A. 2010. Changes in Developmental Trends of Caregiver-Child Interactions among the San: Evidence from the !Xun of Northern Namibia. *Afr. Study Monogr.* **40**:155–177. [6]

———. 2015. Narratives on San Ethnicity: The Cultural and Ecological Foundations of Lifeworld among The !Xun of North-Central Namibia. Kyoto and Melbourne: Kyoto Univ. Press, Trans Pacific Press. [6]

Takahashi, K. 1986. Examining the Strange-Situation Procedure with Japanese Mothers and 12-Month-Old Infants. *Dev. Psychol.* **22**:265–270. [1]

———. 2004. Close Relationships across the Lifespan: Towards a Theory of Relationship Types. In: Growing Together: Personal Relationships across the Lifespan, ed. F. R. Lang and K. L. Fingerman, pp. 130–158. New York: Cambridge Univ. Press. [5]

———. 2005. Toward a Lifespan Theory of Close Relationships: The Affective Relationship Model. *Human Dev.* **48**:48–66. [5]

Takeshita, H., M. Myowa-Yamakoshi, and S. Hirata. 2009. The Supine Position of Postnatal Human Infants: Implications for the Development of Cognitive Intelligence. *Interact. Stud.* **10**:252–269. [3]

Takiguchi, S., T. X. Fujisawa, S. Mizushima, et al. 2015. Ventral Striatum Dysfunction in Children and Adolescents with Reactive Attachment Disorder: Functional MRI Study. *Br, J. Psych. Open* **1**:121–128. [10]

Talhelm, T., X. Zhang, S. Oishi, et al. 2014. Large-Scale Psychological Differences within China Explained by Rice versus Wheat Agriculture. *Science* **344**:603–608. [7]

Tardif, S. D., D. G. Layne, and D. A. Smucny. 2002. Can Marmoset Mothers Count to Three? Effect of Litter Size on Mother–Infant Interactions. *Ethology* **108**:825–836. [3]

Tattersall, I. 2012. Masters of the Planet: The Search for Our Human Origins. New York: Palgrave Macmillan. [2]

Taylor, C. 1989. Sources of the Self: The Making of the Modern Identity. Cambridge, MA: Harvard Univ. Press. [5]

Taylor, J. M. 1996. Cultural Stories: Latina and Portuguese Daughters and Mothers. In: Urban Girls: Resisting Stereotypes, Creating Identities, ed. B. J. R. Leadbeater and N. Way, pp. 117–131. New York: NYU Press. [6]

Teicher, M. H., S. L. Andersen, A. Polcari, et al. 2003. The Neurobiological Consequences of Early Stress and Childhood Maltreatment. *Neurosci. Biobehav. Rev.* **27**:33–44. [10]

Teo, A., E. Carlson, P. Mathieu, B. Egeland, and L. A. Sroufe. 1996. A Prospective Longitudinal Study of Psychosocial Predictors of Achievement. *J. Sch. Psychol.* **34**:285–306. [12]

Thelen, E., D. M. Fisher, and R. Ridley-Johnson. 1984. The Relationship between Physical Growth and a Newborn Reflex. *Infant Behav. Dev.* **7**:479–493. [4]

Thierry, B. 2004. Social Epigenesis. In: Macaque Societies: A Model for the Study of Social Organization, ed. B. Thierry et al., pp. 267–294. Cambridge: Cambridge Univ. Press. [4]

———. 2007. Unity in Diversity: Lessons from Macaque Societies. *Evol. Anthropol.* **16**:224–238. [4]

———. 2008. Primate Socioecology, the Lost Dream of Ecological Determinism. *Evol. Anthropol.* **17**:93–96. [4]

Thompson, R. A. 2000. The Legacy of Early Attachments. *Child Dev.* **71**:145–152. [12]

———. 2006. The Development of the Person: Social Understanding, Relationships, Conscience, Self. In: Handbook on Child Psychology, ed. N. Eisenberg, pp. 24–98, Social, Emotional, and Personality Development, vol. 3, W. Damon and R. Lerner, series ed. Hoboken, NJ: Wiley. [6]

———. 2008a. Attachment-Related Mental Representations: Introduction to the Special Issue. *Attach. Hum. Dev.* **10**:347–358. [12]

———. 2008b. Early Attachment and Later Development: Familiar Questions, New Answers. In: Handbook of Attachment: Theory, Research, and Clinical Applications (2nd ed.), ed. J. Cassidy and P. R. Shaver, pp. 348–365. New York: Guildford. [8]

———. 2013a. Adaptations and Adaptations. In: Evolution, Early Experience and Human Development: From Research to Practice and Policy ed. D. Narvaez et al., pp. 329–336. New York: Oxford Univ. Press. [12]

———. 2013b. Attachment Theory and Research: Précis and Prospect. In: The Oxford Handbook of Developmental Psychology, vol. 2: Self and Other, ed. P. Zelazo, pp. 191–216. New York: Oxford Univ. Press. [8]

———. 2014. Stress and Child Development. *Future Child.* **24**:41–59. [8]

———. 2016. Early Attachment and Later Development: Reframing the Questions. In: Handbook of Attachment (3rd Ed.), ed. J. Cassidy and P. R. Shaver, pp. 330–348. New York: Guilford. [12]

Thompson, R. A., and M. Goodman. 2011. The Architecture of Social Developmental Science: Theoretical and Historical Perspectives. In: Social Development, ed. M. K. Underwood and L. H. Rosen, pp. 3–28. New York: Guilford. [12]

Thompson, R. A., and H. A. Raikes. 2003. Toward the Next Quarter-Century: Conceptual and Methodological Challenges for Attachment Theory. *Dev. Psychopathol.* **15**:691–718. [12]

Tibu, F., M. A. Sheridan, K. A. McLaughlin, et al. 2016. Disruptions of Working Memory and Inhibition Mediate the Association between Exposure to Institutionalization and Symptoms of Attention Deficit Hyperactivity Disorder. *Psychol. Med.* **46**:529–541. [9]

Tiger, L. 1969. Men in Groups. New York: Random House. [2]

Tinbergen, N. 1963. On Aims and Methods in Ethology. *Z. Tierpsychol.* **20**:410–433. [2, 11]

Tizard, B., and J. Rees. 1975. The Effect of Early Institutional Rearing on the Behaviour Problems and Affectional Relationships of Four-Year-Old Children. *J. Child Psychol. Psychiatry* **16**:61–73. [9]

Tobin, J., D. Wu, and D. Davidson, eds. 1989. Preschool in Three Cultures: Japan, China and the United States. New Haven, CT: Yale Univ. Press. [6, 8]

Toda, K., and M. L. Platt. 2015. Animal Cognition: Monkeys Pass the Mirror Test. *Curr. Biol.* **25**:R64–R66. [4]

Tomasello, M. 1999. The Cultural Origins of Human Cognition. Cambridge, MA: Harvard Univ. Press. [4, 11]

Tomasello, M., and J. Call. 1997. Primate Cognition. In: Oxford: Oxford Univ. Press. [4]

Tomasello, M., M. Carpenter, J. Call, T. Behne, and H. Moll. 2005. Understanding and Sharing Intentions: The Origins of Cultural Cognition. *Behav. Brain Sci.* **28**:675–735. [8, 11]

Tomasello, M., and E. Herrmann. 2010. Ape and Human Cognition: What's the Difference? *Curr. Dir. Psychol.* **19**:3–8. [4]

Tomasello, M., S. Savage-Rumbaugh, and A. C. Kruger. 1993. Imitative Learning of Actions on Objects by Children, Chimpanzees, and Enculturated Chimpanzees. *Child Dev.* **64**:1688–1705. [3]

Tomoda, A. 2016. Preliminary Evidence for Impaired Brain Activity of Neural Reward Processing in Children and Adolescents with Reactive Attachment Disorder. *Yakugaku Zasshi* **126**:711–714. [10]

Tooby, J., and L. Cosmides. 1990. On the Universality of Human Nature and the Uniqueness of the Individual: The Role of Genetics and Adaptation. *J. Pers.* **58**:17–67. [5]

Tottenham, N. 2014. The Importance of Early Experiences for Neuro-Affective Development. *Curr. Top. Behav. Neurosci.* **16**:109–129. [9]

Tottenham, N., T. A. Hare, A. Millner, et al. 2011. Elevated Amygdala Response to Faces Following Early Deprivation. *Dev. Sci.* **14**:190–204. [9]

Tottenham, N., T. A. Hare, B. T. Quinn, et al. 2010. Prolonged Institutional Rearing Is Associated with Atypically Larger Amygdala Volume and Difficulties in Emotion Regulation. *Dev. Sci.* **13**:46–61. [9, 11]

Tottenham, N., and M. A. Sheridan. 2010. A Review of Adversity, the Amygdala and the Hippocampus: A Consideration of Developmental Timing. *Front. Hum. Neurosci.* **3**:68. [10]

Townsend, S. W., K. E. Slocombe, M. E. Thompson, and K. Zuberbühler. 2007. Female-Led Infanticide in Wild Chimpanzees. *Curr. Biol.* **17**:R355–R356. [4]

Trawick, M. 1990. Notes on Love in a Tamil Family. Berkeley: Univ. of California Press. [5, 14]

Trevarthen, C. 2009. The Intersubjective Psychobiology of Human Meaning: Learning of Culture Depends on Interest for Co-Operative Practical Work and Affection for the Joyful Art of Good Company. *Psychoanal. Dialog.* **19**:507–518. [11]

———. 2011. What Is It Like to Be a Person Who Knows Nothing? Defining the Active Intersubjective Mind of a Newborn Human Being. *Infant Child Dev.* **20**:119–135. [11]

Trivers, R. L. 1972. Parental Investment and Sexual Selection. In: Sexual Selection and the Descent of Man 1871–1971, ed. B. Campbell, pp. 139–179. Chicago: Aldine. [1, 4]

———. 1974. Parent-Offspring Conflict. *Am. Zool.* **14**:249–264. [2, 4, 6, 11, 12]

Tronick, E. 2003. Of Course All Relationships Are Unique: How Co-Creative Processes Generate Unique Mother–Infant and Patient–Therapist Relationships and Change Other Relationships. *Psychoanal. Inq.* **23**:473–491. [7]

———. 2007. The Neurobehavioral and Social-Emotional Development of Infants and Children. New York: W. W. Norton. [11]

Tronick, E., and M. Beeghly. 2011. Infants' Meaning-Making and the Development of Mental Health Problems. *Am. Psychol.* **66**:107. [8]

Tronick, E., G. A. Morelli, and S. Winn. 1987. Multiple Caretaking of Efe (Pygmy) Infants. *Am. Anthropol.* **89**:96–106. [1, 5, 7]

Tuli, M., and N. Chaudhary. 2010. Elective Interdependence: Understanding Individual Agency and Interpersonal Relationships in Indian Families. *Cult. Psychol.* **16**:477–496. [5]

Turecki, G., and M. J. Meaney. 2016. Effects of the Social Environment and Stress on Glucocor-Ticoid Receptor Gene Methylation: A Systematic Review. *Biol. Psychiatry* **79**:87–96. [11]

Ujiie, T., and K. Miyake. 1985. Responses to the Strange Situation in Japanese Infants. Research and Clinical Center of Child Development Annual Report. Sapporo: Hokkaido Univ. [6]

UNICEF. 2011. Behaviour Change Communication Cells and Village Information Centres. http://unicef.in/Uploads/Publications/Resources/pub_doc81.pdf (accessed April 26, 2017). [14]

———. 2016. Uprooted: The Growing Crisis for Refugee and Migrant Children. Geneva: UNICEF. [6]

United Nations. 1989. Convention on the Rights of the Child, General Assembly of the United Nations. http://www.ohchr.org/en/professionalinterest/pages/crc.aspx (accessed Nov. 9, 2016). [8, 13]

———. 2005. Committee on the Rights of the Child: General Comment 7: Implementing Child Rights in Early Childhood. United Nations: CRC/C/GC/7/Rev.1. http://www2.ohchr.org/english/bodies/crc/docs/AdvanceVersions/GeneralComment-7Rev1.pdf (accessed Nov. 9, 2016). [13]

U.S. Census Bureau. 2013. How Do We Know? Child Care an Important Part of American Life. Washington, D.C.: GPO. [6]

Uvnas-Moberg, K., L. Handlin, and M. Petersson. 2014. Self-Soothing Behaviors with Particular Reference to Oxytocin Release Induced by Non-Noxious Sensory Stimulation. *Front. Psychol.* **5**:1529. [10]

van der Horst, F. C. P. 2011. John Bowlby: From Psychoanalysis to Ethology. Oxford: Wiley-Blackwell. [2]

Vanderwert, R. E., P. J. Marshall, C. A. Nelson III, C. H. Zeanah, and N. A. Fox. 2010. Timing of Intervention Affects Brain Electrical Activity in Children Exposed to Severe Psychosocial Neglect. *PLoS One* **5**:e11415. [9]

van IJzendoorn, M. H. 2005. Attachment in Social Networks: Toward an Evolutionary Social Network Model. *Human Dev.* **48**:85–88. [12]

van IJzendoorn, M. H., M. J. Bakermans-Kranenberg, and F. Juffer. 2008. Video-Feedback Intervention to Promote Positive Parenting: Evidence-Based Intervention for Enhancing Sensitivity and Security. In: Promoting Positive Parenting: An Attachment-Based Intervention, ed. F. Juffer et al., pp. 193–202. New York: Lawrence Erlbaum. [14]

van IJzendoorn, M. H., M. J. Bakermans-Kranenburg, and A. Sagi-Schwartz. 2006. Attachment across Diverse Sociocultural Contexts: The Limits of Universality. In: Parenting Beliefs, Behaviors, and Parent-Child Relations: A Cross-Cultural Perspective, ed. K. H. Rubin and O. B. Chung, pp. 107–142. New York: Psychology Press. [2, 3]

van IJzendoorn, M. H., K. A. Bard, M. J. Bakermans-Kranenburg, and K. Ivan. 2009. Enhancement of Attachment and Cognitive Development of Young Nursery-Reared Chimpanzees in Responsive versus Standard Care. *Dev. Psychobiol.* **51**:173–185. [8, 10]

van IJzendoorn, M. H., and F. Juffer. 2006. Adoption as Intervention: Meta-Analytic Evidence for Massive Catch-up and Plasticity in Physical, Socio-Emotional, and Cognitive Development (the Emanuel Miller Memorial Lecture) *J. Child Psychol. Psychiatry* **47**:1228–1245. [14]

van IJzendoorn, M. H., and P. M. Kroonenberg. 1988. Cross-Cultural Patterns of Attachment: A Meta-Analysis of the Strange Situation. *Child Dev.* **59**:147–156. [7]

van IJzendoorn, M. H., G. Morgan, J. Belsky, et al. 2000. The Similarity of Siblings' Attachment to Their Mother. *Child Dev.* **71**:1086–1098. [5]

van IJzendoorn, M. H., and A. Sagi-Schwartz. 1999. Cross-Cultural Patterns of of Attachment: The Universal and Contextual Dimensions. In: Handbook of Attachment: Theory, Research, and Clinical Applications, ed. J. Cassidy and P. R. Shaver, pp. 713–734. New York: Guilford. [6]

———. 2008. Cross-Cultural Patterns of Attachment: Universal and Contextual Dimensions. In: Handbook of Attachment: Theory, Research, and Clinical Applications (2nd ed.), ed. J. Cassidy and P. R. Shaver, pp. 880–905. New York: Guilford. [5, 6, 7]

van IJzendoorn, M. H., C. M. Vereijken, M. J. Bakermans-Kranenburg, and J. M. Riksen-Walraven. 2004. Assessing Security with the Attachment Q Sort: Meta-Analytic Evidence for the Validity of the Observer Aqs. *Child Dev.* **75**:1188–1213. [5]

van Lawick-Goodall, J. 1968. The Behaviour of Free-Living Chimpanzees of the Gombe Stream Nature Reserve. *Anim. Behav. Monogr.* **1**:161–311. [3]

van Rosmalen, L., M. H. van IJzendoorn, and M. J. Bakermans-Kranenburg. 2014. ABC + D of Attachment Theory: The Strange Situation Procedure as the Gold Standard of Attachment Assessment. In: The Routledge Handbook of Attachment: Theory, ed. P. Holmes and S. Farnfield, pp. 11–30. New York: Routledge. [6]

van Schaik, C. P. 1989. The Ecology of Social Relationships Amongst Female Primates. In: Comparative Socioecology: The Behavioral Ecology of Humans and Other Mammals, ed. V. Standen and R. A. Foley, pp. 195–218. Oxford: Blackwell. [4]

———. 1996. Social Evolution in Primates: The Role of Ecological Factors and Male Behavior. *Proc. Br. Acad.* **88**:9–31. [4]

van Schaik, C. P., and J. M. Burkart. 2010. Mind the Gap: Cooperative Breeding and the Evolution of Our Unique Features. In: Mind the Gap: Tracing the Origins of Human Universals, ed. P. M. Kappeler and J. Silk, pp. 477–496. Berlin: Springer. [5]

van Schaik, C. P., and C. Janson, eds. 2000. Infanticide by Males and Its Implications. Cambridge: Cambridge Univ. Press. [4]

VanTieghem, M. R., and N. Tottenham. 2017. Neurobiological Programming of Early Life Stress: Functional Development of Amygdala-Prefrontal Circuitry and Vulnerability for Stress-Related Psychopathology. *Curr. Top. Behav. Neurosci.* doi: 10.1007/7854_2016_1042. [10]

van Vugt, M., and T. Kameda. 2014. Evolution of the Social Brain: Psychological Adaptations for Group Living. In: Mechanisms of Social Connections: From Brain to Group, ed. M. Mikulincer and P. R. Shaver, pp. 335–355. Washington, D.C.: American Psychological Association. [6]

Varela, F., E. Thompson, and E. Rosch. 1991. The Embodied Mind: Cognitive Science and Human Experience. Cambridge, MA: MIT Press. [11]

Varendi, H., and R. H. Porter. 2001. Breast Odour as the Only Maternal Stimulus Elicits Crawling Towards the Odour Source. *Acta Paediatr* **90**:372–375. [10]

Vasey, N. 2007. The Breeding System of Wild Red Ruffed Lemurs (*Varecia rubra*): A Preliminary Report. *Primates* **48**:41–54. [3]

Vaughn, B. E., and E. Waters. 1990. Attachment Behavior at Home and in the Laboratory: Q-Sort Observations and Strange Situation Classifications of One-Year Olds. *Child Dev.* **61**:1965–1973. [5]

Verhage, M. L., C. Schuengel, S. Madigan, et al. 2015. Narrowing the Transmission Gap: A Synthesis of Three Decades of Research on Intergenerational Transmission of Attachment. *Psychol. Bull.* **142**:337–366. [8]

Vicedo, M. 2009. Mothers, Machines, and Morals: Harry Harlow's Work on Primate Love from Lab to Legend. *J. Hist. Behav. Sci.* **45**:193–218. [6]

———. 2011. The Social Nature of the Mother's Tie to Her Child: John Bowlby's Theory of Attachment in Post-War America. *Br. J. Hist. Sci.* **43**:401–426. [6]

———. 2013. The Nature and Nurture of Love: From Imprinting to Attachment in Cold War America. Chicago: Univ. Chicago Press. [1, 2, 5, 6]

———. 2017. Putting Attachment in Its Place: Disciplinary and Cultural Contexts. *Eur. J. Dev. Psychol.* http://dx.doi.org/10.1080/17405629.2017.1289838 (accessed July 7, 2017). [2, 6]

Vogeley, K., and A. Roepstorff. 2009. Contextualizing Culture and Social Cognition. *Trends Cogn. Sci.* **13**:511–516. [11]

Voland, E. 2000. Contributions of Family Reconstitution Studies to Evolutionary Reproductive Ecology. *Evol. Anthropol.* **9**:134–146. [4]

Voland, E., and J. Beise. 2002. Opposite Effects of Maternal and Paternal Grandmothers on Infant Survival in Historical Krummhörn. *Behav. Ecol. Sociobiol.* **52**:435–443. [4]

Voland, E., W. Schiefenhövel, and A. Chasiotis, eds. 2005. Grandmotherhood. The Evolutionary Significance of the Second Half of Female Life. New York: Rutgers Univ. Press. [5]

Vouloumanos, A., and J. F. Werker. 2007. Listening to Language at Birth: Evidence for a Bias for Speech in Neonates. *Dev. Sci.* **10**:159–164. [6]

Waddington, C. H. 1942. Canalization of Development and the Inheritance of Acquired Characters. *Nature* **150**:563–565. [11]

Wade, M. J., D. S. Wilson, C. Goodnight, et al. 2010. Multilevel and Kin Selection in a Connected World. *Nature* **463**:E8–10. [11]

Wahl, S., and H. Spada. 2000. Children's Reasoning About Intentions, Beliefs and Behaviour. *Cogn. Sci. Q.* **1**:5–34. [11]

Waldmeir, P. 2015. China Migration: Children of a Revolution. *Financial Times*, Dec. 27. https://www.ft.com/content/c4fe010a-a324-11e5-8d70-42b68cfae6e4 (accessed July 7, 2017). [7]

Walker, A. S., C. J. Owsley, J. Megaw-Nyce, E. J. Gibson, and L. E. Bahrick. 1980. Detection of Elasticity as an Invariant Property of Objects by Young Infants. *Perception* **9**:713–718. [4]

Walker, H. 2013. Under a Watchful Eye: Self, Power, and Intimacy in Amazonia. Berkeley: Univ. of California Press. [6]

Walsh, D. M. 2015. Organisms, Agency, and Evolution. Cambridge: Cambridge Univ. Press. [2]

Wang, C. D., and S. S. Young. 2010. Adult Attachment Reconceptualized: A Chinese Perspective. In: Attachment, ed. P. Erdman and K.-M. Ng, pp. 15–33. New York: Routledge. [6]

Wang, Q., and N. Chaudhary. 2005. The Self. In: Psychological Concepts: An International Historical Perspective, ed. K. Pawlik and G. d'Ydewalle, pp. 325–358. Hove, UK: Psychology Press. [6]

Wang, Y. B., and X. L. Wu. 2003. Report on the Case Study of "Left-Behind Children" in Rural China. *J. Youth Stud.* **4**:7–10. [7]

Washabaugh, K., C. T. Snowdon, and T. E. Ziegler. 2002. Variations in Care for Cottontop Tamarin, Saguinus Oedipus, Infants as a Function of Parental Experience and Group Size. *Anim. Behav.* **63**:1163–1174. [3]

Waters, E. 1978. The Reliability and Stability of Individual Differences in Infant-Mother Attachment. *Child Dev.* **49**:483–494. [12]

Waters, E. 1995. Appendix A: The Attachment Q-Set (Version 3.0). *Monogr. Soc. Res. Child Dev.* **60**:234–246. [8]

Waters, E., and E. M. Cummings. 2000. A Secure Base from Which to Explore Close Relationships. *Child Dev.* **71**:164–172. [7]

Waters, E., B. E. Vaughn, and D. M. Teti. 2017. Assessing Secure Base Behavior in Naturalistic Environments: The Attachment Q-Set. In: Measuring Attachment, ed. E. Waters et al. New York: Guilford, in press. [8]

Watson, J. S. 1972. Smiling, Cooing, and "the Game". *Merrill Palmer Q. Behav. Dev.* **18**:323–339. [11]

———. 1985. Contigency Perception in Early Social Development. In: Social Perception of Infants, ed. T. Field and N. Fox, pp. 157–176. Norwood, NJ: Ablex. [6]

———. 1994. Detection of Self: The Perfect Algorithm. In: Self-Awareness in Animals and Humans: Developmental Perspectives, ed. S. Parker et al., pp. 131–149. Cambridge: Cambridge Univ. Press. [11]

———. 2001. Contingency Perception and Misperception in Infancy: Some Implications for Attachment. *Bull. Menninger Clin.* **65**:296–320. [11]

Weaver, I. C. G., N. Cervoni, F. A. Champagne, et al. 2004. Epigenetic Programming by Maternal Behavior. *Nat. Neurosci.* **7**:847–854. [11]

Weinfield, N. S., L. A. Sroufe, B. Egeland, and E. Carlson. 2008. Individual Differences in Infant-Caregiver Attachment: Conceptual and Empirical Aspects of Security. In: Handbook of Attachment, ed. J. Cassidy and P. R. Shaver, pp. 78–101. New York: Guilford. [6]

Weisman, O., O. Zagoory-Sharon, and R. Feldman. 2012. Oxytocin Administration to Parent Enhances Infant Physiological and Behavioral Readiness for Social Engagement. *Biol. Psychiatry* **72**:982–989. [10]

Weisner, T. S. 1984. Ecocultural Niches of Middle Childhood: A Cross-Cultural Perspective. In: Development During Middle Childhood, ed. W. A. Collins, pp. 334–369. Washington, D.C.: National Academy Press. [6]

———. 1997. Support for Children and the African Family Crisis. In: African Families and the Crisis of Social Change, ed. T. S. Weisner et al., pp. 20–44. Westport, CT: Greenwood Press. [5]

———. 2002. Ecocultural Understanding of Children's Developmental Pathways. *Human Dev.* **45**:275–281. [8]

———. 2005. Attachment as a Cultural and Ecological Problem with Pluralistic Solutions. *Human Dev.* **48**:89–94. [2, 5, 14]

———. 2011a. Culture. In: Social Development: Relationships in Infancy, Childhood and Adolescence, ed. M. K. Underwood and L. H. Rosen, pp. 372–399. New York: Guilford. [8]

———. 2011b. The Ecocultural Family Interview: New Conceptualizations and Uses for the Study of Illness. In: Sviluppo E Salute del Bambino: Fattori Individuali, Sociali E Culturali (in Ricordo Di Vanna Axia), ed. S. Bonichini and M. R. Baroni, pp. 166–173. Padova: Libraria Editrice Universitaria di Padova. [8]

———. 2011c. If You Work in This Country You Should Not Be Poor, and Your Kids Should Be Doing Better: Bringing Mixed Methods and Theory in Psychological Anthropology to Improve Research in Policy and Practice. *Ethos* **39**:455–476. [8]

———. 2014. The Socialization of Trust: Plural Caregiving and Diverse Pathways in Human Development across Cultures. In: Different Faces of Attachment: Cultural Variations on a Universal Human Need, ed. H. Otto and H. Keller, pp. 263–277. Cambridge: Cambridge Univ. Press. [2, 5]

————. 2016a. Culture, Context, and the Integration of Qualitative and Quantitative Methods in the Study of Human Development. In: Advances in Culture and Psychology, ed. M. Gelfand et al. New York: Oxford Univ. Press. [8]

————. 2016b. Relationships and Social Trust in Early Childhood Programs: The Importance of Context and Mixed Methods. In: The Culture of Child Care: Attachment, Peers, and Quality in Diverse Communities, ed. K. E. Sanders and A. W. Guerra, pp. 107–120. New York: Oxford Univ. Press. [12]

Weisner, T. S., and G. J. Duncan. 2014. The World Isn't Linear or Additive or Decontextualized: Pluralism and Mixed Methods in Understanding the Effects of Antipoverty Programs on Children and Parenting. In: Societal Contexts of Child Development, ed. E. T. Gershoff et al. New York: Oxford Univ. Press. [8]

Weisner, T. S., and B. H. Fiese. 2011. Introduction to Special Section of the Journal of Family Psychology, Advances in Mixed Methods in Family Psychology: Integrative and Applied Solutions for Family Science. *J. Fam. Psychol.* **25**:795. [8]

Weisner, T. S., and R. Gallimore. 1977. My Brother's Keeper: Child and Sibling Caretaking. *Curr. Anthropol.* **18**:169–190. [1, 2, 6]

Weisner, T. S., and M. C. Hay. 2015. Practice to Research: Integrating Evidence-Based Practices with Culture and Context. *Transcult. Psychiatry* **52**:222–243. [13]

Wen, M., and D. Lin. 2012. Child Development in Rural China: Children Left Behind by Their Migrant Parents and Children of Nonmigrant Families. *Child Dev.* **83**:120–136. [7]

West, S. A., and A. Gardner. 2010. Altruism, Spite, and Greenbeards. *Science* **327**:1341–1344. [4]

West-Eberhard, M. J. 2003. Developmental Plasticity and Evolution. New York: Oxford Univ. Press. [11]

White, G. M., and J. Kirkpatrick. 1985. Person, Self, and Experience: Exploring Pacific Ethnopsychologies. Berkley: Univ. of California Press. [6]

Whiten, A., J. Goodall, W. C. McGrew, et al. 1999. Cultures in Chimpanzees. *Nature* **399**:682–685. [4, 8]

Whiten, A., N. McGuigan, S. Marshall-Pescini, and L. M. Hopper. 2009. Emulation, Imitation, Over-Imitation and the Scope for Culture of Child and Chimpanzee. *Phil. Trans. R. Soc.* **364**:2417–2428. [11]

Whiting, B. B. 1963. Six Cultures: Studies of Child Rearing. New York: Wiley. [1]

Whiting, B. B., and C. P. Edwards. 1988. Children of Different Worlds: The Formation of Social Behavior. Cambridge, MA: Harvard Univ. Press. [6, 8]

Whiting, B. B., and J. W. M. Whiting. 1975. Children of Six Cultures: A Psycho-Cultural Analysis. Cambridge, MA: Harvard Univ. Press. [2, 5, 6]

Wiesel, T. N., and D. H. Hubel. 1965. Extent of Recovery from the Effects of Visual Deprivation in Kittens. *J. Neurophysiol.* **28**:1060–1072. [9]

Williamson, G., I. Pérez, F. Modesto, G. Coilla, and N. Raín. 2012. Infancia y Adolescencia Mapuche en Relatos de la Araucanía. *Cont. Educ. Rev. Educ.* **12**:135–152. [6]

Wilson, D. S., and E. O. Wilson. 2007. Rethinking the Theoretical Foundation of Sociobiology. *Q. Rev. Biol.* **82**:327–348. [11]

Wilson, E. O. 1975. Sociobiology: A New Synthesis. Cambridge, MA: Harvard Univ. Press. [5]

————. 2012. The Social Conquest of Earth. New York: Liveright, Norton. [2]

Wilson, M., and M. Daly. 1985. Competitiveness, Risk-Taking, and Violence: The Young Male Syndrome. *Ethol. Sociobiol.* **6**:59–73. [11]

Wismer Fries, A. B., T. E. Ziegler, J. R. Kurian, S. Jacoris, and S. D. Pollak. 2005. Early Experience in Humans Is Associated with Changes in Neuropeptides Critical for Regulating Social Behavior. *PNAS* **102**:17237–17240. [10]

Wittfoth-Schardt, D., J. Grunding, M. Wittfoth, et al. 2012. Oxytocin Modulates Neural Reactivity to Children's Faces as a Function of Social Salience. *Neuropsychopharmacology* **37**:1799–1807. [10]

Wittig, R. M., and C. Boesch. 2003. Food Competition and Linear Dominance Hierarchy among Female Chimpanzees of the Taı National Park. *Int. J. Primatol.* **24**:847–867. [4]

Witzany, G., ed. 2014. Biocommunication of Animals. New York: Springer. [11]

Wood, B. M., D. P. Watts, J. C. Mitani, and K. E. Langergraber. 2017. Favorable Ecological Circumstances Promote Life Expectancy in Chimpanzees Similar to That of Human Hunter-Gatherers. *J. Hum. Evol.* **105**:41–56. [4]

Workman, A. D., C. J. Charvet, B. Clancy, R. B. Darlington, and B. L. Finlay. 2013. Modeling Transformations of Neurodevelopmental Sequences across Mammalian Species. *J. Neurosci.* **33**:7368–7383. [4]

World Association for Infant Mental Health. 2016. Waimh Position Paper on the Rights of Infants. http://www.waimh.org/files/Perspectives%20in%20IMH/2016_1-2/PositionPaperRightsInfants_%20May_13_2016_1-2_Perspectives_IMH_corr.pdf (accessed Nov. 9, 2016). [13]

World Health Organization. 1962. Deprivation of Maternal Care: A Reassessment of Its Effects. Geneva: World Health Organization. [2]

Worthman, C. M. 2010. The Ecology of Human Development: Evolving Models for Cultural Psychology. *J. Cross-Cult. Psychol.* **41**:546–562. [8]

Wrangham, R. W. 1980. An Ecological Model of Female-Bonded Primate Groups. *Behaviour* **75**:262–300. [4]

Wroblewski, E. E. 2008. An Unusual Incident of Adoption in a Wild Population at Gombe National Park. *Am. J. Primatol.* **70**:995–998. [3]

Xie, X., and Y. Xia. 2011. Grandparenting in Chinese Immigrant Families. *Marriage Fam. Rev.* **47**:383–396. [5]

Yamaguchi, M. K., S. Kanazawa, M. Tomonaga, and C. Murai. 2003. Development of Face Recognition in Humans and Japanese Monkeys. In: Cognitive and Behavioral Development in Chimpanzees, ed. M. Tomonaga et al., pp. 347–352. Kyoto: Kyoto Univ. Press [3]

Yeoh, B. S., S. Huang, and T. Lam. 2005. Transnationalizing the "Asian" Family: Imaginaries, Intimacies and Strategic Intents. *Global Networks* **5**:307–315. [7]

Yopak, K. E., T. J. Lisney, R. B. Darlington, et al. 2010. A Conserved Pattern of Brain Scaling from Sharks to Primates. *PNAS* **107**:12946–12951. [4]

Yoshikawa, H. 2011. Immigrants Raising Citizens: Undocumented Parents and Their Children. New York: Russell Sage Foundation. [7]

Yoshikawa, H., T. S. Weisner, A. Kalil, and N. Way. 2008. Mixing Qualitative and Quantitative Research in Developmental Science: Uses and Methodological Choices. *Dev. Psychol.* **44**:344–354. [8]

Young, L. J. 2002. The Neurobiology of Social Recognition, Approach, and Avoidance. *Biol. Psychiatry* **51**:18–26. [4]

Young, L. J., M. M. Lim, B. Gingrich, and T. R. Insel. 2001. Cellular Mechanisms of Social Attachment. *Horm. Behav.* **40**:133–138. [4]

Yovsi, R., J. Kärtner, H. Keller, and A. Lohaus. 2009. Maternal Interactional Quality in Two Cultural Environments: German Middle Class and Cameroonian Rural Mothers. *J. Cross-Cult. Psychol.* **40**:701–707. [5, 8]

Yovsi, R., and H. Keller. 2003. Breastfeeding: An Adaptive Process. *Ethos* **31**:141–171. [6]

Zahed, S. R., S. L. Prudom, C. T. Snowdon, and T. E. Ziegler. 2008. Male Parenting and Response to Infant Stimuli in the Common Marmoset (*Callithrix jacchus*). *Am. J. Primatol.* **70**:84–92. [3]

Zeanah, C. H., H. L. Egger, A. T. Smyke, et al. 2009. Institutional Rearing and Psychiatric Disorders in Romanian Preschool Children. *Am. J. Psychiatry* **166**:777–778. [9]

Zeanah, C. H., and M. M. Gleason. 2015. Annual Research Review. Attachment Disorders in Early Childhood: Clinical Presentation, Causes, Correlates, and Treatment. *J. Child Psychol. Psychiatry* **56**:207–222. [9]

Zeanah, C. H., S. F. Koga, B. Simion, et al. 2006. Ethical Considerations in International Research Collaboration: The Bucharest Early Intervention Project. *Infant Ment. Health J.* **27**:559–576. [9]

Zeanah, C. H., C. A. Nelson III, N. A. Fox, et al. 2003. Designing Research to Study the Effects of Institutionalization on Brain and Behavioral Development: The Bucharest Early Intervention Project. *Dev. Psychopathol.* **15**:885–907. [9]

Zeanah, C. H., M. Scheeringa, N. W. Boris, et al. 2004. Reactive Attachment Disorder in Maltreated Toddlers. *Child Abuse Neglect* **28**:877–888. [10]

Zeanah, C. H., A. T. Smyke, S. F. Koga, and E. Carlson. 2005. Attachment in Institutionalized and Community Children in Romania. *Child Dev.* **76**:1015–1028. [9]

Zelizer, V. A. 1985. Pricing the Priceless Child: The Changing Social Value of Children. New York: Basic Books. [2]

Zheng, G., and S. Shi. 2004. Intercultural and Intracultural Differences in the Value of Children: Comparisons between Four Countries and the Urban, Rural, and Floating Populations in China. In: Perspectives and Progress in Contemporary Cross-Cultural Psychology: Selected Papers from the Seventeenth International Congress of the International Association for Cross-Cultural Psychology, ed. G. Zheng et al., pp. 129–148. Beijing: China Light Industry Press. [7]

Zontini, E., and T. Reynolds. 2007. Ethnicity, Families and Social Capital: Caring Relationships across Italian and Caribbean Transnational Families. *Int. Rev. Sociol.* **17**:257–277. [7]

Subject Index

Further Titles in the Strüngmann Forum Report Series[1]

Better Than Conscious? Decision Making, the Human Mind, and Implications For Institutions
edited by Christoph Engel and Wolf Singer, ISBN 978-0-262-19580-5

Clouds in the Perturbed Climate System: Their Relationship to Energy Balance, Atmospheric Dynamics, and Precipitation
edited by Jost Heintzenberg and Robert J. Charlson, ISBN 978-0-262-01287-4

Biological Foundations and Origin of Syntax
edited by Derek Bickerton and Eörs Szathmáry, ISBN 978-0-262-01356-7

Linkages of Sustainability
edited by Thomas E. Graedel and Ester van der Voet, ISBN 978-0-262-01358-1

Dynamic Coordination in the Brain: From Neurons to Mind
edited by Christoph von der Malsburg, William A. Phillips and Wolf Singer, ISBN 978-0-262-01471-7

Disease Eradication in the 21st Century: Implications for Global Health
edited by Stephen L. Cochi and Walter R. Dowdle, ISBN 978-0-262-01673-5

Animal Thinking: Contemporary Issues in Comparative Cognition
edited by Randolf Menzel and Julia Fischer, ISBN 978-0-262-01663-6

Cognitive Search: Evolution, Algorithms, and the Brain
edited by Peter M. Todd, Thomas T. Hills and Trevor W. Robbins, ISBN 978-0-262-01809-8

Evolution and the Mechanisms of Decision Making
edited by Peter Hammerstein and Jeffrey R. Stevens, ISBN 978-0-262-01808-1

Language, Music, and the Brain: A Mysterious Relationship
edited by Michael A. Arbib, ISBN 978-0-262-01962-0

Cultural Evolution: Society, Technology, Language, and Religion
edited by Peter J. Richerson and Morten H. Christiansen, ISBN 978-0-262-01975-0

Schizophrenia: Evolution and Synthesis
edited by Steven M. Silverstein, Bita Moghaddam and Til Wykes, ISBN 978-0-262-01962-0

Rethinking Global Land Use in an Urban Era
edited by Karen C. Seto and Anette Reenberg, ISBN 978-0-262-02690-1

Trace Metals and Infectious Diseases
edited by Jerome O. Nriagu and Eric P. Skaar, ISBN 978-0-262-02919-3

Translational Neuroscience: Toward New Therapies
edited by Karoly Nikolich and Steven E. Hyman, ISBN: 9780262029865

---

[1] available at https://mitpress.mit.edu/books/series/str%C3%BCngmann-forum-reports-0

The Pragmatic Turn: Toward Action-Oriented Views in Cognitive Science
edited by Andreas K. Engel, Karl J. Friston and Danica Kragic
ISBN: 978-0-262-03432-6

Complexity and Evolution: Toward a New Synthesis for Economics
edited by David S. Wilson and Alan Kirman, ISBN: 9780262035385

Computational Psychiatry: New Perspectives on Mental Illness
edited by A. David Redish and Joshua A. Gordon, ISBN: 9780262035422

Investors and Exploiters in Ecology and Economics: Principles and Applications
edited by Luc-Alain Giraldeau, Philipp Heeb and Michael Kosfeld
Hardcover: ISBN: 9780262036122, eBook: ISBN: 9780262339797